Cellular Function and Metabolism

Developments in Molecular and Cellular Biochemistry

Series Editor: Naranjan S. Dhalla, Ph.D., FACC

1. V.A. Najjar (ed.): *Biological Effects of Glutamic Acid and Its Derivatives.* 1981 ISBN 90-6193-841-4
2. V.A. Najjar (ed.): *Immunologically Active Peptides.* 1981 ISBN 90-6193-842-2
3. V.A. Najjar (ed.): *Enzyme Induction and Modulation.* 1983 ISBN 0-89838-583-0
4. V.A. Najjar and L. Lorand (eds.): *Transglutaminase.* 1984 ISBN 0-89838-593-8
5. G.J. van der Vusse (ed.): *Lipid Metabolism in Normoxic and Ischemic Heart.* 1989 ISBN 0-7923-0479-9
6. J.F.C. Glatz and G.J. van der Vusse (eds.): *Cellular Fatty Acid-Binding Proteins.* 1990 ISBN 0-7923-0896-4
7. H.E. Morgan (ed.): *Molecular Mechanisms of Cellular Growth.* 1991 ISBN 0-7923-1183-3
8. G.J. van der Vusse and H. Stam (eds.): *Lipid Metabolism in the Healthy and Diseased Heart.* 1992 ISBN 0-7923-1850-1
9. Y. Yazaki and S. Mochizuki (eds.): *Cellular Function and Metabolism.* 1993 ISBN 0-7923-2158-8

SPRINGER-SCIENCE+BUSINESS MEDIA, B.V.

Cellular Function and Metabolism

edited by

YOSHIO YAZAKI

Professor of Medicine
Third Department of Internal Medicine
Faculty of Medicine, University of Tokyo
Bunkyo-ku, Hongo 7-3-1
Tokyo 113, Japan

and

SEIBU MOCHIZUKI

Professor of Medicine
Department of Internal Medicine
Jikei University School of Medicine
Katsushika-ku, Aoto 6-41-2
Tokyo 125, Japan

Springer–Science+Business Media, B.V.

Library of Congress Cataloging-in-Publication Data

ISBN 978-0-7923-2158-3 ISBN 978-1-4615-3078-7 (eBook)
DOI 10.1007/978-1-4615-3078-7

Printed on acid-free paper

Molecular and Cellular Biochemistry:

An International Journal for Chemical Biology in Health and Disease

CONTENTS VOLUME 119, Nos. 1&2, 1993

CELLULAR FUNCTION AND METABOLISM

Professor Makoto Nagano

This book has been compiled to commemorate the retirement of Professor Makoto Nagano, Head of the Department of Internal Medicine, Aoto Hospital, Jikei University, Tokyo, Japan. Professor Nagano is a founder of the Working Group of Cardiac Structure and Metabolism in Japan and has been a strong promoter for exchange of information between basic scientists and clinical cardiologists. In this regard he organized various conferences and symposia and edited several books and monographs. Professor Nagano discovered a growth for cardiac hypertrophy and was successful in starting a new strain of hamsters (J-2-N) for studying the pathogenesis of cardiomyopathy. He has received many awards including a decoration from the Government of Germany for his research achievements in the field of heart metabolism. In addition, Professor Nagano possesses great human qualities and is a dedicated husband to Dr. Brigitte Nagano and a kind father to their three children.

Molecular and Cellular Biochemistry **119**: 1, 1993.

Preface

Cellular function and metabolism

Various metabolic processes are known to be intimately involved in the maintenance of cellular structure and function. It has also become clear that metabolic events involved in the synthesis and hydrolysis of ATP as well as for the synthesis of proteins and phospholipids are essential for cellular health. The regulation of cell function is generally achieved through participation of a wide variety of hormones and different signal transduction mechanisms for the activation/deactivation of some specific metabolic processes. In this regard cyclic AMP and calcium seem to play a crucial role. Various hormones are also known to affect the genetic machinery of the cell; however, the exact signals for genetic control of cellular function are not well defined. In particular, the sequence of events concerned with remodelling of different types of cells under various pathological situations is poorly understood. In this volume we have therefore dealt with some of these issues from biochemical, molecular biological, physiological, and pharmacological viewpoints. Special emphasis has been laid on understanding heart function and metabolism in health and disease in general, and cardiac hypertrophy, heart failure, and ischemic heart disease in particular. It is hoped that this multidisciplinary information will be of value to basic scientists and clinical investigators.

It is with great pride and honour that we dedicate this focussed issue of Molecular and Cellular Biochemistry to Professor Makoto Nagano on the occasion of his retirement as Director of the Department of Internal Medicine, Aoto Hospital of the Jikei University, Tokyo. This monograph on Cellular Function and Metabolism is based on contributions by Professor Nagano's colleagues and friends and is a most befitting tribute to him because of his unparalleled commitment to understanding metabolic derangements under pathological situations. Dr. Nagano obtained his M.D. and Ph.D. degrees from the Jikei University School of Medicine in 1953 and progressed through the ranks to assume the position of Professor and Director of the Department of Internal Medicine in 1976 at the same institution. He had a vision to develop a strong Department with emphasis on blending basic science and clinical practice and now he stands as tangible evidence of strength in fulfilling its objectives of patient care, education and research. Dr. Nagano has been a source of inspiration and wisdom not only to the members of his Department but also to clinical and experimental cardiologists in Japan and other parts of the world. In his capacity as Executive Secretary (1978–1989) and now as President of the Japanese Section of ISHR, Dr. Nagano has played an important role in the development of the International Society for Heart Research. Furthermore, as President of the Japanese Working Group of Cardiac Metabolism since 1975, as well as membership on various professional societies and editorial boards, he has helped in promoting the scientific basis of cardiology throughout the world.

On behalf of the Department of Internal Medicine of the Jikei University and the Japanese Section of ISHR, we want to thank all colleagues and friends for their scientific contributions. We also appreciate the sincere support of Mrs. Brigitte Nagano for the high achievements of her husband. Special thanks are also due to Naranjan S. Dhalla, Editor-in-Chief, Molecular and Cellular Biochemistry, for inviting publication of these papers.

Yoshio Yazaki,
University of Tokyo
Tokyo, Japan

Seibu Mochizuki,
Jikei University
Tokyo, Japan

Molecular and Cellular Biochemistry **119**: 3–9, 1993.

Control of growth in neonatal pig hearts

Cathy J. Beinlich and Howard E. Morgan
Sigfried and Janet Weis Center for Research, Geisinger Clinic, Danville, PA 17822, USA

Abstract

The pig heart grows at a maximal rate in the first 2–3 days of life due to a volume overload imposed on the heart at birth. Rates of ribosome formation and protein synthesis cannot be further accelerated during *in vitro* perfusion with agents that increase cyclic AMP, that bind to α_1-adrenergic receptors or that bind to angiotensin II receptors. Growth of the heart *in vivo* can be restrained by treatment with an angiotensin-converting enzyme inhibitor, enalapril maleate, or an angiotensin receptor antagonist, DuP 753. In the enalapril-treated heart, norepinephrine plus propranolol, but not angiotensin II, accelerated ribosome formation. Rapid growth of the left ventricle of pig heart during the first 10 days of life is due largely to eccentric hypertrophy. (Mol Cell Biochem **119:** 3–9, 1993)

Key words: cardiac hypertrophy, neonatal heart, ribosome formation, protein synthesis, α_1-adrenergic agonist, cardiac myocytes

Introduction

The mechanisms of cardiac hypertrophy can best be studied in an animal model in which a rapid rate of growth is initiated and rigorous measurements of the rates of protein synthesis and ribosome formation are possible. The neonatal pig heart grows rapidly in the first two weeks of life due to the physiological work load imposed on the left ventricle (LV) at birth. In studies of LV output in late gestation sheep fetuses and the same animals 1 and 4 h after birth, LV output rose from 264 ± 23 ml · min^{-1} · kg body wt^{-1} in fetuses to 444 ± 33 ml · min^{-1} · kg body wt^{-1} at 1 and 4 h after delivery [1]. In fetuses, LV output was less than right ventricular (RV) output which averaged 340 ± 33 ml · min^{-1} · kg body wt^{-1}. LV output was 44% and RV output was 56% of combined ventricular output. As a result of the increase in LV output at birth, LV flow contribution to tissues distal to the ductus arteriosus rose from 51 to 226 ml · min^{-1} · kg body wt^{-1}. These tissues had been perfused mainly by the RV *in utero*. However, average blood flow to body tissues was approximately the same (39 ml · min^{-1} · 100 g tissue^{-1}) in fetuses and newborn lambs. The mean systolic aortic pressure was 51 ± 1 mmHg in fetuses and 54 ± 3 mmHg, 4 h after birth. Mean systolic pressure in the pulmonary artery declined from 52 ± 1 to 37 ± 2 mmHg in 4 h lambs compared to fetuses. This with the onset of pulmonary ventilation, RV output perfused the lungs due to a decrease in pulmonary vascular resistance. Teitel *et al.*, [2] found that lung ventilation was accompanied by a decrease in lower body perfusion which was corrected by increased LV output. Thus, the work of the LV is abruptly increased by approximately 60% due primarily to a volume overload. The factors implicated in increased LV output include faster heart rate [1, 3] increased contractile state of the LV [4, 5] related to an increase in circulating catecholamine levels [6–8] and an increase in LV cavity dimensions [5, 9].

In the first 3 days of life of the pig, weight of the left ventricular free wall (LVFW) increased by 71% while the weight of the right ventricular free wall (RVFW) was unchanged [10, 11]. Over the first 10 days of life, weight of the LVFW increased by 314% while the

Address for offprints: H.E. Morgan, Senior Vice President for Research, Weis Center for Research, Geisinger Clinic, Danville, PA 17822-2601, USA

weight of the RVFW increased 180%. The remainder of this paper will focus on the mechanisms by which increased heart rate, increased LV cavity dimensions and elevated catecholamine levels raise rates of protein synthesis. In addition, measurements of cell volume are used to determine whether the increase in ventricular weight is due to greater cell volume or cell division.

Effects of contraction and cell stretch on protein synthesis

In vivo, changes in the release of neurotransmitters or hormones in association with increased heart work makes a rigorous assessment of the direct effect of mechanical parameters on protein synthesis impossible. As a result, effects of mechanical parameters on generation of intracellular signals and cell hypertrophy have been studied *in vitro*.

Cell stretch has been identified as the mechanical parameter most closely associated with faster protein synthesis and ribosome formation (for review, see 12). These studies were carried out in adult heart muscle and led to the hypothesis that stretch of the ventricular wall initiates a signal transduction cascade that results in faster protein synthesis via faster initiation, elongation and termination of peptide chains on preexisting ribosomes and accelerated formation of new ribosomes and other components of the synthetic pathway. When adult feline myocytes were cultured on a deformable membrane, linear stretch by 10% increased the incorporation of [^{3}H] uridine into RNA and [^{3}H] phenylalanine into protein [13]. These studies demonstrated a direct effect of myocyte stretch on protein and RNA synthesis.

Contracting neonatal rat myocytes hypertrophied during 3 days in culture as compared to non-beating cells due to accelerated rates of RNA and protein synthesis [14, 15]. Because growth was not correlated with the rate or efficiency of contraction, it is not possible to assess whether the 30% increase in heart rate that occurs at birth [1] is sufficient to initiate hypertrophy.

Mechanical stretch and cell swelling initiate several signal transduction cascades due to the presence of stretch-activated ion channels, adenylyl cyclase, phospholipase C and Na^+-H^+ exchanger (for review see 12). More recently, contraction and stretch [16–18] were shown to translocate protein kinase C to the nucleus and to increase the formation of inositol phosphates. The effect of contraction to increase RNA and protein synthesis was inhibited by a 5 nM staurosporine, a blocker of protein kinase C activation, but this concentration of the drug reduced the rate of beating [16]. When cultured neonatal rat cardiac myocytes were allowed to adhere to a deformable silicone culture dish, stretching of the myocytes by 10% increased the incorporation of [^{14}C] phenylalanine after 2 h but led to maximal content for c-fos mRNA after 30 min [17]. When the promoter function of the c-fos gene was analyzed, deletion of the sequences between -227 and -404 base pairs abolished the effect of stretch on transcription of the c-fos gene [18]. Since serum and cAMP responsive elements are located in this region, a desensitization study was carried out with phorbol esters, epidermal growth factor and forskolin to down regulate protein kinase C, tyrosine kinase and cyclic AMP-dependent protein kinase, respectively. Only pretreatment with phorbol esters desensitized the myocytes against stretch. This finding suggested that induction of the c-fos gene by myocyte stretch involved the activation of protein kinase C.

Adrenergic effects on cell growth and RNA and protein synthesis in neonatal myocytes

The increase in circulating catecholamines could accelerate growth and macromolecular synthesis by binding to either α- or β-receptors (for review, see 12). Hypertrophic effects of α_1-adrenergic agonists have been clearly demonstrated in cultured neonatal rat myocytes [16, 19–21]. The growth effect was characterized by increased rates of protein synthesis, myocyte surface area and protein content. DNA content was unaffected by α_1-adrenergic agonists. Rapid activation of protein kinase C in culture rat myocytes suggests that protein kinase C activation represents the link between receptor binding and accelerated transcription [16, 22].

Staurosporine, the protein kinase C inhibitor blocked the increase in protein and RNA contents in cultured rat myocytes following exposure to norepinephrine [16]. A preliminary report indicated that cAMP analogues, forskolin or cholera toxin resulted in hypertrophy of cultured rat myocytes [23]. More recently Bishopric and Kedes [24] found that the β-adrenergic agonist, isoproterenol induced expression of the human skeletal α-actin gene in high-density cultures but not at low density. These studies indicate that adrenergic ago nists will accelerate growth of neonatal myocytes and the α_1-adrenergic activation is likely to be the more important route of signal transduction.

Regulation of growth of the neonatal pig heart

In the first 10 days of life when the LVFW of the neonatal pig heart increased in weight by 3.1 fold, RNA content of this heart portion increased 2.9 fold [11]. The 1.8 fold increase in weight of the RVFW was accompanied by a 1.8 fold increase in RNA of this heart portion. During the first 3 days of life the weight of the LVFW increased 1.7 fold and the RNA content by 2.4 fold [10]. Since only 15 to 20% of total RNA was in ribosomal subunits, the major fraction was in polysomes that were actively synthesizing protein. The hypotheses that have been tested in these experiments are that faster growth of the left ventricle as compared to the right ventricle is due to greater capacity for protein synthesis as indicated by content of total RNA, and that the increase in the rate of protein synthesis depends on a preceding synthesis of new ribosomes. As shown in Table 1, the rates of protein synthesis in the LVFW and RVFW were 50 and 57%, respectively, of the values found in 48 h old animals and were not significantly different [10]. However, ribosome formation was 40% faster in the LVFW than RVFW of these hearts. Rates of protein synthesis in hearts from 48 h pigs increased 43% in the LVFW compared to 13% in the RVFW. Rates of ribosome formation decreased in both the LVFW and RVFW as age increased from 18 to 48 h, but the rate of formation remained 40% greater in the LVFW as compared with RVFW. These findings indicate that the formation of new ribosomes precedes the accelerated rate of synthesis of total heart protein. When hearts from 5 day pigs were perfused *in vitro* with buffer that contained insulin, the rate of protein synthesis in the LVFW was 56% greater than in the RVFW. This increase reflected a 43% greater content of RNA [11]. These findings supported the hypotheses stated above.

Table 1. Effect of age on rates of protein synthesis and ribosome formation in pig heart

Heart portion	Age of animal, h	Protein synthesis % of the LVFW value at 48 h	Ribosome formation
LVFW	18	57.3 ± 2.9 (9)	137 ± 9.5 (7)
RVFW	18	50.1 ± 2.1 (9)	98 ± 10.4 (7)*
LVFW	48	100 ± 2.5 (32)	100 ± 4.3 (28)
RVFW	48	63.4 ± 1.6 (32)*	59.1 ± 2.8 (28)*

Values represent the mean ± SE of the number of animals given in parenthesis. The rate of protein synthesis in the LVFW at 48 h was 4350 ± 108 nmol phenylalanine ä g heart protein^{-1} ä h^{-1}. The rate of ribosome formation in this tissue was 2042 ± 87 nmol phenylalanine ä g ribosomal protein^{-1} ä h^{-1}. Hearts were removed from the piglets and perfused as described earlier [10]. LVFW, left ventricular free wall; RVFW, right ventricular free wall. * $P < 0.05$ vs LVFW. Data were calculated from values given in [10].

In the experiments that have been performed in this laboratory [10, 11, 25–27], measurements of the rate of ribosome formation have been relied upon to detect growth effects during *in vitro* perfusion. Because the rates of protein synthesis were determined primarily by ribosome content and even a doubling of the rate of ribosome formation would not increase ribosome content in 1 h, the rates of protein synthesis reflect rates of ribosome formation and degradation that existed in the preceding several days. The next series of experiments were undertaken to identify factors that would accelerate ribosome formation during *in vitro* perfusion.

The substances that were explored included agents that increase adenosine 3', 5'-cyclic monophosphate (cAMP), that translocate and activate protein kinase C, and that bind to the angiotensin II receptor. Each of these agents were found to increase rate of ribosome formation and/or protein synthesis in isolated perfused rat hearts or cultured neonatal myocytes (for review, see 12). In the perfused rat heart, glucagon, forskolin, and the cAMP phosphodiesterase inhibitor 1-methyl-3-isobutylxamthine (IBMX) increased cAMP levels and accelerated rates of protein synthesis and ribosome formation [28, 29]. In the perfused piglet heart, glucagon increased heart rate, but did not alter the content of cAMP, activity of cAMP-dependent protein kinase, or rates of ribosome formation [26]. Similarly, addition of IBMX increased heart rate but had no effect on ribosome formation. In this case, IBMX raised cAMP content and the activity of cAMP-dependent protein kinase.

Exposure of cultured chick cardiac myocytes to angiotensin II (AII) increased protein synthesis and cell growth [30, 31]. In cultured neonatal rat myocytes, AII increased RNA content by 28% over 3 days of culture but had no effect on growth [16]. As shown in Table 2, addition of 1 μM AII to the perfusate did not alter the rate of ribosome formation in either the LVFW or RVFW of control pigs (untreated) [27]. When present in the perfusate, AII increased the rate of beating, and the contents of ATP and creatine phosphate were maintained [25]. High affinity AII receptors were present in the neonatal pig heart [25].

In cultured cardiac myocytes from neonatal and adult rat hearts, α_1-adrenergic agonists accelerated protein synthesis and cell growth [16, 19–21, 32]. Addition of norepinephrine and propranolol (1 μM each) to the perfusate increased heart rate but did not alter ribosome formation in either the LVFW or RVFW (Table 3) [27]. Tissue contents of ATP, creatine phosphate and cAMP were unchanged in hearts exposed to norepinephrine plus propranolol. These results with IBMX, AII and norepinephrine plus propranolol indicated that at 2–3 days of age rates of ribosome formation in the LVFW and RVFW are unresponsive to stimulation by these agents. A hypothesis was proposed that the lack of effect on ribosome formation was due to the fact that the rates were already maximal.

Because enalapril maleate, an angiotension-converting enzyme inhibitor blocked the pressure overload hypertrophy in rats that followed coarctation of the abdominal aorta [33], the drug was administered to newborn pigs in an attempt to slow the growth of the LV (LVFW plus septum). Treatment of piglets for 3 days with enalapril maleate decreased LV/body weight ratio (19%), LV RNA content (20%), LV total RNA (36%) and LV total protein (23%), [27]. DuP 753, an AII receptor antagonist decreased LV/body wt ratio (16%), LV RNA content (11%), LV total RNA (21%) and LV RNA/DNA (20%) [27]. These parameters were unaffected in the right ventricle. Because both an angiotensin-converting enzyme inhibitor and an AII receptor antagonist restrained growth, AII may be required for rapid growth of the LV in the neonatal pig heart. The major reservations in accepting this conclusion are 1) enalapril lowered mean carotid blood pressure from 70 ± 3 mmHg in untreated pigs to 55 ± 4 mmHg in enalapril-treated pigs and 2) enalapril, but not DuP 753 impaired renal function and resulted in a 4.5–5 fold elevation of serum urea nitrogen, creatinine and uric acid [27]. Serum K^+ was unchanged by drug treatment. Mean serum urea nitrogen was slightly increased in DuP 753-treated pigs. These findings raise the possibility that enalapril may have restrained LV growth indirectly by reducing the work of the heart and modifying the function of other organs. Impairment of renal function did not appear to be required for decreased left ventricular growth by the angiotensin receptor antagonist.

The aim of the next experiments was to determine whether restraint of growth of the LV in enalapril-treated pigs was accompanied by a reduction in the rates of ribosome formation and protein synthesis in the perfused neonatal pig heart (Table 4). After 3 days of enalapril treatment, ribosome formation and protein synthesis in the LVFW were reduced 32.9 and 24.4%, respectively. Protein synthesis was also decreased in the RVFW by 11.3%, but ribosome formation was unchanged. These data indicate that the reduction of LV

Table 2. Effect of angiotension II on ribosome formation in hearts from control and enalapril-treated pigs

Heart portion	Angio-tensin II	Ribosome formation % of value of untreated LVFW	
		Untreated	Treated
LVFW	0	100 ± 4.7 (5)	70.9 ± 8.6* (6)
	+	86.9 ± 13.5	73.1 ± 5.9
RVFW	0	72.1 ± 5.4	64.2 ± 6.5
	+	66.4 ± 8.5	63.7 ± 6.4

Values represent the mean ± SE of the number of animals given in parenthesis. The rate of ribosome formation in the 3 day untreated pigs was 1727 ± 81 nmol phenylalanine ä g protein^{-1} ä h^{-1}. Treated piglets were given 5 mg enalapril maleate daily for 3 days beginning the day of birth. Hearts were removed from untreated or treated piglets and perfused in the absence or presence of AII (1 1M) as descrived earlier [25]. * P < 0.01 vs corresponding values from untreated piglets. Data were calculated from values given in [25].

Table 3. Effect of norepinephrine/propranolol on ribosome formation in hearts from control and enalapril-treated pigs

Heart portion	Norepinephrine/ propranolol	Ribosome formation % of value in untreated LVFW	
		Untreated	Treated
LVFW	0	100 ± 11.3 (10)	82.3 ± 4.4 (22)
	+	109 ± 8.8 (14)	103 ± 5.3 (20)*
RVFW	0	72.2 ± 6.1 (8)	69.8 ± 2.9 (22)
	+	81.8 ± 7.7 (15)	75.2 ± 3.2 (19)

Values represent the mean ± SE of the number of animals given in parenthesis. The rate of ribosome formation in the LVFW of 2 day untreated pigs was 1541 ± 174 nmol phenylalanine ä g protein^{-1} ä h^{-1}. Treated piglets were given 5 mg enalapril maleate daily for 2 days. Hearts from treated or untreated piglets were perfused in the presence or absence of 1 1M norepinephrine and 1 1M propranolol as described earlier [25]. P < 0.02 as compared to corresponding value with no addition to perfusate. Data were calculated from values given in [25].

mass was largely accounted for by reduced rates of ribosome formation and protein synthesis.

The hearts from enalapril-treated pigs were used to reassess the ability of angiotensin II (Table 2) and norepinephrine plus propranolol (Table 3) to accelerate ribosome formation [25]. AII continued to be ineffective in accelerating ribosome formation in either the LVFW or RVFW. As in untreated pigs, AII accelerated heart rate over the hour of perfusion but did not alter tissue contents of ATP and creatine phosphate [25]. Enalapril treatment increased the number of AII receptors by 67% but had no effect on the dissociation constant [25]. These results suggest either that AII does not have a direct effect on ribosome formation or that the onset of the effect is delayed so that it is not apparent during the first hour of exposure. In contrast to AII, addition of norepinephrine (α-adrenergic receptor agonist) and propranolol (β-adrenergic receptor antagonist) restored the rate of ribosome formation in the LVFW to that observed in hearts from untreated pigs (Table 3). There was no effect of these drugs in the RVFW. The stimulatory effect of norepinephrine plus propranolol was completely inhibited by the addition of 100 nM prazosin, an α_1-adrenergic receptor antagonist, and mimicked by 1 μM phenylephrine, a more specific α_1-adrenergic receptor agonist [25]. Enalapril-treatment had no effect on α_1-adrenergic receptor number or affinity as assessed by [^{3}H]prazosin binding to membrane preparations from hearts of treated and untreated pigs [25]. Treatment of piglets for 3 d with prazosin (1 mg, twice daily) decreased LV/body weight ratio (10%) and LV total RNA (14%), but also lowered the mean arterial blood pressure from 62.6 ± 3.3 to 45.7 ± 4.9 mmHg [25]. To assess whether the slower growth of the LV in prazosin-treated piglets was a direct effect of α_1-adrenergic receptor blockade or an indirect effect of decreased heart work, other neonatal pigs were treated with hydralazine, a vasodilator. Hydralazine decreased mean arterial pressure from 64.8 ± 2.5 to 55.0 ± 2.3 mmHg, but had no effect on growth or RNA or DNA contents of either the LV or RV. These data support the hypothesis that norepinephrine is directly involved in growth of the neonatal pig heart by binding to α_1-adrenergic receptors. Similar experiments with agents that increase cAMP content or stretch the ventricular wall have not been carried out in hearts from enalapril-treated pigs in which growth of the left ventricle occurs at submaximal rates.

Table 4. Effect of *in vivo* pretreatment with enalapril maleate on cytoplasmic ribosome formation and total protein synthesis

Heart portion	Enalapril treatment	Ribosome formation	Protein synthesis
		% of value in untreated LVFW	
LVFW	0	100 ± 8.5	100 ± 4.4
	+	67.1 ± 6.0*	75.6 ± 0.4*
RVFW	0	73.4 ± 7.5	59.2 ± 2.0
	+	63.5 ± 5.8	52.5 ± 1.8$^{\delta}$

Values represent the mean ± SE for 10 to 21 observations. The rate of ribosome formation in the LVFW of 3 day untreated pigs was 1609 ± 137 nmol phenylalanine ä g protein^{-1} ä h^{-1}. The rate of protein synthesis was 3805 ± 167 nmol phenylalanine ä g protein^{-1} ä h^{-1} in this same heart portion. Piglets were treated with 5 mg enalapril maleate daily for 3 days beginning with the day of birth. Hearts were removed and perfused as described earlier [27]. * $P < 0.01$; † $P < 0.05$ as compared to values from heart of untreated piglets. Data were calculated from values given in [27].

Contributions of cell division and hypertrophy to growth of the neonatal pig heart

Rapid growth of the LVFW of newborn pigs (3.1 fold) in the first 10 days of life could be due primarily to division of heart cells or to enlargement of the cells present at birth. In the neonatal rat heart, three phases of growth have been described:

a) a hyperplastic phase,
b) a transitional phase and
c) a hypertrophic phase [34, 35].

The hyperplastic phase lasted 4–5 days; the transitional phase with accompanying binucleation of myocytes occurs between days 6 and 14. Subsequent growth was by hypertrophy. Even though heart weight doubled during the first 6 days of age, myocyte volume was unchanged [36]. These findings indicated that growth of the rat heart in the early neonatal period was entirely by an increase in cell number. In man, however, more than 90% of human postnatal growth of the heart can be accounted for by hypertrophy [37]. As shown in Fig. 1, pig myocyte volume increased progressively in the LV and RV. The rate of increase in the LV was 418 ± 28 μm/day compared to 127 ± 20 μm^3 in the RV, a 3.3 fold faster rate of cell hypertrophy. During the first 10 days of life the LVFW of the pig heart increased in mass at a rate that was 3,2 fold faster than the RVFW [11]. At birth, the dimensions of LVFW myocytes were 79.5 ± 1.7 μm by 7.20 ± 0.41 μm; at 2 weeks of age the dimensions were 119 ± 5.4 μm and 9.31 ± 0.44 μm. Myocytes

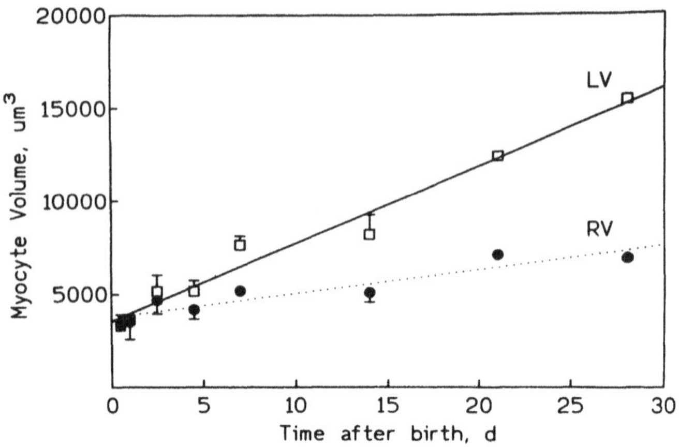

Fig. 1. Myocyte volumes in the left and right ventricular free walls of piglet hearts as a function of age.

Littermate piglets ranging in age from 4 h to 4 weeks were obtained from a local commercial breeder. Before the hearts were excised, piglets were heparinized with sodium heparin (40 mg/kg body weight, i.p.) and anesthetized with sodium pentobarbital (40 mg/kg body weight, i.p.).

Myocytes were isolated as descrived by Kozlouskis *et al.* [38] with slight modification. The heart was flushed for 3 min at a flow rate of 24 ml/min by retrograde coronary perfusion with Joklik media containing 35 mM $NaHCO_3$, and 0.1 mM EGTA, pH 7.4 at 37° C. This was followed by a 20–30 min perfusion with Joklik media containing 0.1% bovine serum albumin (fraction V) and 0.1% collagenase B (Boehinger Mannheim), pH 7.4 at 37° C. The flaccid heart was sectioned into RVFW and LVFW portions. Each tissue section was minced using a sterile razor blade and the tissue digest was transferred to a flask containing Joklik media, 0.1 mM EGTA and 1% bovine serum albumin. The cell suspension was drawn up and down in a wide bore plastic 25 ml pipette twice and filtered through 250 μm nylon mesh. After centrifugation at 70 g for 3 min the resultant cell pellet was resuspended in the same media that had been equilibrated with 95% O_2/5% CO_2 and allowed to sediment at 37° C for 15 min with continuous gassing of the fluid surface. Aliquots of the cell suspension were fixed in 1.5% glutaraldehyde in 0.08 M phosphate buffer, pH 7.4 Cells prepared by this method were generally 60–70% rod shaped. Aliquots of fresh cells were used to prepare cytospins (Shannon Cytospin 2) which were subsequently stained for assessment of number of nuclei.

The length and width of the fixed isolated myocytes from the right and left ventricle were determined using an imaging system (Technology Resources, Nashville, TN). These measurements were used to calculate cellular volume assuming that the myocytes were cylinderical in shape. Measurements were made on 45–50 rod shaped cells, with no evidence of membrane alterations, from each cell preparation. Values during the first 14 days after birth represent the mean ± SE of the average values from 3–9 hearts. The values at 21 days are the average of 2 hearts and the value at 28 days is a single observation. Linear regression of the LV data gave a y intercept of 3538 μm^3 and a slope of 418 μm^3/day. The r value was 0.9867. Linear regression of the RV data gave a y intercept of 3756 μm^3 and a slope of 127 μm^3/day. The r value was 0.9317.

from the RVFW increased in length over this period (77.1 ± 1.86 μm to 104 ± 6.8 μm) but the width remained approximately the same (7.46 ± 0.39 and 7.73 ± 0.33 at birth and 2 weeks respectively). The myocytes of both ventricles were approximately 90% mononucleated up to 4–5 days of age, by 2 weeks approximately half of the cells were binucleated in the LVFW and 30% in the RVFW. Thus, rapid growth of the left ventricle of the pig heart during the first 10 days of life appears to be caused primarily by eccentric hypertrophy, with increased cell width and length, as is typical of volume overload hypertrophy [39].

The neonatal pig heart is an excellent model in which to study the mechanisms of a very rapid rate of cardiac growth both *in vivo* and during *in vitro* perfusion of the heart.

References

1. Smolich JJ, Soust M, Berger PJ, Walker AM: Indirect relation between rises in oxygen consumption and left ventricular output at birth in lambs. Circ Res 71: 443–450, 1992
2. Teitel DF, Iwamoto HS, Rudolph AM: Effects of birth-related events on central blood flow patterns. Pediatr Res 22: 557–566, 1987
3. Klopfenstein HS, Rudolph AM: Postnatal changes in the circulation and responses to volume loading in sheep. Circ Res 42: 839–845, 1978
4. Reimenscheider TA, Brenner RA, Mason DT: Maturational changes in myocardial contractile state of newborn lambs. Pediatr Res 15: 349–356, 1981
5. Anderson PAW, Glick KL, Manring A, Crenshaw C Jr: Developmental changes in cardiac contractility in fetal and postnatal sheep: *In vitro* and *in vivo*. Am J Physiol 247: H371–H379, 1984
6. Comline RS, Silver M: The composition of fetal and maternal blood during parturition in the ewe. J Physiol (London) 222: 233–256, 1972
7. Davidson D: Circulating vasoactive substances and hemodynamic adjustments at birth in lambs. J Appl Physiol 63: 676–684, 1987
8. Padbury JF, Martinez AM: Sympathoadrenal system activity at birth: Integration of postnatal adaptation. Semin Perinatol 12: 163–172, 1988
9. Kirkpatrick SE, Covell JM, Friedman WF: A new technique for the continuous assessment of fetal and neonatal cardiac performance. Am J Obstet Gynecol 116: 963–972, 1973
10. Camacho JA, Peterson CJ, White GJ, Morgan HE: Accelerated ribosome formation and growth in neonatal pig hearts. Am J Physiol 258: C86–C91, 1990
11. Peterson CJ, Whitman V, Watson PA, Schuler HG, Morgan HE: Mechanisms of differential growth of heart ventricles in newborn pigs. Circ Res 64: 360–369, 1989
12. Morgan HE, Baker KM: Cardiac hypertrophy: Mechanical, neural and endocrine dependence. Circulation 83: 13–25, 1991
13. Cooper G IV, Mercer WE, Hoober JK, Gordon PR, Kent RL,

Lauva IK, Marino TA: Load regulation of the properties of adult feline cardiocytes. Circ Res 58: 692–705, 1986
14. McDermott PJ, Morgan HE: Contraction modulates the capacity for protein synthesis during growth in neonatal heart cells in culture. Circ Res 64: 540–553, 1989
15. McDermott PJ, Rothblum LI, Smith SD, Morgan HE: Accelerated rates of ribosomal RNA synthesis during growth of contracting heart cells in culture. J Biol Chem 246: 18220–18227, 1989
16. Allo SN, Carl LL, Morgan HE: Acceleration of growth of cultured cardiomyocytes after translocation of protein kinase C. Am J Physiol 263: C319–C325, 1992
17. Komuro I, Kaida T, Shibazaki Y, Kurabayashi M, Katoh Y, Hoh E, Takaku F, Yazaki Y: Stretching cardiac myocytes stimulates protooncogene expression. J Biol Chem 265: 3595–3598, 1990
18. Komuro I, Katoh Y, Kaida T, Shibazaki Y, Kurabayashi M, Hoh E, Takaku F, Yazaki Y: Mechanical loading stimulates cell hypertrophy and specific gene expression in cultured rate cardiac myocytes. J Biol Chem 266: 1265–1268, 1991
19. Simpson P: Norepinephrine stimulated hypertrophy of cultured rat myocardial cells in an α_1 adrenergic response. J Clin Invest 72: 732–738, 1983
20. Meidell RS, Sen A, Henderson SA, Slahetka MF, Chein KR: Alpha $_1$-adrenergic stimulation of rat myocardial cells increases protein synthesis. Am J Physiol 251: H1076–H1084, 1986
21. Simpson P: Stimulation of hypertrophy of cultured neonatal rat heart cells through an α_1-receptor and induction of beating through an α_1 and β_1-adrenergic receptor interaction. Circ Res 56: 884–894, 1985
22. Henrich CJ, Simpson PC: Differential acute and chronic response of protein kinase C in cultured neonatal rat heart myocytes to α_1-adrenergic and phorbol ester stimulation. J Mol Cell Cardiol 20: 1081–1085, 1988
23. Rocha-Singh KJ, Simpson PC, Karliner JS: Cyclic AMP induces hypertrophy of cultured rat cardiac myocytes. Circulation 80 (Suppl II): II-93, 1989
24. Bishopric NH, Kedes L: Adrenergic regulation of the skeletal α-actin gene promoter during myocardial cell hypertrophy. Proc Natl Acad Sci USA 88: 2132–2136, 1991
25. Beinlich CJ, Baker KM, Morgan HE: α-Adrenergic agonists stimulate ribosome formation in hearts from enalapril-treated piglets. J Mol Cell Cardiol, in press
26. Beinlich CJ, Baker KM, White GJ, Morgan HE: Control of growth in the neonatal pig heart. Am J Physiol Suppl (Oct) 261: 3–7, 1991
27. Beinlich CJ, White GJ, Baker KM, Morgan HE: Angiotensin II and left ventricular growth in newborn pig heart. J Mol Cell Cardiol 23: 1031–1038, 1991
28. Xenophontos XP, Watson PA, Chua BHL, Haneda T, Morgan HE: Increased cAMP content accelerated protein synthesis in rat heart. Circ Res 65: 647–656, 1989
29. Watson PA, Haneda T, Morgan HE: Effect of higher aortic pressure on ribosome formation and cAMP content in rat heart. Am J Physiol 256: C1257–C1261, 1989
30. Aceto JF, Baker KM: [Sar[1]]Angiotensin II receptor-mediated stimulation of protein synthesis in chick heart cells. Am J Physiol 258: H806–H813, 1990
31. Baker KM, Aceto JF: Angiotensin II stimulation of protein synthesis and cell growth in chick heart cells. Am J Physiol 259: H610–H618, 1990
32. Fuller SJ, Gaitanaki CJ, Sugden PH: Effects of catecholamines on protein synthesis in cardiac myocytes and perfused hearts isolated from adult rats. Biochem J 266: 727–736, 1990
33. Baker KM, Chernin MI, Wixson SK, Aceto JF: Renin-angiotensin system involvement in pressure overload cardiac hypertrophy in rats. Am J Physiol 259: H324–H332, 1990
34. Anversa P, Olivetti G, Loud AV: Morphometric study of early postnatal development in the left and right ventricular myocardium of the rat. 1. Hypertrophy, hyperplasia and binucleation of myocytes. Circ Res 46: 495–502, 1980
35. Clubb FJ Jr, Bishop SP: Formation of binucleated myocardial cells in the neonatal rat. Lab Invest 50: 571–577, 1984
36. Clubb FJ Jr, Penney DG, Baylerian MS, Bishop SP: Cardiomegaly due to myocyte hyperplasia in perinatal rats exposed to 200 ppm carbon monoxide. J Mol Cell Cardiol 18: 477–486, 1986
37. Zak R: Development and proliferative capacity of cardiac muscle cells. Circ Res 35(suppl II): 17–26, 1974
38. Kozlovskis PJ, Gerdes AM, Smets M, Moore JA, Bassett AL, Myerburg RJ: Regional increase in isolated myocyte volume in chronic myocardial infarction in cats. J Mol Cell Cardiol 23: 1459–1466, 1991
39. Gerdes AM: The use of isolated myocytes to evaluate myocardial remodeling. Trends Cardiovasc Med 2: 152–155, 1992

Molecular and Cellular Biochemistry **119**: 11–16, 1993.

Role of protein kinase system in the signal transduction of stretch-mediated protooncogene expression and hypertrophy of cardiac myocytes

Yoshio Yazaki, Issei Komuro, Tsutomu Yamazaki, Kazuyuki Tobe, Kouji Maemura, Takashi Kadowaki and Ryozo Nagai
Third Department of Internal Medicine, The Faculty of Medicine, University of Tokyo, 7-3-1 Hongo, Bunkyo, Tokyo 113, Japan

Abstract

To examine the molecular mechanisms by which mechanical stimuli induce protooncogene expression and hypertrophy of cardiac myocytes directly, we cultured rat neonatal cardiac myocytes in deformable dishes and imposed mechanical load on adherent cultured cardiac myocytes by stretching the dishes. Myocyte stretching increased total cell RNA content and mRNA levels of c-fos and skeletal α-actin followed by the aminoacid incorporation into cardiac proteins. CAT assay analysis indicated that the sequences containing serum response element were required for the efficient transcription of c-fos gene by stretching. This accumulation of c-fos mRNA by myocyte stretching was inhibited markedly by down-regulation of protein kinase C. Moreover, myocyte stretching increased inositol phosphate levels. These findings suggests that mechanical stimuli might directly induce protooncogene expression possibly via protein kinase C activation. Furthermore, we observed the activation of MAP kinase by myocytes stretching. This result suggests that MAP kinase activation induced by mechanical stimuli might increase the efficiency of protein synthesis on ribosomes induced by mechanical stimuli. (Mol Cell Biochem **119:** 11–16, 1993)

Key words: protooncogenes, cardiac hypertrophy, mechanical stimulus, protein kinase C, MAP kinase

Introduction

During the process of cardiac hypertrophy, the expression of specific genes, such as protooncogenes and fetal-type genes of contractile proteins, was induced as well as an increase in protein synthesis. The 'immediate early genes' such as protooncogenes and heat shock genes are induced as an early response to pressure overload, and 'late responsive genes' such as fetal contractile protein genes and the atrial natriuretic peptide gene are reexpressed as a later event. Recently, many hormones and growth factors have been reported to induce cardiac hypertrophy and specific gene expression in cultured cardiac myocytes. However, whether hemodynamic overload directly stimulates cellular hypertrophy and specific gene expression in cardiac myocytes without the participation of humoral factors remains unknown.

Address for offprints: Y. Yazaki, Third Department of Internal Medicine, The Faculty of Medicine, University of Tokyo, 7-3-1 Hongo, Bunkyo, Tokyo 113, Japan

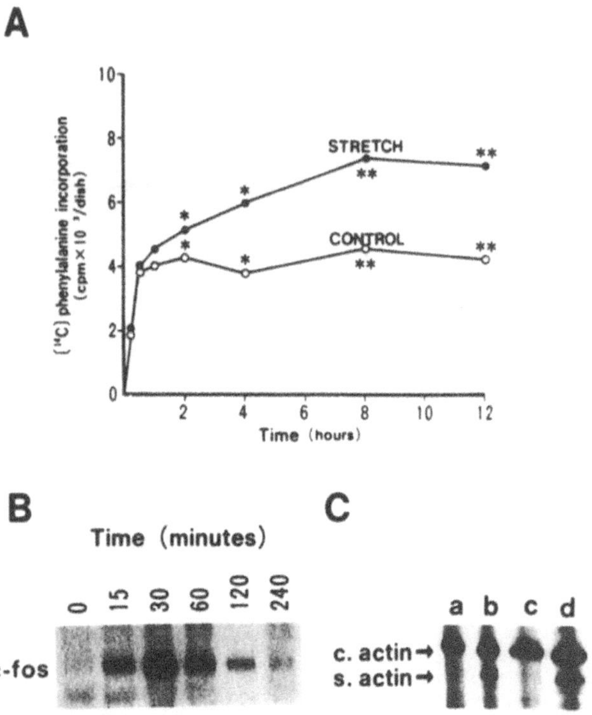

Fig. 1. Stimulation of amino acid uptake and c-fos and skeletal α-actin gene expression by myocyte stretching. A, after 2 days maintained in a serum-free medium, culture dishes were stretched by 10% in length along a single axis and 1 μCi/ml [^{14}C]phenylalanine was added for 30 min prior to processing the cells for intracellular trichloroacetic acid-insoluble radioactivity. Each point represents the mean ± SE from three experiments performed in duplicate. *, $P < 0.05$, **, $P < 0.01$. B, cardiac myocytes were stretched by 10% for 30 min. RNA was extracted and 10 or 20 μg of total RNA (indicated in parentheses) was analyzed by Northern blot hybridization using a 0.8-kb AccI fragment of human c-fos as a probe. C, cardiac α-actin (c. actin) and skeletal α-actin (s. actin) were separated by a primer extension technique. RNA was extracted from neonatal cardiocytes cultured for one (a, b) or two (c, d) days with (b, d) or without (a, c) stretching.

In the present study, to examine whether mechanical stimuli are directly coupled to specific gene expression, we cultured neonatal rat cardiac myocytes in deformable silicone culture dishes with defined serum-free medium, imposed mechanical stimuli by stretching adherent myocytes and examined protooncogene expression. Furthermore, using protooncogene induction by stretching, we studied the signal transduction pathway of mechanical stimuli on cardiac myocytes. We also observed the activation of MAP kinase which is proposed to activate S6 kinase of rebosomes resulting in increased efficiency of protein synthesis.

Mechanical load stimulates cell hypertrophy and specific gene expression

We devised deformable culture dishes to impose mechanical stimuli directly on cardiac myocytes. Whole culture dishes were made of silicone. The bottom of the dish is 1 mm-thick, and it is highly transparent because of no inorganic filler in either component. We mechanically expanded the dishes with the plastic frame and increased their length uniaxially. Following expanding the length of the dishes, attached cardiac myocytes were stretched. The resting length of the myocytes was increased parallel to the axis of expansion by the same percent length as the dish [1]. This method allowed us to do more detailed analysis, such as quantitative assessment of mRNA levels, because we could obtain larger scaled samples.

We prepared primary cultures of cardiac myocytes from the ventricles of 1-day-old Wistar rats. A cardiac myocyte-rich fraction was obtained by preplating method. Myocytes not attached to the preplated dishes were plated into laminine-coated silicone dishes at a field density of 1×10^5 cells/cm^2. A nonmuscle cell-rich fraction was obtained by preplating the cells into silicone dishes for the first hour.

The effect of myocyte stretch on amino acid incorporation into cardiac proteins was shown in Fig. 2A. To avoid the effect of serum, we performed this experiment after 2 days in the serum-free, chemically defined medium. Myocytes were stimulated by 10% increase in the length of the attached dishes. At this point, more than 90% of cells were beating. The incorporation of [^{14}C]phenylalanine was significantly increased 2 h after stretch and the stimulation was maintained for over 12 h (Fig. 1A). This finding suggests that mechanical stress stimulates cardiac cellular hypertrophy.

To ascertain whether mechanical stress induces specific genes, such as protooncogenes and fetal-type isogenes of contractile proteins as observed in the heart *in vivo,* we examined the expression of c-fos and skeletal α-actin gene. Northern blot analysis revealed that c-fos was rapidly and transiently expressed by stretching myocytes. The level of c-fos mRNA was increased as early as 15 min after the passive stretch of myocytes, and reached the maximum level at 30 min, followed by a decline to the undetectable level (Fig. 1B). The kinetics of this c-fos expression by stretching is the same as those when cells are stimulated with serum or growth factors. This protooncogene expression was observed abundantly in cardiac myocyte-rich fraction rather than in

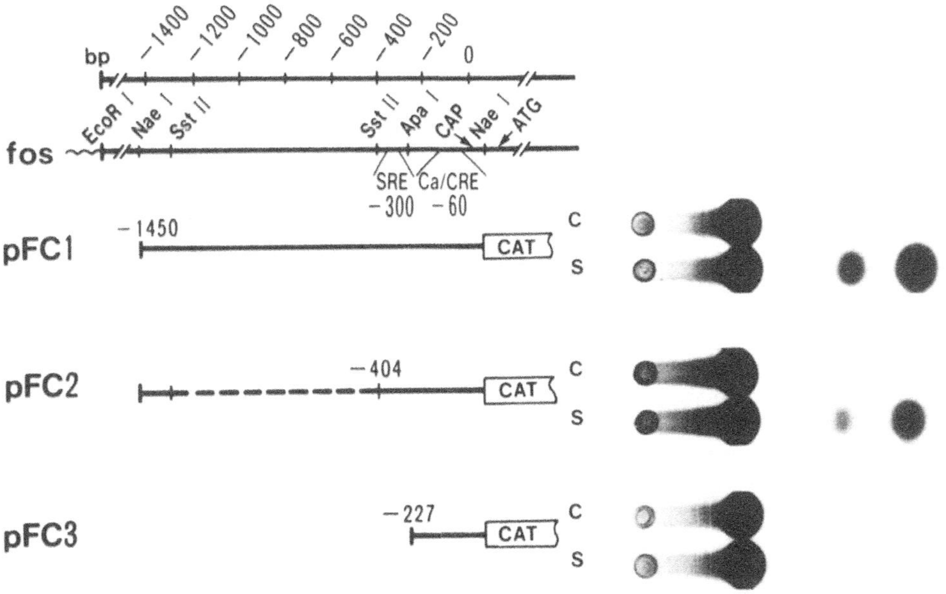

Fig. 2. CAT activity of c-fos recombinants containing deletions in the 5'-flanking region [2]. Cardiac myocytes were transfected with 1 μg of DNA per dish and stretched for 2 h. Cell extracts were prepared and assayed for CAT activity. C, control cells; S, stretched cells; SRE, serum response element; Ca/CRE, calcium/cAMP response element.

nonmuscle cell-rich fraction. This result confirmed that the stimulation of c-fos gene expression by stretching occurred in cardiac myocytes. The induction of c-fos mRNA depended on the extent of expansion of the dishes. The stimulation of c-fos gene expression was recognized by 5% increase in the length of the dishes. The maximum stimulation was obtained by 20% of stretch.

The level of skeletal α-actin mRNA was also accumulated after the passive stretch of myocytes. Skeletal α-actin mRNA was significantly increased 4 h after stretching, and gradually accumulated up to 2 days during stimulation (Fig. 1C). Because it has been known that acute pressure overload induces cardiac hypertrophy and the gene expression, such as protooncogenes and fetal-type isogenes of contractile proteins in the heart *in vivo*, our observations revealing the expression of c-fos and skeletal α-actin gene by myocytes stretching, suggested that stretching cardiac myocytes *in vitro* could substitute for hemodynamic overload *in vivo*.

Transcriptional activation of c-fos gene by myocyte stretching

To examine whether the c-fos gene expression by myocytes stretching was regulated at the transcriptional level or post-transcriptional level, we analyzed its promoter function by CAT assay method [2]. We linked the 5' flanking region of the fos gene including its promoter, to the 5' end of the chloramphenicol acetyltransferase (CAT) encoding sequences in the plasmid and transfected this plasmid into primary cultures of neonatal rat cardiac myocytes and measured CAT activity of the cell extracts. The pSVO CAT construct containing the entire coding sequences of the procaryotic CAT gene minus its promoter showed very little CAT activity in either the absence or presence of stretch stimulation. By contrast when pSVO fos CAT, which contained the 5' c-fos flanking region, was introduced into the system, myocytes stretching for 48 h reproducibly caused more than a 7-fold increase in CAT activity but there was little activity without stretching. When the pSV2CAT construct, which contained SV40 enhancer and early promoter sequences, was introduced into myocytes, a large amount of CAT activity was observed. However, additional activity was not obtained after stretching. Furthermore, the run-on study using myocyte nuclei also revealed the accumulation of c-fos mRNA by stretching. These results suggested that the c-fos gene expression by stretching was regulated at the transcriptional level and that the stretch response element was located in the 5' flanking region of the c-fos gene.

These results revealed that mechanical stress markedly induced the expression of c-fos protooncogene

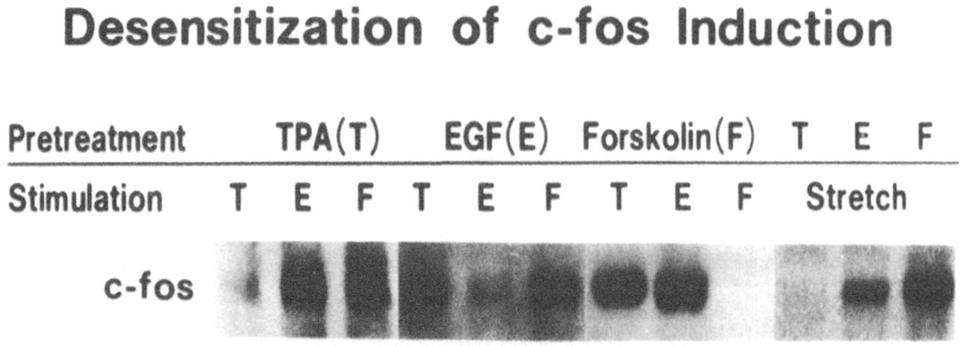

Fig. 3. Desensitization, down-regulation, and pharmacological study of c-fos induction [2]. A, myocytes were pretreated with 10 ng/ml TPA (T), 10 ng/ml epidermal growth factor (EGF E), or 1 μM forskolin (F) for 4 h, then stimulated by the same doses of these inducers or stretching for 30 min. B, cells were stretched in the presence and absence of the indicated inhibitors for 30 min. RNA was extracted, run on a gel, and transferred to the filters. The filters were hybridized with the c-fos probe. S. stretched cells.

without the participation of humoral factors. Therefore, hemodynamic overload itself seems to be one of the main factors to stimulate the expression of the c-fos gene in the heart *in vivo*. Recently, Fos, the protein which the fos gene encodes, was elucidated to be localized in the nucleus and to bind to the 12-0-tetradecanoylphorbol-13-acetate-responsive elements of some genes, followed by the activation of their gene transcription in cooperation with the transcription factor AP-1. These observations suggest that some early responsive gene products like Fos may stimulate other subsequent gene expressions in the heart under conditions of hemodynamic overload.

To identify the sequences essential for the transcription of the fos gene induced by stretching, we analyzed effects of deletion of the 5' flanking region of the gene on CAT activity.

Mechanism of c-fos gene expression by myocyte stretching

Deletion mutagenesis of the 5'-flanking region of the c-fos gene indicated that the sequences between − 227 and − 404 base pairs were required for the efficient transcription of the fos gene by myocyte stretching (Fig. 2) [10]. Since it has been known that there were serum and cAMP responsive elements in this region, we hypothesized that three known factors, cAMP, protein kinase C and tyrosine kinase, are involved in c-fos gene stimulation by stretching.

Thus, we carried out a desensitization study to determine which protein kinase system plays a central role in the signal transduction induced by mechanical stress. We pretreated myocytes with either phorbol esters (TPA), epidermal growth factor (EGF) or forskolin for 24 h to down-regulate individual protein kinases, C kinase, tyrosine kinase or A kinase, respectively, and then treated again with one of these inducers. Thirty minutes after the treatment, we assessed the mRNA level of c-fos with Northern blot analysis. Our results showed that, after the pretreatment with TPA, c-fos stimulation was not obtained only with TPA. Again, the pretreatment with EGF or forskolin induced desensitization against the stimulation with EGF or forskolin, respectively. The most important observation was that only the pretreatment with TPA desensitized myocytes against stretch (Fig. 3A). These results suggest that the induction of c-fos gene by myocyte stretching might be caused by the activation of protein kinase C.

To confirm this possibility, we examined the effect of protein kinase C inhibitors on the expression of c-fos by myocytes stretching. H-7 strongly inhibited c-fos mRNA induction by stretching, whereas H-1004 inhibited it weakly, depending on their Ki value for protein kinase C. Staurosporin also strongly inhibited c-fos induction (Fig. 3B). Furthermore, the treatment of myocytes with TPA induced both c-fos and skeletal α actin mRNA.

Finally, to examine the mechanism for the activation of protein kinase C by myocyte stretching, we measured phosphatidyl inositol turnover after myocyte stretching. Immediately after stretching, the activation of phosphatidyl inositol turnover was observed in myocytes. One minute after stretching, inositol monophosphate and biphosphate significantly increased and reached about two-fold levels of control after 5 min (Fig. 4). We could not detect the elevation of inositol triphosphate levels at either time point. However, these results suggest that mechanical stress might stimulate

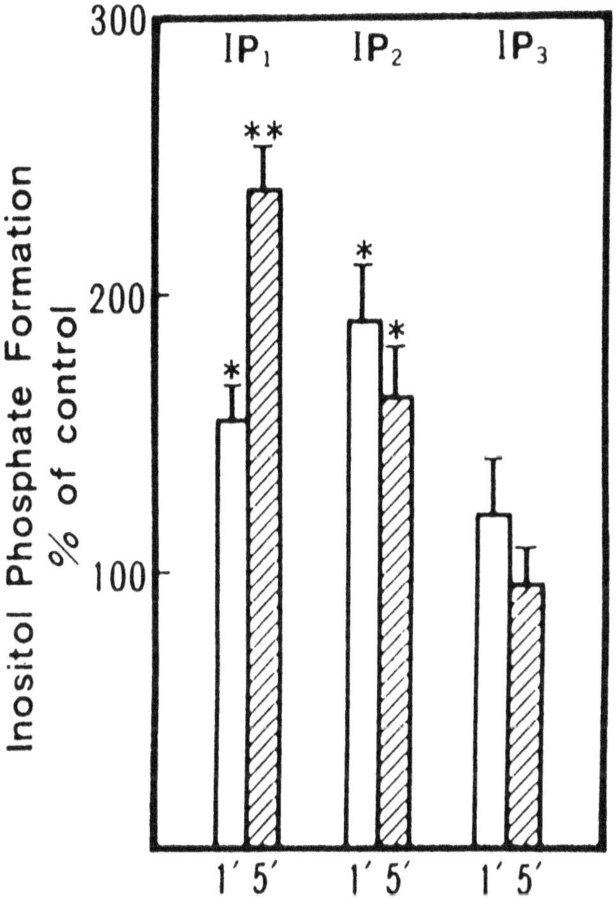

Fig. 4. Effect of stretching on the levels of inositol phosphates [2]. Myocytes were incubated for 24 h in Dulbecco's modified Eagle's medium containing myo [2 ^{3}H]inositol and stretched 10% for the indicated times. The water-soluble inositol phosphates were separated by column chromatography. Each histogram represents the average percentage of control from five experiments performed in duplicate (mean ± S.E.). Control values (cpm × 10^{-2}/dish) were as follows: inositol monophosphate (IP_1), 160–350; inositol bisphosphate (IP_2), 30–42; inositol triphosphate (IP_3), 62–93. Statistical significance was determined by analysis of variance [39]. *$p < 0.05$, **$p < 0.01$.

protein kinase C activity via phospholipase C activation in cardiac myocytes. Recently, mechanical stress has been reported to induce prostagrandin production in skeletal and endothelial cells via phospholipase C pathway. These reports might support our hypothesis. However, further investigation is necessary to elucidate the precise molecular mechanisms how mechanical load activates phospholipase C or protein kinase C [2].

Activation of MAP kinase by myocyte stretching

Mitogen-activated protein kinase, MAP kinase is a serine/threonine protein kinase, which can be activated by a variety of stimuli such as growth factors and TPA, and is proposed to be a general intracellular signaling molecule to link between events at cell surface and those in the nucleus and ribosome. Recent reports indicated that MAP kinase can phosphorylate and activate S6 kinase of ribosome resulting in increased efficiency of protein synthesis [3, 4]. Thus, in order to further define the intracellular signals by which mechanical stimuli result in increased protein synthesis, we examined whether mechanical stimuli can activate MAP kinase.

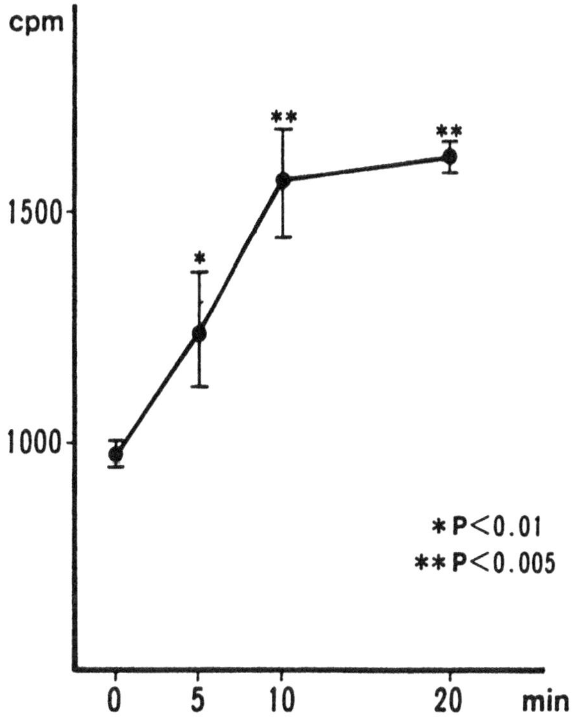

Fig. 5. Activation of MAP kinase by myocyte stretching. Activities of MAP kinase were measured in cell lysates by the phosphorylation of MBP as substrate. The phosphorylation of MBP was stimulated maximally 10–20 min after stretching by approximately 1.8 fold.

We labeled cardiac myocytes with P^{32} orthophosphate and stretched, solubilipd and immunoprecipitated with specific antibodies against MAP kinase. As shown in this slide, stretching myocytes increased phosphorylation of the protein of 43 KDa by approximately 6-fold. Other bands were not increased. This protein was identified to the 43 KDa MAP kinase. MAP II kinase by western immunoblotting.

To address the question, whether increased phosphorylation of MAP II kinase by stretching is indeed associated with increased kinase activity, we measured kinase activity of cell lysate by the phosphorylation of myelin basic protein, MBP as substrate. The phosphorylation of MBP was stimulated maximally 10–20 min after stretching by approximately 1.8 fold (Fig. 5).

To directly determine whether the phosphorylation of MBP is induced by MAP kinase itself, cell lysates were electrophoresed on SDS-polyacrylamide gels containing MBP. After renaturation, the gels were incubated with γ-P^{32} ATP and magnesium. We observed an increased phosphorylation of MBP at 43 KDa after stretching. This increased phosphorylation was also observed when myocytes were treated with TPA. These data demonstrate that stretching myocytes increased 43 KDa MAP kinase, MAP II kinase activity. We also observed the phosphorylation of MBP at 70 KDa, but the identity of this kinase activity is unknown at present.

Our results strongly suggested that MAP kinase is stimulated by mechanical stress Via PKC activation, and this activated MAP kinase probably phosphorylates and stimulates S6 kinase of ribosome accompanied by an increased protein synthesis. Recently, MAP kinase has been shown to phosphorylate and activate the protooncogene product, c-Jun. Therefore, the induction of c-fos and possibly activation of c-jun Via MAP kinase activation by stretching, can synergetically activate the function of AP 1 comples.

Acknowledgement

This investigation was supported in part by a Grant-in-aid for Scientific Research from the Ministry of Education, Science and Culture of Japan, Grants from the Ministry of Welfare and Uehara Memorial Foundation, and for Basic Reserach on Cardiac Hypertrophy from the Vehicle Racing Commemorative Foundation.

References

1. Komuro I, Kaida T, Shibazaki Y, Kurabayashi M, Katoh Y, Hoh E, Takahu F, Yazaki Y: Stretching cardiac myocytes stimulates protooncogene expression. J. Biol. Chem. 265: 3595–3598, 1990
2. Komuro I, Katoh Y, Kaida T, Shibazaki T, Kurbayashi M, Hoh E, Takaku F, Yazaki Y: Mechanical loading stimulates cell hypertrophy and specific gene expression in cultured rat cardiac myocytes. J. Biol. Chem. 266: 1265–1268, 1991
3. Ahn NG, Krebs EG: Evidence for an epidermal growth factor-stimulated protein kinase cascade in swiss 3T3 cells. J. Biol. Chem. 265: 11495–11501, 1990
4. Ahn NG, Seger R, Bratlien RL, Diltz CD, Tonks NK, Krebs EG: Multiple components in an epidermal growth factor-stimulated protein kinase cascade. J. Biol. Chem. 266: 4220–4227, 1991.

Molecular and Cellular Biochemistry **119**: 17–22, 1993.

Growth factor for cardiac hypertrophy

Makoto Nagano, Tadanari Ohkubo, Tohru Arino, Masahito Tsuchiya and Makoto Nagai
Department of Internal Medicine, Jikei University School of Medicine, Katsushika-ku, Tokyo, Japan

Abstract

Cardiac size can be regulated by the balance in activity between cardiac growth factors and inhibiting factors, chalones. This study was undertaken to verify the role of the cardiac growth factor and its purification from hypertrophied hearts. For this propose the hypertrophied hearts of renovascular hypertensive rats were used. The purification was made by using an isoelectric focusing chromatography and the HPLC method. We examined the cardiac growth effect of the isolated fractions with cultured chicken embryonic cardiac myocytes. Simultaneously, the influence of these fractions on the cardiac cell cycle was examined by DNA analysis with the flow cytometric method. If the hearts were overloaded due to hypertension, the growth of the cardiac size could be induced by increased the level of five proteins with different molecular weight and with an isoelectric point of 8.3. The significant growth activities were observed at these five proteins compared to the absence of the fractions. For the appearance of these growth effect, it is necessary that the structure of the protein is there fundamentally as a form with a molecular weight of 27 k dalton. After addition of these isolated fractions, BrdU content is S and G2 phases by flow cytometry was increased. This change indicates that the cardiac myocytes are stimulated in form DNA synthesis. (Mol Cell Biochem **119:** 17–22, 1993)

Key words: cardiac growth factor, cardiac myocyte, chicken embryonic heart, cell culture, cell cycle

Introduction

Cardiac myocytes proliferate throughout fetal and early postnatal development and the proliferation ceases at approximately 2 weeks of age. Cardiac growth after 2 weeks of age primarily occurs through an increase in myocytes size rather than number.

It is of very high interest but unknown, what substance in the embryonic heart induces the proliferation of the cardiac myocytes and what substance that first appears in the postnatal heart stops the hyperplasia of cardiac myocytes.

On the other side, cardiac hypertrophy is induced by pressure and volume overloads. The increase in cardiac size is essentially cell enlargement due to an increase in the rate of protein synthesis at a molecular level.

But it is not known which biological substance induces the growth of the cardiac myocytes.

In our previous study [1–3] we have demonstrated that the protein with molecular weight of 43 k dalton and with isoelectric point of 8.3 appears in hypertrophied hearts and stimulates cardiac size in chicken embryonic hearts *in vivo* and in cultured rat cardiac myocytes *in vitro*.

The present report describes the purification of the cardiac growth factor to homogeneity and effects of the

Address for offprints: M. Nagano, Department of Internal Medicine, Jikei University School of Medicine, Katsushika-ku, Aoto 6-41-2, Tokyo 125, Japan

cardiac growth factor, including the effect on the cell cycles.

Material and method

The examined growth factors were obtained from hypertrophied rat hearts. Cardiac hypertrophy was induced by stenosis of the right renal artery and by nephrectomy of left kidney of male Wistar rats (G-B-rat). As control rats, normal Wistar rats with sham operation were used. The heart weights of the G-B rats (above 0.45 g/body weight 100 g) were markedly greater than those of the control rats (under 0.33 g/b.w. 100 g).

Table 1. Purification of cardiac growth factor from the hypertrophied heart muscle. The ventricles were homogenized in 10 volume of HEPES buffer (20 mMol HEPES, 0.54 Mol NaCl, pH 7.35 adjusted with Tris). The homogenate was centrifuged. The supernatant was dialyzed in persaturated (NH4)2SO4 solution. All precipitate was dialyzed in aqua dest. The resoluble proteins after dialysis were fractionated with isoelectric focusing chromatography.

Purification of Protein

Heart (ca. 15 g/routine)
Wash-out of blood by coronary perfusion
↓
Homogenate by Polytron after mince
10 sec x 2 time at dial 6
(1g/10ml tissue)
↓
x 100,000 g 1hr
(supernatant fluid)
↓
dialysis in persaturated $(NH_4)_2SO_4$ solution
(pH7.4) 2 days
↓
after the desalting and solvent change
(by PD-10 column)
1st purification by Rotofor
Separate to 20 fractions
↓
dialysis in persaturated $(NH_4)_2SO_4$ solution
(pH7.4) 2 days and pooled
↓
after the desalting and solvent change
Fraction No.17
↓
2nd purification by Superdex 75 column
Separate to 6 fractions
↓
dialysis in persaturated $(NH_4)_2SO_4$ solution
(pH7.4) 2 days

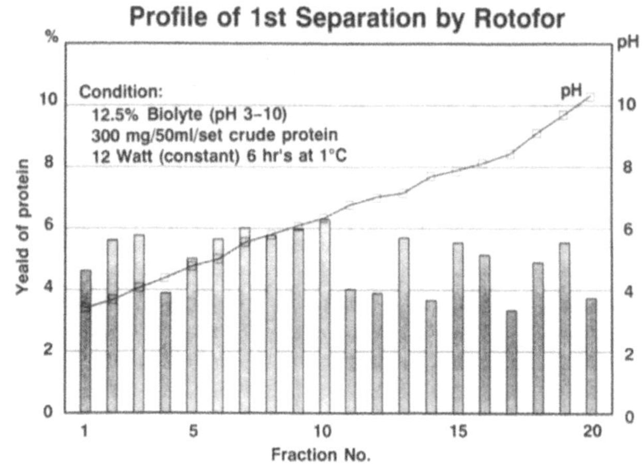

Fig. 1. The isoelectric focusing chromatography by Rotofor. Voltage: started with 400 V. Power: 12 w constant. Cooling: A refrigerated circulation water bath to keep the coolant temperature at 3° C. The run was completed in 5.5 hr. The P.I. (isoelectric point) can be determined by running the sample in the Rotofor cell using a broad range of 3–10 ampholyte.

Purification of cardiac growth factor from the heart muscle

The hypertrophied hearts were obtained from G-B rats, 5–6 weeks after operation. The hearts were removed under pentobarbital anesthesia and perfused with cold saline through coronary perfusion for washout of blood. The ventricles were homogenized in 10 volumes of buffer. As shown in Table 1, the homogenate was centrifuged at 100.000 × g for 60 min and the white lipids layer of the supernatant was filtrated off. The supernatant was dialyzed in persaturated ammonium sulfate solution (pH 7.4). The precipitate was dialyzed against water. The resoluble samples after dialysis were fractionated with the method of isoelectric focusing chromatography by Rotofor (Bio-Rad). Figure 1 shows the profile of the 1st separation by Rotofor chromatography. The fraction numbered 17, I.P. (isoelectric point) 8.3 was applied to Superdex 75 column of HPLC.

Bioassay system for the cardiac growth effect [4]

For the determination of the cardiac growth effect, a bioassay system of cultured chicken cardiac myocytes were used. The cardiac myocytes were isolated from 9–11 days old chicken embryos and cultured in HAM-F12 medium containing 10% calf serum at 37° C in a condition of 95% O_2 and 5% CO_2. After 24 hr the

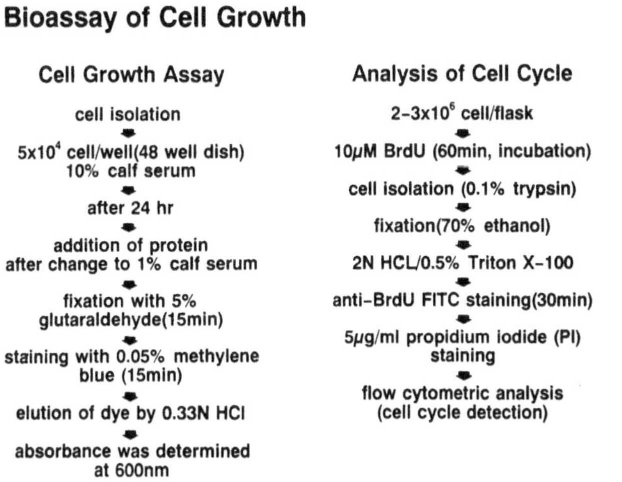

Fig. 2. Bioassay of cell growth. Cell growth assay and analysis of cell cycle (for details, see text).

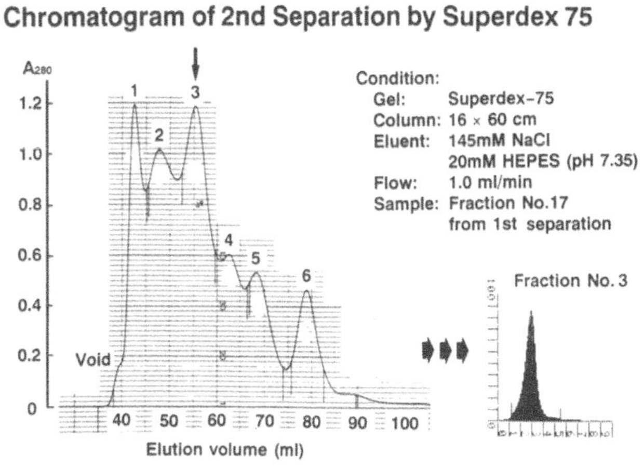

Fig. 3. A separation pattern of fraction with I.P. of 8.3 by high performance liquid chromatography (HPLC). Right bottom shows monoclonal fraction of fraction 3.

medium was removed and fresh medium containing 1% calf serum was added.

The cells were incubated in the presence or absence of the isolated fractions by HPLC for 72 hr at 37° C.

At the end of the incubation, the cells were fixed with glutaraldehyde and stained with methylen blue, then this indicator eluted from cells was determined at 600 nm (Fig. 2).

Analysis of the cell cycle with bromodesoxyuridine (BrdU) [5]

The chicken embryonic cardiac myocytes were cultured in medium added 10 μM BrdU for 60 min. Then the isolated cells after trypsin treatment were fixed with 70% ethanol. As shown in Fig. 2, the cells in G1, G2, S and M phases were determined by flow cytometric method.

Statistical analysis

As statistical analysis the Student's t test was used. Values are mean + SD. Significant: $p < 0.05$.

Results

Separation of the monoclonal protein fractions

Figure 1 shows the pattern of the first separation by Rotofor chromatography. The fraction numbered 17, corresponding to I.P. 8.3, was applied to Superdex 75 column of HPLC, Pharmacia, HPLC system.

Figure 3 shows a separation pattern of I.P. 8.3 fraction by HPLC. The vertical axis shows UV absorption at 280 nm. By HPLC the fraction with I.P. 8.3 was separated 6 subfractions.

The molecular weights were the following

Fraction-1(F-1): over 150 k dalton, fraction-2(F-2): 117 k dalton, fraction-3(F-3): 72 k dalton, fraction-4(F-4): 45 k dalton, fraction-5(F-5): 27 k dalton and fraction-6(F-6): 16 k dalton. Each subfraction was isolated as a monoclonal spot. For example, Fig. 3 shows a monoclonal pattern of F-3.

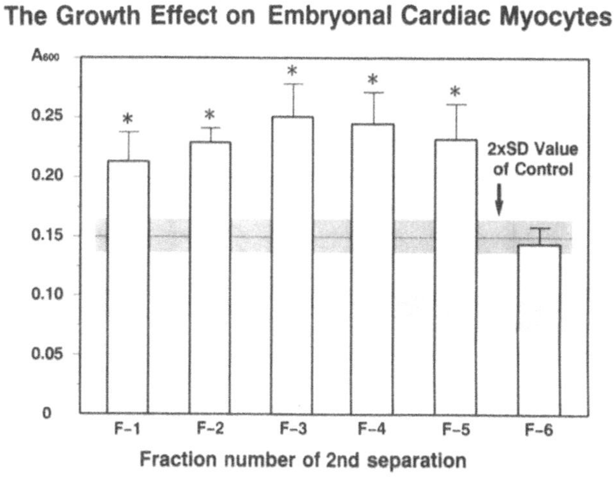

Fig. 4. The growth effect of each fraction on the embryonic chicken cardic myocytes. Number of each fraction, n = 6. Mean + SD, * $p < 0.05$.

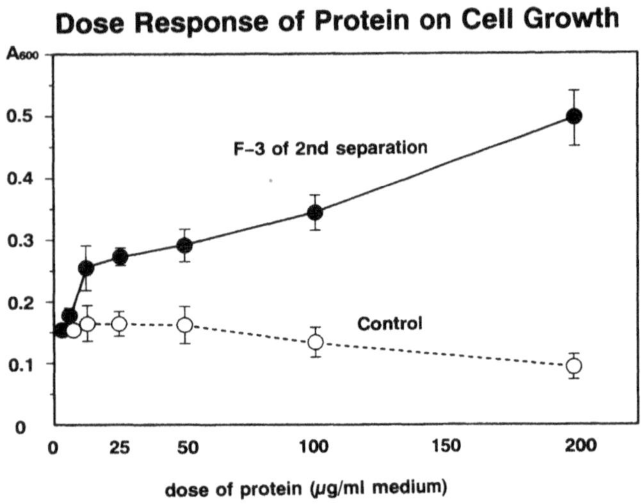

Fig. 5. Dose response curve of fraction 3. The vertical line shows dose of protein (μg/ml) and the horizontal axis shows absorption of methylene blue at 600 nm, corresponding to cell numbers.

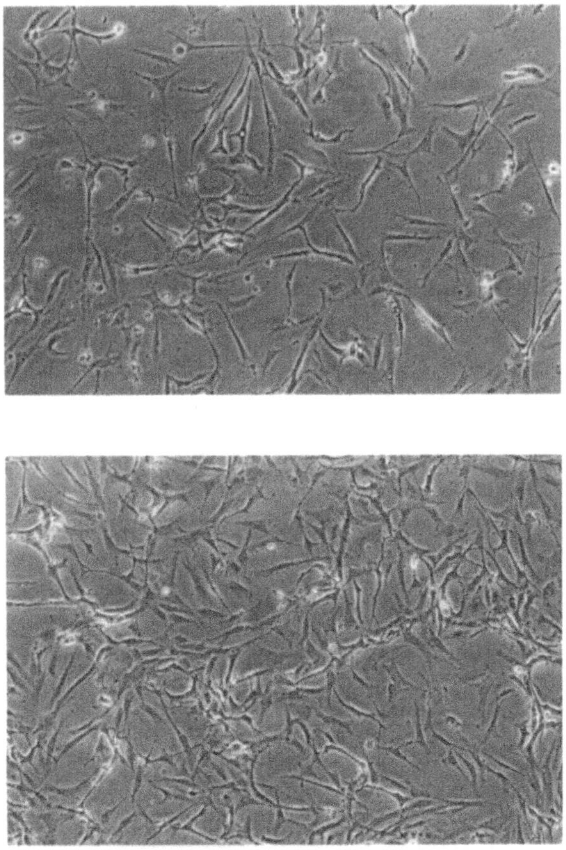

Fig. 6. The phase contrast microscopic pictures of chicken embryonic cardiac myocytes. Fraction 3 stimulated the growth of cardiac cells: the top shows cell growth in the absence of F-3 and the bottom in the presence.

Growth effect on chicken embryonic myocyte

After 72 hr the indicator eluted from cultured cells in the presence of absence of the fractions from no. 1 to no. 6 was determined. The significant growth activities were observed at F-1, 2, 3, 4 and 5 compared to the rate in a control assay system without addition of the fractions (Fig. 4).

Only the F-6 has shown non-effect on the stimulation of cell growth. At this bioassay, 100 μg/ml of each fraction was added to incubated medium.

For the determination of minimum doses, F-3 was tested for the dose-response curve. As shown in Fig. 5, F-3 already stimulated cell growth with 10 μg/ml corresponding to 0.13 μmol.

Figure 6 shows that F-3 stimulated the growth of cardiac myocytes of the chicken embryo. A significant stimulation in the increasing rate of cell numbers was observed compared to the rate in absence of the fraction.

Figure 7 shows the change of DNA synthesis by flow cytometric method. After adding F-3 into the culture medium, an increase of BrdU content in S and G2

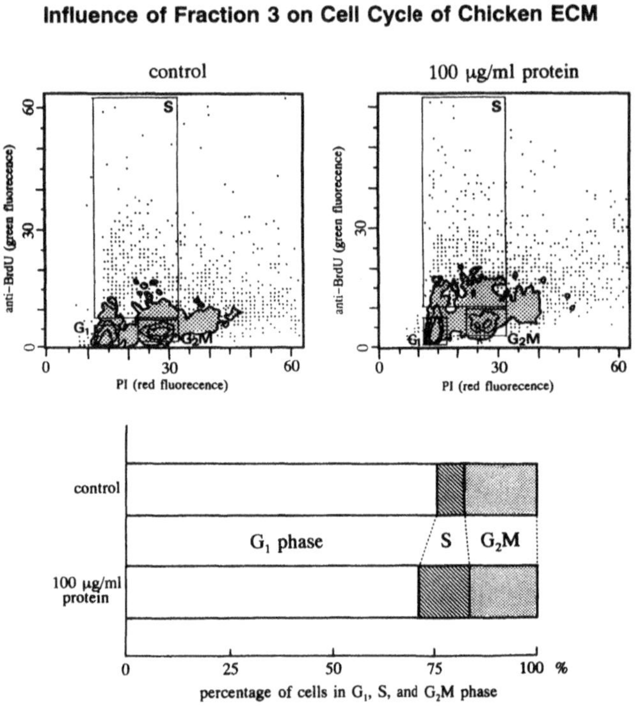

Fig. 7. The effect of fraction 3 on the cell cycle of the chicken embryonic cardiac myocytes. The left top shows control and the right top shows the stimulating effect of F-3 on the cell cycle. The bottom shows percent change of cells in each phase.

phases was observed compared to that in absence of this fraction.

Discussion

Cardiac mass can be regulated under the interaction between the stimuli and the inhibits.

In fetal and early postnatal development, cardiac myocytes proliferate and this proliferation ceases at approximately 2 weeks of age. Cardiac growth after 2 weeks of age primarily occurs through an increase in myocyte size rather than number [6].

It is of very high interest but unknown, what substance in the embryonic heart induces the growth of the cardiac myocytes and what substance that first appears in the postnatal heart stops the hyperplasia of cardiac myocytes. On the other side, the cardiac hypertrophy is induced by pressure and volume overloads. The increase in cardiac size is essentially cell enlargement due to an increase in the rate of protein synthesis at a molecular level. In our previous studies [1, 2], we have demonstrated that a factor which we have named 'cardiac growth factor' appears in the hypertrophied hearts and stimulates cardiac size in chicken embryonic hearts *in vivo* and in cultured rat cardiac myocytes *in vitro*.

From hypertrophied rat hearts, we have isolated five fractions with molecular weights 27, 45, 72, 117 and over 150 k daltons. These fractions with I.P. 8.3 stimulate cardiac cell growth with 10 μg/ml corresponding to 0.1 μmol levels. These fractions were analyzed through SDS-polyacrylamide Gel electrophoresis and two protein fractions with lower molecular weights were fundamentally detected in each chromatogram. These details will be discussed at a later date. For this reason, it is quite likely that the growth factor is fundamentally constructed from these two protein components, and it is necessary for the appearance of the growth effect that the structure of the protein is there fundamentally as a form with a molecular weight of 27 k dalton. The other fractions with a molecular weight over 72 k dalton also have growth activities. But these proteins may be constructed as combinations of the growth factor with the other structure proteins in cells.

In our previous study [3], we have reported that the cardiac growth factor is a protein with molecular weight close to 43 k dalton. From these new findings, we concluded that the growth factor may have a molecular weight of 45 k dalton *in vivo*.

In the first study [3], we observed that the myocardial DNA and RNA contents and the DNA/RNA ratio of the chicken embryo injected with a fraction of a MW above 30 k dalton from G-B hypertensive rat hearts increased compared with the extract from normal control rat hearts. The absolute increase of cardiac DNA content suggests that this cardiac growth effect appears in form of cardiac enlargement with polyploid cardiac myocytes and cardiac hyperplasia. As shown in Fig. 6, F-3 remarkably stimulated the growth of cardiac myocytes in the chicken embryo. A significant stimulation in the increasing rate of cell numbers was observed compared to the rate in absence of the fraction. This cardiac growth factor induced the hyperplasia of cardiac myocytes.

In a separate experiment, this cardiac growth factor also induced cell enlargement and hyperplasia of H9C2.

In the normal rat hearts, S and G2 phases of the cardiac cell exist only minimally. If the heart is overloaded, S and G2 phases in myocytes increase 3–4 days after cardiac overload. The changes are seen later than the appearance of proto-oncogenes in the overloaded hearts.

It is very interesting in what time after addition of F-3 fraction, the cell cycle of chicken embryonic myocytes is stimulated. Figure 7 shows the change of DNA synthesis by flowcytometry. S and G2 phases in the cell cycle do not appear within 42 hr, only 42 hr after addition of F-3. After addition of F-3, BrdU contents in S and G2 phases were increased. This change indicates that cardiac myocytes of the chicken embryo are stimulated in form DNA synthesis. It is also very interesting in what stage after cardiac overload the cardiac growth factor will appear. The proteins with I.P. of 8.3 reaches its maximum within the fifth to sixth postoperative week and then gradually decreased. Contrary to these findings, the I.P. 7.1 fraction, which can also be found in the normal rat heart, decreases around the fourth to sixth week when the heart is hypertrophied. This fraction can inhibit the growth of cardiac myocytes. According to the presence of the cardiac growth and inhibiting factors, we conclude that cardiac size can be regulated by the balance in activity between the cardiac growth and inhibiting factors.

In previous studies, there are several examples of humoral factors involved in the regulation of the cardiac size. These factors are insulin, insulin-like growth factor, IGF-1 [7], growth hormone [8], thyroid hormone [9], angiotensin II [4], norepinephrine [10], PDGF [11].

Recently, myotrophine isolated from SHR-rats has been reported [12]. These above substances inducing cardiac hypertrophy are substances with other molecular weight and other I.P. than our protein.

In our first report, this protein could neither be identified in the normal rat and in the old SHR hearts. According to the HPLC method, however, it is present to a certain extent. In the young SHR and the young cardiomyopathic Syrian hamster hearts, this protein can be identified. This protein is therefore not exclusive property of the hypertrophied heart of Goldblatt hypertension rats.

Acknowledgements

This work was supported by the Grant of Japanese Ministry of Education, no. 02670409, 1990/1991.

References

1. Nagano M, Ohkubo T, Suzuki H: Growth factor and its role for cardiac hypertrophy. In: Jacob, Seipel, Zucher (eds.) Cardiac Dilatation. Gustav Fischer Verl, Stuttgart, New York, 1990, pp 81–88
2. Suzuki H, Nagai M, Ohkubo T, Nagano M: Cardiac growth factor in the hypertrophied rat heart. Jikeikai Med J 37 (Suppl): 97–106, 1990
3. Suzuki H, Ohkubo T, Nagano M: Cardiac growth factor. Jikei Med J (Japanese) 103: 1441–1448, 1988
4. Baker KM, Aceto JF: Angiotensin II stimulation of protein synthesis and cell growth in chick heart cells. Am J Physiol 259: H610–H618, 1990
5. Dolbearà F, Gratzner N, Pallaricini MG, Gray WJ: Flow cytometric measurement of total DNA content and incorporated bromodesoxyuridine. Proc Natl Acad Sci 80: 5573–5577, 1983
6. Bugaisky L, Zak R: Biological mechanisms of hypertrophy. In: Fozzard, Haber, Jennings, Katz, Morgan (eds.) The Heart and Cardiovascular System-Scientific Foundations. Raven Press, New York, 1986, pp 1491–1506
7. Tsushima T: Insulin-like growth factor. In: Ohno, Umeda, Kaji, Kasuya, Toshima (eds.) Cellular and Molecular Biology. Part II. Asakua-shoten, Tokyo, 1987, pp 25–30
8. Clemmons DR, van Wyk JJ: Somatomedin, Physiological control and effect on cell proliferation. In: Basegra (ed.) Tissue Growth Factors. Springer Verl, Berlin, Heidelberg, New York, 1981, pp 161–198
9. Morkin E, Flink IL, Goldman S: Biochemical and physiological effects of thyroid hormone on the cardiac performance. Proc Cardiovas 25: 435–464, 1983
10. Simpson P: Stimulation of hypertrophy of cultured neonatal rat heart cells through an a1 adrenergic receptor and induction of beating throug an a1 and β1 adrenergic receptor interaction: Evidence for independent regulation of growth and beating. Circ Res 56: 884–894, 1985
11. Ross R: The platelet-derived growth factor. In: Basegra (ed.) Tissue Growth Factors. Springer Verl, Berlin, Heidelberg, New York, 1989, pp 133–157
12. Sen S, Kundu G, Mekhail N, Castel J, Misono K, Healy B: Myotrophin: Purification of a novel peptide from spontaneously hypertensive rat heart that influences myocardial growth. J Biol Chem 265: 16635–16643, 1990

Molecular and Cellular Biochemistry **119**: 23–28, 1993.

Effect of angiotensin coverting enzyme inhibitor on regression in cardiac hypertrophy

Naoki Makino, Hirosuke Matsui, Kazuhiro Masutomo, Tomoji Hata and Takashi Yanaga
Department of Bioclimatology and Medicine, Medical Institute of Bioregulation, Kyushu University, Beppu, Japan

Abstract

Cardiac hypertrophy in rats was produced by aortic banding for 6 weeks and regression of hypertrophy in these experimental animals was induced by administration of angiotensin converting enzyme inhibitor, enalapril (10 mg/kg/day) for 6 weeks. The left ventricular muscle mass and systolic pressure were decrease upon treating the hypertrophied rats with enalapril. This drug also decreased the number of α1-adrenoceptors in hypertrophyied myocardium without any changes in β-adrenoceptors. The regression of cardiac hypertrophy in spontaneously hypertensive rats by enalapril for 10 weeks was not associated with any alterations in α1-adrenoceptors in hypertrophied myocardium, but was decreased in β-adrenoceptors. Effects of enalapril on extracellular matrix in the myocardium was also observed in regression of hypertrophy in which the type III collagen mRNA expression and collagen contents were reduced in comparison with those of hypertrophied myocardium. These results indicate that regression of cardiac hypertrophy is not always associated with a decrease in the number of α1-adrenergic receptors and that the beneficial effects of enalapril in the hypertrophied heart in aortic banding animals may be of some specific nature. (Mol Cell Biochem **119**: 23–28, 1993.

Key words: cardiac hypertrophy, cardiac regression, enalapril, collagen expression, adrenergic receptor, rat heart

Introduction

Myocardial hypertrophy serves a compensatory function to maintain wall stress within the normal range. Usually, it occurs in hemodynamic overload but it may also be initiated by nonhemodynamic stimuli. Numerous experimental studies in hypertensive animals adressed the regression of left ventricular (LV) hypertrophy under various conditions. In spontaneously hypertensive rat (SHR) captoril and alpha-metyldopa reserved LV hypertrophy, the latter regardless of blood pressure control. Hydralazine controlled blood pressure without affecting LV hypertrophy and minoxidil actually increase cardiac weight despite lowering blood pressure [1]. These findings give strong evidence that hemodynamic effects are not the only determinants for regression of LV hypertrophy. Other factors such as neurohumoral systems may be of additional importance. The involvement of the sympathetic system nervous system in mediating development and regression of LV hypertrophy varies in different animal models of hypertension. In view of the suspected role of adrenergic mechanisms in cardiac hypertrophy [2, 3], several studies have examined the status of both alpha and beta adrenergic receptors during the development of hypertrophy [4–6]. Since an angiotensin converting en-

Address for offprints: N. Makino, Department of Bioclimatology and Medicine, Medical Institute of Bioregulation, Kyushu University, 4546 Tsurumihara, Beppu 874, Japan

zyme inhibitor (ACE) has been reported to reverse the LV hypertrophy due to reduction in blood pressure and peripheral resistance [7–9], it was thought of interest to study changes in both alpha and beta adrenergic receptors in hypertrophied heart upon treating the aortic banding rats with ACE inhibitor, enalapril. In contrast to systemic arterial hypertension, LV afterload in this experimental model is determined predominantly by the fixed obstruction of the aorta and therefore cannot be reduced significantly by pharmacologic intervention. A possible reduction of LV mass would therefore reflect nonhemodynamic effects of enalapril. In addition, SHR treated with enalapril were used for the purpose of comparison.

Methods

Animal model

LV hypertrophy was produced by banding of the abdominal aorta in rats anesthetized with pentobarbital sodium (40 mg/kg, i.p.). The abdominal aorta was surgically isolated above the renal vessels and constricted by using the ligature-needle technique as described previously [10]. Sham-operated rats underwent a similar surgical procedure except for the aortic banding. Male SHR and normotensive Wister Kyoto (WKY) rats (22 weeks old) were also used in this study.

Experimental protocol

Cardiac regression was induced 6 weeks after aortic banding in rats. ACE inhibitor, enalapril was added in drinking water and given ad libitum for an additional 6 weeks. Tap water was given to the sham control or untreated aortic banding rats. It was estimated that treated rats received about 10 mg/kg of enalapril daily by oral intake. Cardiac regression for SHR was studied by the administration of enalapril at the same concentration from 22 weeks old of age for 10 weeks as described for the aortic banding rats. All animals were kept under similar conditions and fed rat chow.

Hemodinamics

The left ventricular pressure as well as positive and negative dP/dt were measured by introducing a transducer-tippled No. 2 catheter (Miller, Houston, Tx) in the LV before killing rat. Systolic and end-diastolic pressure as well as positive and negative dP/dt max were determined from recording made at a paper speed of 125 mm/sec. To eliminate differences caused solely by differences in developed pressure, both of positive and negative dP/dt max were normalized with respect to the peak developed pressure to reveal intrinsic attributes of the LV muscle. After hemodynamical measurements, rats were killed by decapitation, and the LV was immediately removed, cleaned, weighed and prepared for the determination of adrenoceptors, collagen contents and collagen mRNA expressions.

Preparation of membrane

The membrane preparation from cardiac muscle was isolated by the method described previously [10]. The final pellet was resuspended in 50 mM Tris-HCl (pH 7.4), at a concentration between 1.0 and 2.0 mg of protein per ml. The concentration of protein was measured by the method of Lowry *et al.* [11]. The membrane protein yield from control hearts varied from 1.16 to 1.20 mg/g heart wet weight and this was not different from that for the experimental group. Furthermore, no significant ($p > 0.05$) difference was seen between the ouabain sensitive Na^+-K^+ ATPase activity [10] between the control ($24.6 \pm 0.5\ \mu$mol pi/mg/hr, $n = 6$) and experimental ($23.1 \pm 0.8\ \mu$mol pi/mg/hr) preparations.

Radioligand binding assay for α1 and β-adrenoceptors

To determine β-receptor binding, membrane were incubated with increasing concentration of [125I]cyanopindolol in total volume 0.25 ml for 60 min at 30° C in the presence and absence of 10 μM/L 1-propranolol [12]. α1-adrenoceptor binding was analyzed in a manner similar to the preceding one except [3H]prazosin was used in the presence and absence of 10 μM/L phentolamine hydrocloride [13]. All incubations were terminated by rapid vacum filtration through Whatman GF/C glass filters (Whatman Inc., Crifton NJ). The filter was rinsed a three aliquots (5 ml) of ice cold 50 mM Tris-HCl buffer. Counting was done in a ligand scintillation spectrometer with triton/toluen based scintillation cocktail for α1-adrenoceptor binding, while a gamma-counter was used for β-adrenoceptor. Saturation data were calculated by Scatchard analysis using linear regression of bound/free versus bound ligand in order to obtain the maximum number of binding sites (Bmax) and the dissociation constant (Kd).

RNA Isolation and Northern blotting

Total cellular RNA was isolated by the methods outlined by Chirgwin *et al.* [14]. RNA was quantitated by absorbance at 260 nm and 310 nm. Quantification was confirmed by size determination of ribosomal bands that had been fractionated on agarose-formaldehyde gel stained with ethidium bromide. RNA (30–50 μg) was electrophoresed on a 1% agarose, 6% formaldehyde gel and transferred to a nitrocellulose filter. The filter was prehybridized for 4–8 hours at 42° C. Hybridization was performed using [32P]d-ATP (0.5–1.0 × 10 cpm per lane) of labelled probe (specific activity; 4–1.0 × 10 cpm/μg). The cDNA probes in this experiment were α2 (I) collagen and α1 (III) collagen (kindly provided by Dr. de Crombrugghe) and a β actin probe. After hybridization, the nitrocellulose filter was washed according to methods described by Thomas [15]. The filter was autoradiographed on to X-ray film and densitometric analysis was performed.

Determination of collagen

Separation of collagen from dry defatted tissue and hydrolysis of collagen to hydroxyproline was carried out by the method of Bergman and Loxley [16]. Hydroxyproline content was multiplied by a factor of 7.46 to estimate the content of collagen.

Reagents

[3H]prazosin, as 7-methy-[3H]prazosin (specific activity, 80.9 Ci/mmol), was obtained from New England Nuclear. Phentolamin hydrochloride (regitine) was obtained from Ciba-Geigy, Japan Ltd. [125I]iodocyanopindolol was obtained from New England Nuclear and had a specific activity of ~2.250 count per min. All other chemicals were purchases from Sigma Chemical (St. Louis, MO).

Statistics analysis

The Student's t-test employed to determine the statistical significance of the differenced between two groups of data. For comparison of more than two groups, multiple analysis of variance was carried out, and Duncan's new multiple range test was used to determine the significance. All values were expressed as mean (SEM), and $P < 0.05$ was considered significant.

Results

The data in Table 1 indicate that aortic banding for 6 weeks resulted in a significant increase in LV weight ($p < 0.05$) as well as LV to body weight ratio. Although systolic blood pressure in these experimental animals was increased significantly, no change in heart rate was evident (data were not shown). When enalapril was administrated to aortic banding rats from 6 week after surgery for 6 weeks, LV weight and LV to body weight ratio in aortic banding animals (12 weeks) were significantly decrease ($p < 0.05$), compared with those in rats without the treatment. Blood pressure in treated group was also significantly decreased ($p < 0.05$). The both of LV weight and LV to body weight ratio in SHR revealed to decrease in the group treated for 12 weeks with enalapril for compared with untreated group of SHR. Effect of enalapril on blood pressure also showed to be evident in group of the treatment as well as results of experiments in aortic banding rats.

Adrenoceptor binding studies was indicated in Table 2. The maximum number (Bmax) of α1-adrenoceptor, as measured by [3Hprazosin] specific binding, was in-

Table 1. Left ventricular weight, left ventricule/body weight ratio and systolic blood pressure in experimental groups

Group		LV weight (g)	LV/body weight (10^{-3})	SBP (mmHg)
Sham	6 w	0.73 ± 0.02	2.23 ± 0.03	131 ± 7
AB	6 w	0.98 ± 0.02*	3.03 ± 0.11*	192 ± 8*
Sham	12 w	0.75 ± 0.04⁺	2.38 ± 0.05⁺	167 ± 10⁺
AB	12 w	1.13 ± 0.06*	2.97 ± 0.12*	216 ± 14*
AB (E)	12 w	0.98 ± 0.03*⁺	2.61 ± 0.11*⁺	181 ± 11⁺
WKY	20 w	0.88 ± 0.08	2.08 ± 0.09	121 ± 11
SHR	20 w	0.98 ± 0.06	2.18 ± 0.10	186 ± 10*
WKY	32 w	0.93 ± 0.07⁺	2.13 ± 0.08⁺	146 ± 15⁺
SHR	32 w	1.03 ± 0.07*	2.64 ± 0.08*	191 ± 13*
SHR (E)	32 w	0.89 ± 0.07⁺	2.15 ± 0.09⁺	167 ± 10⁺

Values are mean ± SE of 6 experiments. Enalapril (E) (10 mg/Kg/day) was administered for 6 weeks to aortic banding (AB) rats from 6 weeks after operation, whereas in spontaneously hypertensive rats (SHR) was administered from 22 weeks of age to 32 weeks of age. * $p < 0.05$: compared with sham-operated rats or Wistar Kyoto (WKY) rats at the same weeks of age. ⁺ $p < 0.05$: compared with untreated AB rats or untreated SHR at the same weeks of age. SBP: systolic blood pressure.

creased in membrane of aortic banding rats at 6 weeks after surgery. On the contrary, Bmax of α1-adrenoceptor was significantly decreased ($p < 0.05$) by the treatment with enalapril for 6 weeks. Similar data was obtained from Bmax in membrane isolated from hearts of SHR treated with enalapril. Kd was not significantly different from any groups. Although β-adrenoceptor binding study was no difference between treated and untreated rats undertaken by aortic banding, Bmax of β-adrenoceptor in SHR revealed to decrease by the administration of enalapril for 12 weeks in spite of no change of Kd.

The effects of enalapril on hemodynamics were also studied aortic banding animals treated with enalapril for 6 weeks. It can be seen from Table 3 that positive and negative dP/dt increased in the hearts of untreated aortic banding rats compared to the sham-operated rats. Furthermore, these functional parameters were not significantly altered in rat hearts treated with enalapril. When normalized to the developed pressure, positive and negative dP/dt values were similar in the sham-operated and enalapril treated rats.

Northern blot analysis was performed on total RNA in hearts obtained from sham-operated rats, untreated aortic banding rats, and treated aortic banding rats. As seen in Fig. 1, steady state mRNA levels for cardiac type III (α1) collagen were greater in hearts of untreated banding rats than that of sham-operated rats. This expression was slightly decreased in hearts treated with enalapril in comparison with that of banding rats without the treatment. However, there was not different in type I collagen (α2) of hearts among sham-operated, untreated banding and treated banding rats (data not shown). Myocardial collagen concentration, defined by the amount of collagen in milligram per gram of heart tissue estimated from the quantification of hydroxyprolines, significantly increased ($p < 0.05$) in hearts of rats 12 weeks after aortic banding (1.84 ± 0.25 μg/mg wet wt. n = 4), compared with collagen contents in hearts of sham-operated rats (1.06 ± 0.12 μg/mg wet wt. n = 4). By the way, the administration of enalapril for 6 weeks significantly decreased ($p < 0.05$) collage contents in heart with aortic banding (1.14 ± 0.12 μg/mg n = 4). That was no difference from those of sham-operated rats.

Discussion

This study demonstrates that enalapril leads to moderate blood pressure control with a reduction in myocardial hypertrophy. The obvious conclusion is that cardiac α1-adrenergic receptors may play an important role in the response of ventricular mass to the control of blood pressure in experimental cardiac hypertrophy produced by aortic banding. These observations in aortic banded rats are similar to those of other investigators [7, 9]. Although different findings have been reported by β-adrenoceptors in cardiac hypertrophy induced by aortic banding [2, 17], our results have been shown no alteration in the β-adrenergic system in either development or the regression of cardiac hypertrophy through a drug intervention.

In the present study we have shown that α1-adrenergic receptors were altered in cardiac hypertrophy at 6 and 12 weeks of aortic banding in rats whereas β-

Table 2. Left ventricular α- and β-adrenoceptors in experimental animals

Group		α1-adrenoceptor		β-adrenoceptor	
		Kd (pmol)	Bmax (fmol/mg)	Kd (pmol)	Bmax (fmol/mg)
Sham	6 w	69 ± 4	108 ± 4	87 ± 5	108 ± 5
AB	6 w	70 ± 6	125 ± 5*	86 ± 4	109 ± 6
Sham	12 w	77 ± 4	112 ± 5	92 ± 4	104 ± 6
AB	12 w	88 ± 5	132 ± 3*	90 ± 4	114 ± 7
AB (E)	12 w	71 ± 4	118 ± 4⁺	88 ± 4	120 ± 6
WKY	20 w	118 ± 7	118 ± 5	87 ± 4	78 ± 5
SHR	20 w	98 ± 6	194 ± 7*	91 ± 3	118 ± 5*
WKY	32 w	108 ± 6	121 ± 6	92 ± 4	74 ± 4
SHR	32 w	99 ± 5	152 ± 9*	90 ± 5	98 ± 4*
SHR (E)	32 w	104 ± 8	148 ± 6*	94 ± 6	76 ± 4⁺

Values are mean ± SE of 6 experiments. Experimental conditions are the same as in Table 1. * $p < 0.05$: compared with sham-operated rats or WKY rats at the same weeks of age. ⁺ $p < 0.05$: compared with untreated AB rats (12 w) or untreated SHR (32 w) at the same weeks of age.

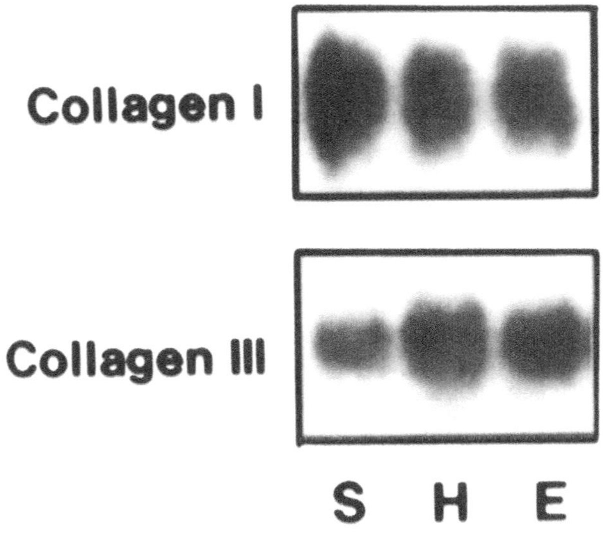

Fig. 1. Data of Norther blotting for collagen type I (α2) and type III (α1) in the myocardium. S: sham-operated rat heart 12 weeks after operation. H: aortic banding rat hearts 12 weeks after operation. E: the administration of enalapril to aortic banding rat hearts for 6 weeks.

adrenergic receptors were increased in hearts from 20 and 32 weeks old SHR. An increase in α1-adrenergic receptors without any change in β-adrenoceptors was seen in prehypertensive stage whereas a decrease in α1-adrenergic receptors and an increase in β-adrenergic receptors was seen in early cardiac hypertrophy due to aortic banding in rats [2]. On the other hand, cardiac hypertrophy due to thyroid hormones has been shown to be associated with decreased α-adrenoceptors and increased β-adrenoceptors [18, 19]. Thus it appears that alterations in adrenergic receptors in the myocardium are dependent upon the stage as well as the types of stimulus leading to the development of cardiac hypertrophy.

The results in this study reveal that regression of cardiac hypertrophy in aortic banded rats upon treatment with enalapril was associated with a decrease in the number of α1-adrenergic receptors in the cardiac membrane. Such an alteration in α1-adrenergic receptors was not due to any artifact because no change in β-adrenergic receptors was evident under similar conditions. The depression of the α1-adrenergic receptors in aortic banded hearts by enalapril does not seem to be due to the decrease in cardiac mass per se because the regression of cardiac hypertrophy was not associated with any change in α1-adrenergic receptors when aortic banded animals were treated with other agents (data not shown). Since treatment of SHR with enalapril decreased cardiac mass without any change in α1-adrenergic receptors in the myocardium, it appears that the reduction in α1-adrenergic receptors in aortic banded rats by enalapril is specific with respect to this experimental model. These findings are consistent with previous reports from several laboratories that the reversal of increased blood pressure and cardiac mass as well as changes in other hemodynamic parameters by different interventions is dependent upon the type of experimental model of cardiac hypertrophy [3, 9].

Results of Northern blot analysis presented here demonstrated that type III collagen (α1) mRNA expression was seen in hearts at 12 weeks after aortic banding operation. Administration of enalapril for 6 weeks also decreased type III collagen expression in hearts of aortic banding rats. This suggests that the rearrangement of the cardiac extracellular matrix was underwent in hypertrophied hearts, and mRNA expres-

Table 3. Effects of enalapril on cardiac function in experimentally induced hypertrophy of rat hearts

	Sham	Untreated	Enalapril
Heart rate (bpm)	394 ± 19	410 ± 20	397 17
LVSP (mmHg)	157 ± 11	218 ± 9*	176 ± 9**
LVEDP (mmHg)	2.2 ± 0.8	4.6 ± 0.7	3.5 ± 0.7
+dP/dt max (mmHg)	4420 ± 253	5413 ± 224*	5200 ± 480
+dP/dt/DP	26.6 ± 1.2	25.2 ± 1.9	29.4 ± 1.7
−dP/dt (mmHg)	2540 ± 222	3313368*	3100 ± 246
−dP/dt/DP	16.0 ± 0.8	15.2 ± 0.8	17.5 ± 0.4

Each value represents mean ± SE (n = 6). Concentration of enalapril was 10 mg/Kg/day. Drug was administered daily for 6 weeks after inducing hypertrophy by aortic banding for 6 weeks. Sham (control) and untreated (hypertrophied) rats were given tap water and maintained for 12 weeks under the conditions similar to the treated rats with enalapril. LVSP: left ventricular systolic pressure; LVEDP: left ventricular end- diastolic pressure; DP: developed pressure. Values of dP/dt/DP represent the percent of peak systolic pressure $\times 10^{-2}$. * p F< 0.05 vs sham-operated rats. ** p < 0.05 vs untreated rats.

sion of collage type III may be contributed to be the development hypertrophy. Our results here were supported by Weber *et al.* [20]. In addition, the data presented here also illustrated that the proportion of type I and type III collagen may be at least partially regulated at the level of the collagen mRNA because this expression was relevant to the function of the heart [20]. Furthermore, since enalapril decreased expression of type III collagen mRNA of aortic banding rats, renin-angiotensin system may be important role in the regression of cardiac hypertrophy.

References

1. Sen S: Regression of cardiac hypertrophy: experimental animal model. Am J Med 75 (suppl 3A): 87–93, 1983
2. Ganguly PK, Lee SL, Beamish RE, Dhalla NS: Altered sympathetic system and adrenoceptors during the development of cardiac hypertrophy. Am Heart J 118: 520–528, 1989
3. Sen S, Trazi RC: Regression of myocardial hypertrophy and influence of adrenergic system. Am J Physiol 244 (Heart Circ Physiol 13): H97–H101, 1983
4. Limas CJ: Increased number of β-adrenergic receptors in the hypertrophied myocardium. Biochem Biophy Acta 588: 174–178, 1979
5. Limas CJ, Limas C: Reduced number of β-adrenergic receptors in the myocardium of spontaneously hypertensive rats. Biochem Biophys Res Commun 83: 710–714, 1978
6. Ayobe MH, Tarazi RC: Reversal of changes in myocardial β-receptors and inotropic responsiveness with regression of cardiac hypertrophy in renal hypertensive rat (RHR). Cir Res 54: 125–134, 1984
7. Kromer EP, Riegger GA: Effects of long-term angiotensin converting enzyme inhibition on myocardial hypertrophy in experimental aortic stenosis in the rat. Am J Cardiol 62: 161–163, 1988
8. Asmer RG, Dannier B, Santoni JPH, Laurent ST, London GM, Levy BI, Safar ME: Reversion of cardiac hypertrophy and reduced arterial compliance after converting enzyme inhibitor in essential hypertension. Cir Res 78: 941–950, 1988
9. Nakashima Y, Fouad FM, Tarazi RC: Regression of left ventricular hypertrophy from systemic hypertension by enalapril. Am J Cardiol 53: 1044–1049, 1984
10. Nakanishi H, Makino N, Hata T, Matsui H, Yano K, Yanaga T: Sarcolemmal Ca transport activities in cardiac hypertrophy caused by pressure overload. Am J Physiol 257 (Heart Circ Physiol 26): H349–H356, 1989
11. Lowry OH, Rosenbrough NJ, Farr AL, Randall RJ: Protein measurement with the Folin phenol reagent. J Biol Chem 193: 256–275, 1951
12. Snavely MD, Mahan LC, O'Connor T, Insel A: Selective down-regulation of adrenergic receptor subtypes in tissues from rats with pheochromocytoma. Endocrinology 113: 354–361, 1983
13. Corr PB, Crafford WA: Enhanced α-adrenergic responsiveness in the myocardium: Role of α-adrenergic blocker. Am Heart J 102: 605–614, 1981
14. Chirgwin JJ, Przbyla AE, MacDonald RJ, Rutter WJ: Isolation of biologically active ribonucleic acid from sources enriched in ribonuclease. Biochemistry 18: 5294–5302, 1979
15. Thomas PS. Hybridization of denaturated RNA transferred or dotted to nitrocellulose paper. Method Enzymol 100: 255–266, 1983
16. Bergman J, Loxley R: Two improved and simplified methods for the spectrophotometric determination of hydroxyproline. Anal Chem 36: 1961–1965, 1963
17. Tamai J, Hori M, Kagiya T, Inoue M, Kamada T: Role of α1-adrenoceptor activity in progression of cardiac hypertrophy in guinea pig hearts with pressure overload. Cardiovasc Res 23: 315–322, 1989
18. Fox AW, Juberg EN, May JM, Johnson RD, Abel PW: Thyroid status and adrenergic receptor subtypes in the rat: Compari son of receptor density and responsiveness. J Pharmacol Exp Therap 235: 715–723, 1985
19. William LT, Lefkowitz RJ, Watanabe AM, Hathaway KR, Besch HR Jr: Thyroid hormone regulation of α-adrenergic receptor number. J Biol Chem 252: 2787–2789, 1977
20. Weber KT: Cardiac interstitum in health and disease: the fibrillar collagen network. J Am Coll Cardiol 13: 1639–1952

Molecular and Cellular Biochemistry **119**: 29–34, 1993.

Differing responses of G2-related genes during differentiation of HL60 cells induced by TPA or DMSO

Junko Horiguchi-Yamada and Hisashi Yamada
Department of Internal Medicine, Aoto Hospital, The Jikei University School of Medicine, 6-41-2 Aoto, Katsushika-ku, Tokyo 125, Japan

Abstract

Differentiation leads to the cessation of cellular proliferation, but little is known about the molecular mechanisms of growth arrest. We compared the effect of two differentiation inducers, 12-o-tetradecanoyl 13-acetate (TPA) and dimethyl sulfoxide (DMSO) on both the cell-cycle and the modulation of G2-related genes in synchronized HL60 cells. TPA treatment of HL60 cells resulted in G1 arrest within 24 h. In contrast, the cell cycling of DMSO-treated cells was initially accelerated and they progressed to the second cycle before accumulating in the G1 phase. Expression of cyclin B, cdc25, weel and cdc2 was studied during cell cycle arrest by Northern blot hybridization. Expression of cyclin B, cdc25 and cdc2 fluctuated in association with cell cycle progression towards the G2/M phase, while weel expression remained constant in untreated cells. These four genes were highly expressed in TPA-treated cells for the first 12 h, but drastic down-regulation was seen at 18 h and expression became undetectable after 24 h. In contrast, no remarked changes of gene expression were seen in DMSO-treated cells. These findings suggest that cell cycle progression along with the initial process of differentiation in response to TPA differs from the response to DMSO and that the down-regulation of cdc2 expression by TPA-treated HL60 cells contributes to endorsement of G1 arrest. (Mol Cell Biochem **119:** 29–34, 1993)

Key words: cell cycle, differentiation, G2 phase, acute leukemia, differentiation inducer

Introduction

Terminal differentiation is usually associated with the cessation of cellular proliferation. Thus, the induction of differentiation represents a possible therapeutic modality for controlling malignant cellular growth. Although the mechanisms of growth arrest and differentiation still remain unclear, modulation of proto-oncogenes during the differentiation process may play an important role. In HL60 cells, the down-modulation of c-myc expression appears to be a key issue in cellular growth arrest [1], while the up-regulation of c-jun is reported to be responsible for lineage specificity [2]. However, the modulation of proto-oncogenes alone is not sufficient for the achievement of differentiation.

Molecular analysis of cell cycle regulation has advanced greatly in recent years and various aspects of the cell cycle have been disclosed. For example, it has been found that retinoblastoma (RB) gene products are phosphorylated as cells progress from G1 to S phase,

Address for offprints: H. Yamada, Department of Internal Medicine, Aoto Hospital, The Jikei University School of Medicine, 6-41-2 Aota, Katsushika-ku, Tokyo 125, Japan

while this phosphorylation does not occur during G1/G0 arrest (see ref. 3 for review). Accumulation of the underphosphorylated form of RB protein is therefore one of the candidates for participating in the G1 arrest of cells. Both TPA and DMSO have been reported to cause underphosphorylation of RB protein [4, 5]. However, it remains unclear how phosphorylation of RB protein is regulated in the machinery of cell cycle. Cell cycle progression involves the sequential activation of certain phase-specific proteins. For instance, activation of mitosis promoting factor (MPF) plays a major role in the progression from G2 to M phase (see ref. 6 for review). MPF consists of the cdc2 product and cyclin B [7–9], and possesses protein kinase activity which is regulated by the phosphorylation status of cdc2 [10–12]. wee1 causes the phosphorylation of cdc2 and inactivates MPF [13], while cdc25 dephosphorylates cdc2 and activates the kinase function of MPF.

Thus, the modulation of G2 genes could cause disarrangement of the G2 phase and distortion of the subsequent phases of the cell cycle. In this study, we compared the effects of the chemical inducers (TPA and DMSO) on the cell cycle and G2 related genes during the differentiation of HL60 cells (see ref. 19 for review). We found that these inducers exerted different effects on both cellular growth arrest and the kinetics of G2-related genes.

Methods

Cell culture and cell cycle analysis

HL60 cells (Japanese Cancer Research Resource Bank) were cultured in RPMI 1640 medium supplemented with 10% fetal bovine serum (Hyclone, USA). All experiments were done using cells passaged less than thirty times.

Synchronization of the cell cycle was achieved by the double-block method described by Heintz *et al.* [20] with a slight modification (i.e., hydroxyurea was used instead of aphidicolin). HL60 cells were first cultured for 12 h in the presence of 2 mM thymidine, which was added to the complete medium. The cells were then washed three times with serum-free medium, and were released by incubation with 20 μM thymidine and 20 μM deoxycytidine for 9 h. This was followed by the second period of nucleotide synthesis inhibition through incubation with 1 mM hydroxyurea for 12 h, after which cells were washed again and the time chasing experiments were initiated. The viability of the synchronized cells at the time of release was 80%. Synchronized cells were recultured in medium without additives (untreated), in medium with 33 nM of TPA or in medium with 1.3% DMSO. The cells were subsequently harvested at the designated times and subjected to cell cycle analysis and RNA extraction.

The distribution of cells in each phase of cell cycle was monitored by flowcytometry using a FACscan (Becton-Dickinson, USA). Cellular DNA content was measured after propidium iodide staining and the percentage of cells in each phase was calculated by the sum of broadened rectangles method.

Analysis of G2-related genes

Total RNA was extracted by the guanidinium/cesium chloride centrifugation method [21]. Fifteen micrograms of total RNA was applied to formaldehyde-denatured agarose gel electrophoresis. After size separation, the RNA was transferred onto nylon filters by the capillary method and baked for two hours at 80° C in a vacuum. Filters were hybridized as previously described [22] with a slight modification. The following probes were labelled by the multiprimer method (Amarsham, USA): 1) cdc2 [23], a full length cDNA fragment; 2) cyclin B [8], a full length cDNA fragment; 3) cdc25Hu2 [24], a NaeI/SacI 0.7 kbp cDNA fragment; 4) wee1Hu [25], a NaeI/PstI 1.6 kbp cDNA fragment. Briefly, the filters were hybridized overnight at 42° C with α-^{32}P dCTP-labeled probes in a hybridization solution without dextran, followed by two washes at room temperature and two washes at 55° C under highly stringent conditions. The filters were then exposed to Kodak XAR films at − 80° C for 24 h or 7 days for longer exposure.

Results

Cell cycle analysis

The cells began to synthesize DNA from immediately after release and progressed through the cell cycle from the G1/S boundary, as shown in Fig. 1 (untreated cells indicated by closed circles). More than 70% of the cells were in G1 phase at the time of release of the blockade.

The majority of the cells then rapidly shifted to the S phase, causing a sharp decrease in the G1 phase population and a sudden increase in the S phase population. The S phase population peaked at 6h at which time about 70% of the cells were in this phase. Then the synchronized cells gradually moved to the G2/M phase, where the maximum accumulation was observed after 12h. Cells returned to the G1 phase at 18h after dividing in M phase. The peak of the second S phase was noted at 24h, although the synchronization was less marked. Due to loss of synchronization, the following G2/M peak was barely detectable at 36h.

TPA treatment did not affect the cell cycle until 18h, when the cells entered the G1 phase (Fig. 1, squares). Then progression towards the second S phase was completely blocked by TPA treatment and the cells remained arrested in the G1 phase. The G2 population was 15 to 20% during this period.

DMSO treatment had different effect from TPA treatment (Fig. 1, triangles). The S phase was observed at 6h, similarly to the case for untreated and TPA-treated cells. The consequent G1 phase, however, was already seen at 12h in DMSO-treated cells, earlier than in both untreated at TPA-treated cells. Corresponding to this earlier G1 phase, the percentage of G2/M phase cells at 12h was significantly lower in DMSO-treated cultures than in the other cultures. A second cell cycle was observed in DMSO-treated cultures with the second S phase peak at 24h, but afterwards the cells tended to accumulate in G1 phase and G1 cells rose to 70%. Thus, DMSO-treated cells showed more rapid progression through the cell cycle for the first 24h, but longer exposure led the cells to accumulate in G1 phase along with differentiation.

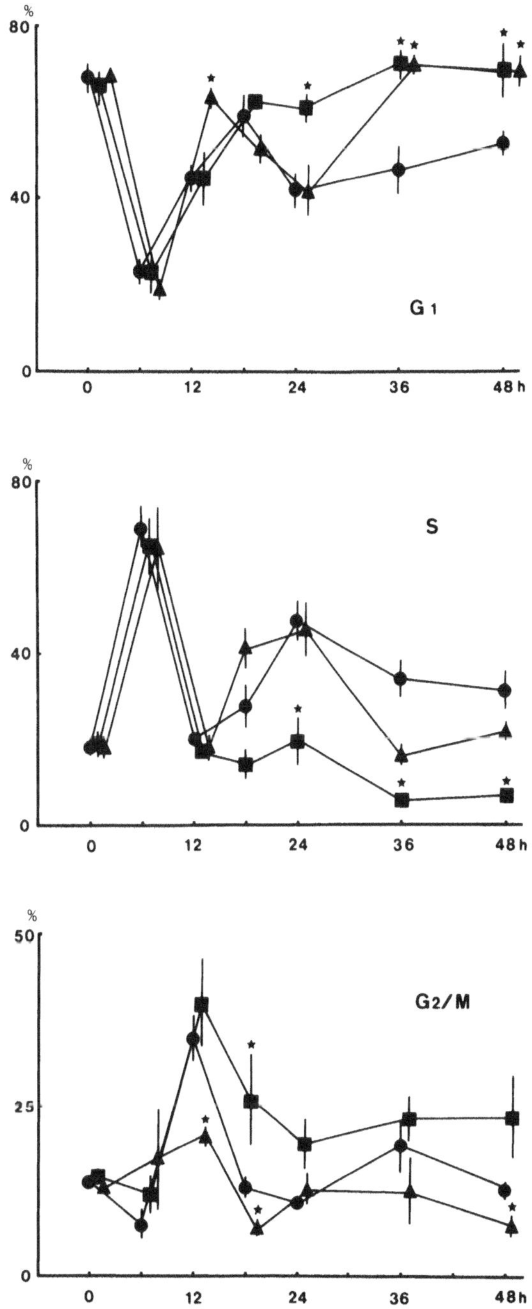

Fig. 1. Percentage of G1 cells (the top panel), S cells (the center panel) and G2/M cells (the bottom panel). Closed circles represent untreated HL60 cells, squares show TPA-treated and triangles show DMSO-treated cells. Time 0 is the point when synchronized cells were released by incubation in medium with or without agents. Cells were harvested at six-hour intervals for the initial 24h and then at twelve-hour intervals until 48h. The plots and bars show the mean ± SE. Experiments were repeated four times for untreated and TPA-treated cells, and five times with untreated and DMSO-treated cells. The stars indicate a significant difference ($p < 0.05$) between the untreated and treated groups.

Modulation of G2-related gene expression

The expression of G2-related genes (cyclin B, cdc25, weel, and cdc2) was studied by Northern blot hybridization (Figs 2 and 3). In untreated cells, the expression of cyclin B, cdc25, and cdc2 was associated with cell cycle progression. The maximum expression of these genes was found at 6–12h and it was decreased at 18h, consistently with cell cycle progression towards G2/M phase. On the other hand, expression of weel was constant and was not affected by the cell cycle stage.

TPA treatment caused marked down-modulation of G2-related genes after 18h. Initially, the expression of

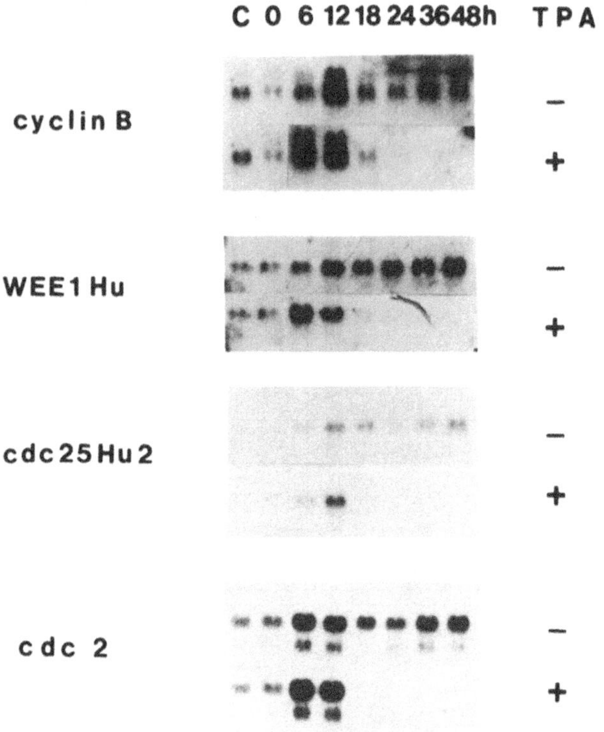

Fig. 2. Expression of cyclin B, cdc25Hu2, wee1 and cdc2 by untreated (upper column) and TPA-treated (lower column) cells. Fifteen micrograms of RNA extracted from cells harvested at the indicated times was hybridized with the labeled probes. C represents the RNA from unsynchronized cells in the log growth phase.

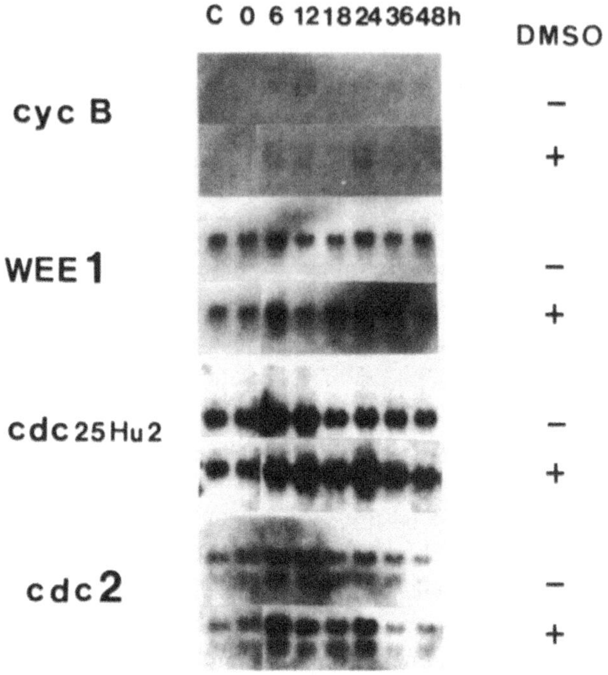

Fig. 3. Expression of cyclin B, cdc25Hu2, wee1 and cdc2 by untreated (upper column) and DMSO-treated (lower column) HL60 cells. The experimental protocol was identical to that described in Fig. 2.

these four genes was actually accentuated (see the bands at 6 and 12 h), but this was followed by sudden down-regulation at 18 h, and no expression was found after 24 h.

DMSO treatment did not elicit a marked change in the expression of these four genes and comparison between the untreated and DMSO treated cells revealed no obvious down- or up-regulation.

Discussion

The G1 and G2 phases represent the time between DNA synthesis and mitosis during the cell cycle and it is suspected that various regulatory mechanisms act during these periods (see ref. 26 and 27 for review). These include systems to recognize the completion of DNA replication, to inspect its accuracy, and to repair any misreading of the newly synthesized DNA chains, all processes that are crucial to provide normal descendants for a cell. Recent advances in the understanding of these mechanisms have enabled us to examine in more detail the phenomena occurring during cellular differentiation. In this study, we examined the alteration of G2-related genes during the chemically induced differentiation of a leukemic cell line HL60. TPA and DMSO cause the differentiation of HL60 cells into two distinct lineages and both cause growth arrest in the G1 phase. However, we found that the time courses of growth arrest differed in each case. TPA arrested cellular growth in the first G1 phase, while DMSO caused accumulation of cells in the G1 phase after two cell divisions. Moreover, the time to complete a cell cycle was shortened by DMSO treatment. This acceleration of progression through the cell cycle was observed within the first cycle, so the G2 phase was probably affected by DMSO treatment. The G2 phase is the period of preparation for entry into mitosis after the replication of DNA has been accomplished. Feedback mechanisms exist to prevent a cell undergoing premature mitosis while DNA replication is still taking place and the main factor in this regulation is the control of MPF activation. Thus, we investigated the alterations of G2-related genes in an attempt to clarify the molecular level of growth arrest during differentiation. Expression of cyclin B, cdc25, and cdc2 oscillated in association with the progression of the cell cycle towards G2 phase, while no obvious changes were observed in wee1 expression.

Little is known about the regulation of weel transcription, but our data suggest that it is not cycle-dependent at the transcriptional level. Although the levels of both cyclin B and cdc2 transcripts increased cycle-dependently, the protein products of these two genes differ in their behavior. The cdc2 protein content is reported to be constant throughout the cell cycle [28], while the cyclin B product increases in close association with G2 progression and is degraded immediately once mitosis occurs [8]. In addition, cdc2 is present at a more than ten-fold excess when compared to cyclin B. Thus, we speculate that cyclin B and cdc25 play key issues to enter into mitosis in untreated cells.

Down-modulation of cdc2 could also affect the cycle regulatory systems in the G1 phase. Since the cdc2 product forms complexes with cyclin A [29], the marked down-regulation of cdc2 would also affect the regulation of the G1 phase in TPA-treated cells. The retinoblastoma gene product is one of key candidates for participating in the G1 to S transition and TPA-induced differentiation is reported to be associated with the accumulation of underphosphorylated RB protein [5]. cdc2 kinase or cdc2-related kinase could contribute to cell cycle regulation by phosphorylating RB protein (see ref. 3 for review). Therefore, the down-regulation of cdc2 would result in distortion of the cycle at the G1-S boundary and cause G1 arrest. However, provided that the half-life of cdc2 protein is reasonably longer, the down-regulation of cdc2 mRNA seen in our experiments was not sufficient to block the cell cycle, since a considerable amount of cdc2 protein would probably remain in the nucleus. But, the cell cycle dependent cdc2 production is also reported [28] that the drastic down-regulation of cdc2 in G2 phase may contribute to subsequent G1 arrest in TPA-treated cells.

The cell cycle advancement seen in DMSO-treated cells was an interesting phenomenon which may have depended on the early entry into mitosis. The early entry into mitosis has been reported to be induced by the modification of weel of cdc25 [30–32]. Three types of cdc25 are known to exist and the timing of their function during G2 phase is different [17]. However, the actual roles of these three genes still remain unclear. The expression of cdc25C was too low to detect by Northern blot hybridization (data not known). We demonstrated only cdc25B (also known as cdc25Hu2), thus cdc25A might be helpful to understand this phenomenon.

In conclusion, our data demonstrated clear differences in cell cycle modulation between the two inducers. Further studies are required to elucidate the mechanisms of cellular differentiation and the upstream regulation of deviation from the cell cycle into G1/G0 arrest. The initial overexpression of G2-related genes in TPA-treated cells suggests that preceding modulation of G2 phase might contribute to subsequent G1 arrest, and further experiments involving modulation of these genes would provide the answer about these issues.

Acknowledgements

We are grateful to Drs. D. Kufe, T. Hunter, and A Nagata for providing probes, and to the Japanese Cancer Research Resources Bank (JCRB) for providing HL60 cells. We also thank Yukihiro Takeuchi for excellent technical assistance in flowcytometric measurement.

References

1. Grosso LE, Pitot HC: Modulation of c-myc expression in HL60 cell line. Biochem Biophys Res Commun 119: 473–480, 1984
2. Gaynor R, Simon K, Koeffler P: Expression of c-jun during macrophage differentiation of HL-60 cells. Blood 77: 2618–2623, 1991
3. Cobrinik D, Dowdy SF, Hinds PW, Mittnacht S, Weinberg RA: The retinoblastoma protein and the regulation of cell cycling. TIBS 17: 312–315, 1992
4. Hihara K, Cao X, Yen A, Chandler S, Driscoll B, Murphree AL, T'Ang A, Fung YT: Cell cycle-dependent regulation of phosphorylation of the human retinoblastoma gene product. Science 246: 1300–1303, 1989
5. Akiyama T, Toyoshima K: Marked alteration in phosphorylation of the RB protein during differentiation of human promyelocytic HL60 cells. Oncogene 5: 179–183, 1990
6. Nurse P: Universal control mechanism regulating onset of M-phase. Nature 344: 503–508, 1990
7. Riabowol K, Draetta G, Brizuela L, Vandre D, Beach D: The cdc2 kinase is a nuclear protein that is essential for mitosis in mammalian cells. Cell 57: 393–401, 1989
8. Pines J, Hunter T: Isolation of a human cyclin cDNA: Evidence for cyclin mRNA and protein regulation in the cell cycle and for interaction with $p34^{cdc2}$. Cell 58: 833–846, 1989
9. Gautier J, Minshull J, Lohka M, Glotzer M, Hunt T, Maller JL: Cyclin is a component of maturation-promoting factor from Xenopus. Cell 60: 487–494, 1990
10. Gautier J, Matsukawa T, Nurse P, Maller J: Dephosphorylation and activation of Xenopus $p34^{cdc2}$ protein kinase during the cell cycle. Nature 339: 626–629, 1989
11. Morla AO, Draetta G, Beach D, Wang JYJ: Reversible tyrosine phosphorylation accompanies activation during entry into mitosis. Cell 58: 193–203, 1989
12. Krek W, Nigg EA: Differential phosphorylation of vertebrate

p34^{cdc2} kinase at the G1/S and G2/M transitions of the cell cycle: Identification of major phosphorylation sites. EMBO J 10: 305–316, 1991
13. Lundgren K, Walworth N, Booher R, Dembski M, Kirschner M, Beach D: mik1 and wee1 cooperate in the inhibitory tyrosine phosphorylation of cdc2: Cell 64: 1111–1122, 1991
14. Kumagai A, Dunphy WG: The cdc25 protein controls tyrosine dephosphorylation of the cdc2 protein in a cell-free system. Cell 64: 903–914, 1991
15. Dunphy WG, Kumagai A: The cdc25 protein contains an intrinsic phosphatase activity. Cell 67: 189–196, 1991
16. Gautier J, Solomon MJ, Booher RN, Bazan JF, Kirschner MW: cdc25 is a specific tyrosine phosphatase that directly activates p34^{cdc2}. Cell 67: 197–211, 1991
17. Galaktionov K, Beach D: Specific activation of cdc25 tyrosine phosphatases by B-type cyclins: Evidence for multiple roles of mitotic cyclins. Cell 67: 1181–1194, 1991
18. Jessus C, Beach D: Occilation of MPF is accompanied by periodic association between cdc25 and cdc2-cyclin B. Cell 68: 323–332, 1992
19. Lübbert M, Harrmann F, Koeffler HP: Expression and regulation of myeloid-specific genes in normal and leukemic myeloid cells. Blood 77: 909–924, 1991
20. Heintz N, Sive HL, Roeder RG: Regulation of human histone gene expression: Kinetics of accumulation and changes in the rate of synthesis and in the half-lives of individual histone mRNAs during the HeLa cell cycle. Mol Cell Biol 3: 539–550, 1983
21. Maniatis T, Fritsch EF, Sambrook J: Molecular cloning: A laboratory manual. Cold Spring Harbor Laboratory, New York, 1982, pp 187–210
22. Yamada H, Sakamoto H, Taira M, Nishimura S, Shimosato Y, Terada M, Sugimura T: Amplifications of both c-Ki-ras with a point mutation and c-myc in a primary pancreatic cancer and its metastatic tumors in lymph nodes. Jpn J Cancer Res 77: 370–375, 1986
23. Lee MG, Nurse P: Complementation used to clone a human homologue of the fission yeast cell cycle control gene cdc2. Nature 327: 31–35, 1987
24. Nagata A, Igarashi M, Jinno S, Suto K, Okayama H: An additional homolog of the fission yeast cdc25$^+$ gene occurs in humans and is highly expressed in some cancer cells. The new biologist 3: 959–968, 1991
25. Igarashi M, Nagata A, Jinno S, Suto K, Okayama H: Wee1$^+$-like gene in human cells. Nature 353: 80–83, 1991
26. Pardee AB: G1 events and regulation of cell proliferation. Science 246: 603–608, 1989
27. Murray AW, Kirschner MW: Dominoes and clocks: The union of two views of the cell cycle. Science 246: 614–621, 1989
28. McGowan CH, Russell P, Reed SI: Periodic biosynthesis of the human M-phase promoting factor catalytic subunit p34 during the cell cycle. Mol Cell Biol 10: 3847–3851, 1990
29. Draetta G, Luca F, Westendorf J, Brizuela L, Ruderman J, Beach D: cdc2 protein kinase is complexed with both cyclin A and B: Evidence for proteolytic inactivation of MPF. Cell 56: 829–838, 1989
30. Enoch T, Nurse P: Mutation of fission yeast cell cycle control genes abolishes dependence of mitosis on DNA replication. Cell 60: 665–673, 1990
31. Rowley R, Hudson J, Young PG: The wee1 protein kinase is required for radiation-induced mitotic delay. Nature 356: 353–355, 1992
32. Smythe C, Newport JW: Coupling of mitosis to the completion of S phase in Xenopus occurs via modulation of the tyrosine kinase that phosphorylates p34^{cdc2}. Cell 68: 787–797, 1992

Molecular and Cellular Biochemistry **119**: 35–41, 1993.

Mechanistic effects of kijimicin on inhibition of human immunodeficiency virus replication

Teruo Yamauchi, Mariko Nakamura, Hitoshi Honma, Motoaki Ikeda, Kazunari Kawashima and Tsuneya Ohno
Department of Microbiology, Jikei University School of Medicine, 3-25-8, Nishi-shinbashi Minato-ku, Tokyo 105, Japan

Abstract

Kijimicin represents an important type of ionophore compound. In veterinary medicine, it is becoming important as anticoccidiostatic agent and feed supplement. We examined Kijimicin for its HIV inhibitory activity. The compound exhibited concentration-dependent inhibition of HIV replication in primary infected cultures of human T-lymphoblastoid H9 cells. Substantial inhibition of viral replication was observed at concentrations of Kijimicin that showed little cytotoxicity. The ratio of IC_{50} values for the MTT to RT assays was 40. Anti HIV activity was also observed in cultures of monocytic lineage U937 cells chronically infected with HIV. Moreover, in attempting to define the inhibitory mechanism of Kijimicin, we investigated its effect on each step of HIV replication. The infectivity of progeny viral particles was reduced by Kijimicin treatment. This decrease may be due to incompletely glycosylated forms of gp120. (Mol Cell Biochem **119:** 35–41, 1993)

Key words: viral replication, anticoccidiostatic agents, human immunodeficiency virus

Introduction

With the identification of HIV as the etiological agent of AIDS, a number of approaches are being explored to develop therapeutic agents for treating this disease. One therapeutic strategy for the treatment of HIV infection has been to develop compounds that interfere with replication of the virus. Kijimicin, one of the polyether antibiotics, exhibits selective toxicity against coccidia, and presently is used as prophylaxis of chicken coccidiosis [1].

We have recently reported the inhibitory effect of Kijimicin on the ability of HIV-1 to replicate in both lymphoid lineage H9 cells and mononuclear phagocyte lineage U937 cells [2]. The inhibition of HIV replication was shown in primary infected cultures with little cytotoxicity, and anti viral activity was also observed in cultures chronically infected with HIV.

In this present communication, we investigated the effects of Kijimicin at various phases of the HIV replication cycle to analyze the mechanism of its inhibitory action. Specifically, virus adsorption was studied in comparison to dextran sulfate, an agent which has been shown to inhibit HIV binding to target cells *in vivo* [3, 4]. Moreover, we examined the process of assembly and the infectivity of progeny viruses to assess the action of Kijimicin on later stage of HIV replication.

The identification of this new class of anti-HIV agent and studying this mode of inhibitory action may lead us

Address for offprints: T.Ohno, Department of Microbiology, Jikei University School of Medicine, 3-25-8, Nishi-shinbashi Minato-ku, Tokyo 105, Japan

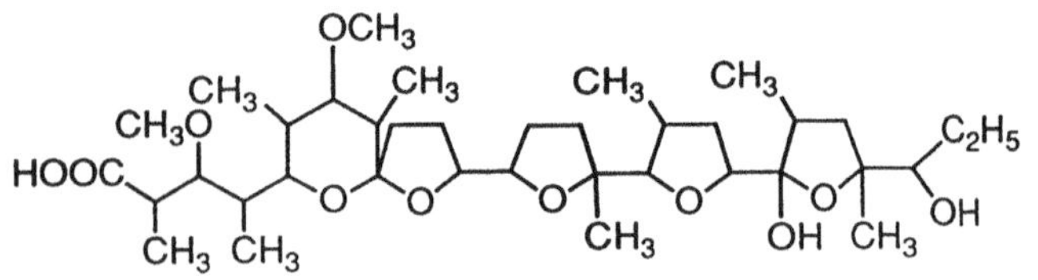

Fig. 1. Structure of Kijimicin.

to a further development of HIV therapy and elucidate the mechanism of viral replication.

Materials and methods

Compounds

Kijimicin was prepared as sodium salt from culture broth of the strain MI215-NF3 in the Institute of Microbial Chemistry (kindly provided by Dr. Tomio Takeuchi, structure is shown in Fig. 1). Dextran sulfate (M.W.8000) was purchased as a sodium salt from Sigma Chemical Company, U.S.A.

Cell lines and preparation of virus stocks

H9 [5] and U937 [6] cell cultures were maintained at densities between 10^5 and 10^6 cells/ml in RPMI 1640 supplemented with 10% heat inactivated fetal bovine serum and 2 mM L-glutamine. HIV-1 was obtained from the culture supernatant of a H9 cell line infected with HIV-1, and stocked at $-80°C$. The titer of the virus stock was 8.2×10^4 $TCID_{50}$ per ml.

Antiviral assays

Acutely infected cells
H9 cells were suspended at 5×10^6 cells per ml in the presence of various concentrations of the antibiotic. Following 30 min incubation at 37° C, the cells were inoculated with HIV-1 IIIB at 1.6×10^4 $TCID_{50}$/ml. Cells were incubated for 90 min at 37° C to permit adsorption of viral particles and then expanded 1 : 10 with fresh media. The cultures were split 1 : 5 on day 4 and on day 7 samples were harvested for reverse transcriptase (RT) and MTT assays.

Chronically infected cells
Chronically HIV-1 IIIB infected U937 (kindly supplied by Dr. Groopman, New England Deaconess Hospital) were suspended at 2.5×10^5 cells/ml and 200 μl of the suspension was added to each well of a 96 well flat bottom plate. 10 μl of serially diluted Kijimicin was then added. Supernatant (10 μl) was used for RT assay, and the cell suspension (100 μl) was examined for viability after 72 h by the MTT assay.

Reverse transcriptase assay [7]

10 μl of supernatant containing HIV particles was disrupted with 10 μl of detergent solution (50 mM Tris-HCl pH 8.0, 10 mM DTT, 300 mM KCl, 0.5% Triton X-100) in a 96 well round bottom plate. Following 15 min incubation at 4° C, 25 μl of RT buffer containing 50 mM Tris-HCl pH 8.0, 10 mM $MgCl_2$, 5 mM DTT, 0.25 U/ml poly(rA)oligo$(dT)_{12\text{-}18}$ (Pharmacia), 15 μCi/ml $[^3H]$-dTTP (American Radiolabeled Chemicals, Inc.) was added. The reaction mixtures were incubated at 37° C for 18 h and 15 μl of mixture was then spotted onto DE81 Whatman filter paper, dried, washed with 2 × SSC 6 times, and rinsed with ethanol. The dried filters were placed in mini vials with 5 ml of Aquasol-2 and radioactivities were analyzed in a scintillation counter (Beckman).

MTT(3-4,5 dimethylthiazol-2,5 diphenyl tetrazolium bromide) assay [8, 9]

Cytotoxicity of compounds was examined by MTT assay. At the end of the culture, 100 μl of cell suspension was transfered to a new 96 well flat bottom plate, 10 μl of 5 mg/ml MTT/PBS was added to each well and the suspension was incubated at 37° C for 4 h. 100 μl of 0.04 N HCl/isopropanol was then added and mixed thoroughly by repeated pipetting. Absorbance was determined on an ELISA plate reader at a test wavelength of 550 nm and a reference wavelength of 630 nm.

Detection of HIV-1 proviral DNA by PCR

HIV-1 infected cells (1×10^6 cells) were incubated in 300 μl of lysis buffer (100 mM NaCl, 10 mM Tris-HCl pH 8.0, 25 mM EDTA pH 8.0, 0.5% SDS, 0.1 mg/ml proteinase K) at 37° C for 18 h. The lysate was then deproteinized by successive phenol/chloroform/isoamyl alcohol extractions. The sequences of the oligonucleotide primers

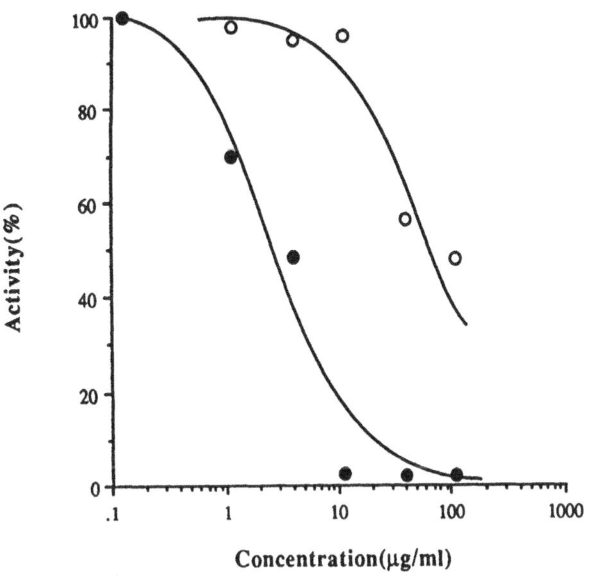

Fig. 2. Inhibitory effect of Kijimicin against primary infection. H9 cells were preincubated with various concentrations of Kijimicin for 30 min at 37° C, and exposed to HIV-1 IIIB. On day 7, viable cells were counted by MTT assay (○) and concentrations of viruses were estimated by RT assay (●).

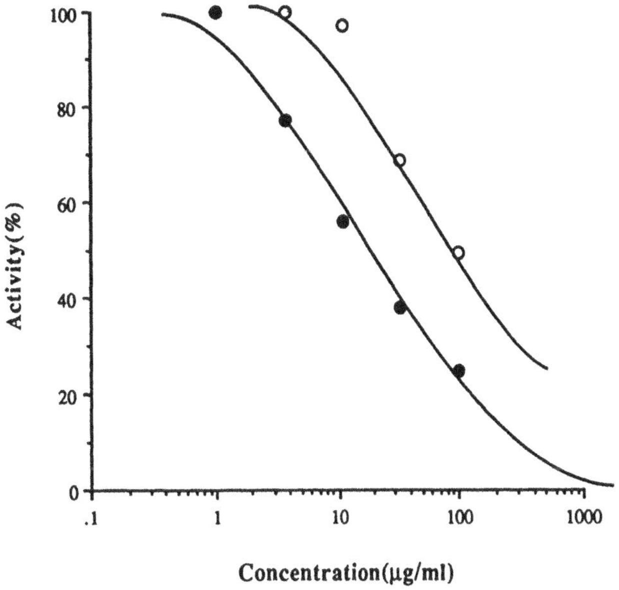

Fig. 3. Inhibitory effect of Kijimicin on chronically infected U937. Chronically HIV infected U937 were treated with Kijimicin. After 72 h, the number of virions was counted by RT assay (●) and viability was examined by MTT assay ○).

used for HIV-1 detection were derived from the sequence of HIV-1 IIIB and were as follows [10]: 5′-TGTACA-CATGGAATTAGGCCAG-3′, nucleotides 6982 to 7014; and 5′ ATTTCTGGGTCCCCTCCTGAGG-3′, nucleotides 7338 to 7359. The predicted size of the amplified region is 378 bp. A β-globin specific primer was used for normalization to the internal cellular DNA control, as decribed previously [11]. β-globin gene primer pair, 5′-GAAGAGCCAAGGACAGGTAC-3′ and 5′-CAACTTCATCCACGTTCACC-3′, flank a 268-bp fragment. To detect HIV-1 specific sequences, 100 ng of oligonucleotide primers was used for each sample. Each reaction mixture contained 0.25 mM of each deoxynucleotide triphosphate, 50 mM KCl, 10 mM Tris-HCl pH 8.0, 1.5 mM $MgCl_2$, 0.1% gelatin, and 2.5 U/μl of Taq DNA polymerase (Boehringer). The reaction mixture was overlaid with one drop of Ampli wax (Perkin Elmer Cetus) and then subjected to 30 cycles of denaturation of 1 min at 94° C, annealing for 1 min at 60° C, and finally polimerization of 2 min at 72° C. The amplified products resulting from PCR were analyzed by electrophoresis in 2% agarose gels and visualized by ethidium bromide staining.

Inhibitory mechanism of Kijimicin on acute infection

In order to define the effect of Kijimicin in early stages of HIV infection, we performed a time of addition experiment. H9 cells were exposed to HIV-1 IIIB at a high multiplicity to ensure that the virus replication steps were synchronized in the whole cell population and then incubated at 37° C. 1 μg/ml of Kijimicin was added at different times (– 30, 0, 30, 60 min) before adsorption or after HIV-1 IIIB exposure to H9 cells. Proviral DNA was analyzed at 24 h postinfection by PCR to ascertain the inhibitory effects on HIV replication from adsorption to integration.

Inhibitory effect on maturation

H9 cells were exposed to HIV-1 IIIB at 8.2×10^4 $TCID_{50}$/ml and incubated at 37° C for 105 h until the progeny viruses could be detected by RT assay.

After 105 h of primary infection, the cells were shaken and washed twice to remove free virus particles. The cells were then resuspended in fresh media at 5×10^5 c/ml, and seeded in a 96 well plate in the presence of serially diluted Kijimicin. Culture supernatant was harvested at 6, 22, 30, and 58 h for RT assay to quantifiy the number of virus particles.

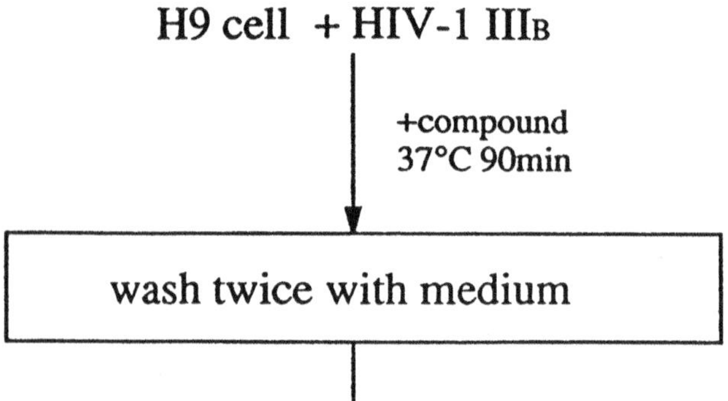

Fig. 4. Schematic outline of the strategy to assess the inhibitory mechanism of Kijimicin on adsorption period.

Infectivity of progeny virus particles

H9 cells were exposed to HIV-1 IIIB at 8.2×10^4 $TCID_{50}$/ml and incubated at 4° C for 60 min. After adsorption, the cells were washed twice and suspended at 1.3×10^5 c/ml with fresh media. The cells were seeded in the presence of serial dilutions of Kijimicin (1.0, 0.1, 0.01, 0 μg/ml) and incubated at 37° C. Culture supernatant was harvested on day 7 and the infectivity of progeny viruses in the original fluid was determined by the $TCID_{50}$ procedure.

Results

Anti-HIV activity on acute infection

H9 cells were pretreated with Kijimicin and inoculated with HIV. Infection and viral replication were monitored by measuring the amount of HIV particles by reverse transcriptase assay in culture supernatants. As shown in Fig. 2, Kijimicin treatment resulted in a concentration dependent inhibition of HIV replication in this acute primary assay system. HIV production was completely abolished at concentrations higher than 10 μg/ml, and IC_{50} was observed at 2.2 μg/ml that produced no cytotoxicity. The IC_{50} of MTT was at 88 μg/ml, consequently the ratio of IC_{50} values for MTT to RT was 40. Identically treated uninfected cultures studied in parallel showed no change in cellular viability as compared to infected cultures (data not shown).

Anti-HIV activity on chronic infection

Kijimicin treatment also inhibited HIV replication in chronically infected U937 (Fig. 3), but it had less difference of IC_{50} ratio between RT and MTT. It should be

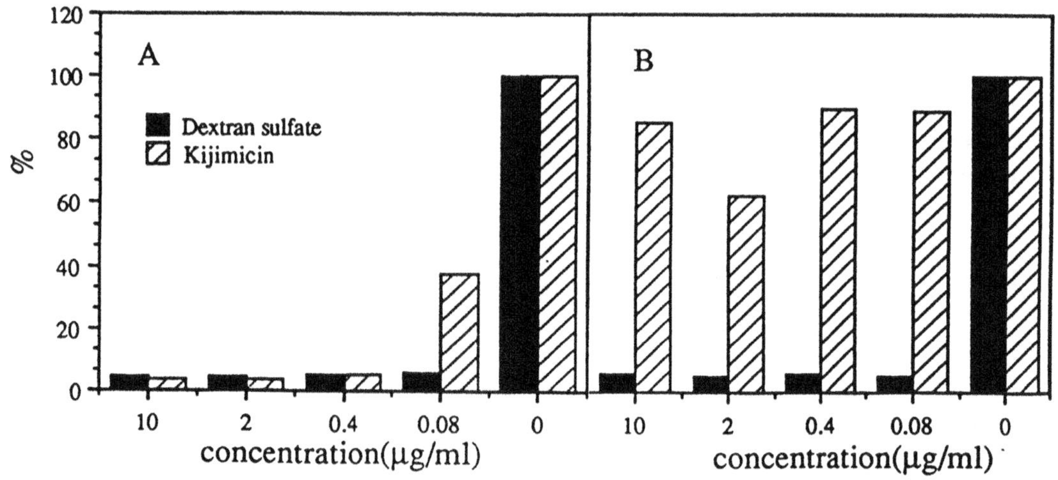

Fig. 5. Inhibitory mechanism of Kijimicin on acute infection.
H9 cells were inoculated with HIV with compounds and incubated at 37° C for 90 min. Cells were then washed and resuspended with fresh media to be 1.5×10^5 c/ml. In treatment A, the cells were exposed to the compounds again in further incubation. In treatment B, they were incubated in the absence of compounds. Each value is the average of 4 replicates experiments. RT activity (cpm) is expressed as percent of the control (no agents).

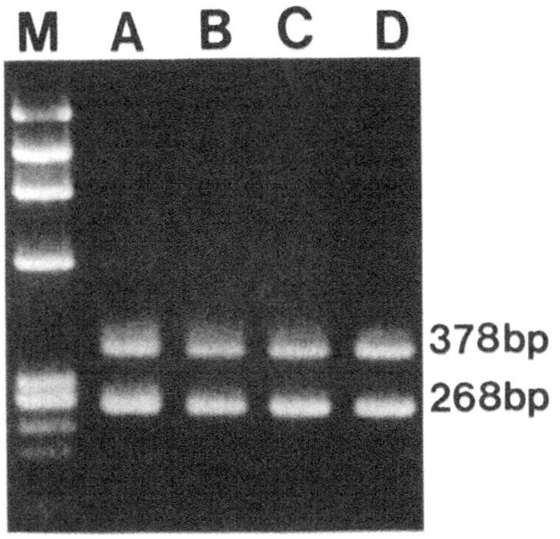

Fig. 6. PCR analysis of HIV-1 proviral DNA in H9 cells. Kijimicin was added to H9 cells at the indicated times (A: − 30 min, B: 0 min, C: 30 min, D: 60 min) after HIV inoculation. HIV proviral DNA was amplified from the genomic DNA of infected H9 cells as described in Materials and methods. The band derived from the HIV-1 corresponds to 378 bp and the β-globin band (internal cellular DNA control) corresponds to 268 bp. M indicates a øX174/Hae III molecular weight marker.

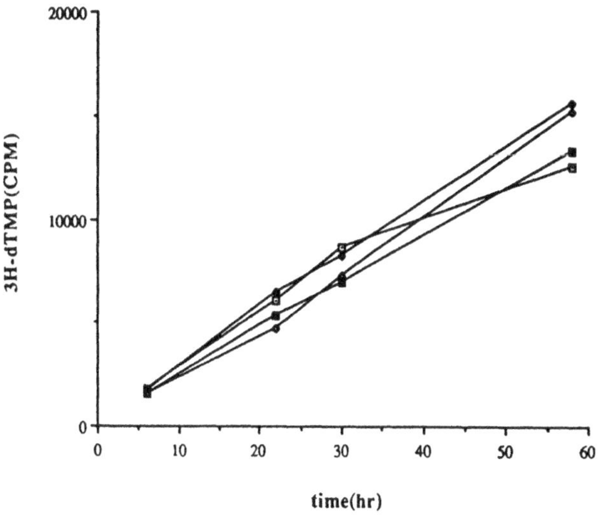

Fig. 7. Inhibitory effect on maturation. After 105 h on primary infection, serially diluted Kijimicin was added to the culture, and the number of progeny viruses was monitored by RT assay at 6, 22, 30, and 58 h. Each mark indicates the concentration of Kijimicin as follows; —□— 12.5 μg/ml —◆— 6.25 μg/ml –■— 1.57 μg/ml —◇— 0 μg/ml.

considerable that the higher cytotoxicity was caused by continuous exposure of antibiotic.

Influence of various treatment periods on anti HIV activity

In order to determine the mechanism of action of Kijimicin on HIV replication, various treatment periods with Kijimicin were examined. We tested dextran sulfate as a control since it is known to block virus adsorption to cells. Schematic outline of Kijimicin treatment was shown in Fig. 4. Briefly, H9 cells were inoculated with HIV at 1.6×10^4 $TCID_{50}$/ml with 10 times concentrated compounds (500, 100, 20, 4.0, 0.8, 0 μg/ml). After incubation at 37° C for 90 min, cells were washed twice to remove compounds and unadsorbed viruses, resuspended in fresh media at 1.5×10^5 c/ml, and 200 μl of culture was added to a new 96 well plate. In further incubations, each compound was added again in treatment A, and no more agents were added in treatment B.

Figure 5 shows that dextran sulfate was highly effective and completely abolished HIV replication at even less than 0.08 μg/ml regardless of the presence of the compound after the adsorption period. In marked contrast, Kijimicin exhibited inhibitory activity only with treatment A, and treatment B with this agent had no effect on HIV replication.

We next examined proviral DNA with various treatment periods to confirm the ineffectiveness of Kijimicin on adsorption steps. As shown in Fig. 6, no inhibition was observed in all these studies since the amount of DNA appeared identically and the density of the bands was equal to that without Kijimicin treatment.

Inhibitory effect on maturation

After 105 h of primary infection, the progeny viruses started to release from host cells. Kijimicin was added

Table 1. Infectivity of progeny viruses

Concentration of Kijimicin (μg/ml)	1.0	0.1	0.01	0
Production of progeny virus (cpm/ml)	2.8×10^5	8.3×10^5	1.4×10^6	1.0×10^6
$TCID_{50}$/ml of progeny viruses	6.4×10^2	5.1×10^3	$>4.1 \times 10^4$	$>4.1 \times 10^4$

The production of progeny viruses in the presence of Kijimicin was estimated by reverse transcriptase activity (cpm/ml). And the infectivity of the progeny viruses was determined by $TCID_{50}$.

to these freshly infected cells and culture supernatant was subsequently harvested to determine whether Kijimicin acts on the maturation step in the HIV replication cycle. Figure 7 shows that the amount of the progeny viruses was increased linearly and the slope of the replication curves was similar at any concentration of Kijimicin.

Infectivity of progeny HIV from Kijimicin-treated cells

In order to determine the degree of infectivity of progeny viruses from Kijimicin-treated infected H9 cells, culture fluids were added to uninfected H9 cells to examine the $TCID_{50}$. As listed in Table 1, Kijimicin treated cells produce substantial amount of viral particles by RT assay. However, marked reduction was observed by $TCID_{50}$. The observed reduction in viral infectivity was of the order of more than 1 log. In contrast, culture supernants of infected H9 cells not exposed to Kijimicin were able to successfully establish infection.

Discussion

A variety of potentially vulnerable steps in the HIV replication cycle could theoretically serve as targets for therapeutic intervention. CD4-expressing lymphiod cells are susceptible to HIV [12, 13] and cells of the mononuclear phagocyte lineage actually appear to be a major reservoir for HIV *in vivo* [14–16]. Furthermore, the importance of chronic infection is anticipated from the pathobiology of HIV infection *in vivo*.

The results of the present studies indicate that Kijimicin inhibits HIV replication in both T-cell and monocyte lineage cell lines, and shows activity in *in vitro* assays of both acute and chronic infections.

We examined antiviral activity in culture supernatants after several days of incubation by determining production of progeny viruses and generation of reverse transcriptase activity. The findings suggest that Kijimicin could inhibit every step during HIV infection. Therefore, additional experiments were performed to elucidate the mechanism of action of Kijimicin. Kijimicin has no detectable effect if the cells are washed free of agent after the virus penetration period even if these cells are pretreated with Kijimicin before HIV inoculation or when it is added to H9 cells during virus adsorption. This result suggests that Kijimicin interferes with events after penetration as the virus goes through its successive replication cycle, such as internalization, uncoating, reverse transcription, integration to genomic DNA, transcription, translation and maturation. Moreover, there was no detectable inhibitory effect of Kijimicin on reverse transcriptase, however a possibility still exists that these agents might, as in the case of AZT, inhibit reverse transcriptase after some modification of their structure in host cells [17]. The presence of proviral HIV DNA was demonstrated in these treatment periods, and the HIV genome is integrated within 1 h after virus binding to the CD4 receptor (preliminary findings). Therefore Kijimicin has no effect on early stages of HIV infection in terms of integration.

Moreover, the data suggest that Kijimicin is also effective in chronically infected U937 cells. Consequently, it affects viral replication, especially during the later stages of the HIV replication cycle, such as assembly and/or maturation. The maturation experiment indicates that viral particles are assembled normally under conditions of Kijimicin treatment since the number of viral particles is not suppressed. The production of infectious viral particles into culture supernatant decreased, despite the finding that RT activity was not affected. Monencin, one of the polyethers, has recently been shown to block endoproteolytic cleavage and the secondary glycosylation steps of gp 160 [18], Monencin treatment also results in a reduction in syncytium formation. Because the envelope glycoproteins of HIV are highly glycosylated, compounds that interfere with post translational processing of glycoprotein gp 120 may prevent viral entry into the cells. Terminal glycosylation of N-linked oligosaccharides and synthesis of O-linked oligosaccharides of viral glycoproteins may be important for viral spread and pathogenesis [19]. A decrease in infectivity by Kijimicin might be due to incompletely glycosylated forms of gp 120 that have a lower affinity for the CD4 receptor.

The identification of this new class of anti-HIV agent proceeds an opportunity to gain a better understanding of the HIV replication cycle and to further the development of HIV treatment.

Acknowledgement

This research was supported by a Grant-in-Aid from the Ministry of Education, Science and Culture.

References

1. Takahashi Y, Nakamura H, Ogata R, Matsuda N, Naganawa H, Taki-ta T, Iitaka Y, Sato K, Takeuchi T: Kijimicin, a polyether antibiotic. J Antibiotics 43: 441–443, 1990
2. Nakamura M, Ohno T, Kunimoto S, Naganawa H, Takeuchi T: Kijimicin: An inhibitor of human immunodeficiency virus in acutely and chronically infected cells. J Antibiotics 44: 569–571, 1991
3. Ito M, Baba M, Sato A, Panwels R, De Clercq E, Shigeta S: Inhibition effect of dextran sulfate and heparin on the replication of human immunodeficiency virus (HIV) *in vitro*. Antiviral Res 7: 361–367, 1987
4. Mitsuya H, Looney DJ, Kuno S, Ueno R, Wong-Staal F, Broder S: Dextran sulfate suppression of viruses in the HIV family: Inhibition binding to CD4+ cells. Science 240: 646–649, 1988
5. Popovic M, Sarngadharan MG, Read E, Gallo RC: Detection, isolation, and continuous production of cytopathic retroviruses (HTLV-III) from patients with AIDS and pre-AIDS. Science 224: 497–550, 1984
6. Chantal Petit AJ, Terpstra FG, Meidema F: Human immunodeficiency virus infection down-regulates HLA class II expression and induces differentiation in promonocytic U937 cells. J Clin Invest 79: 361–367, 1987
7. Hoffman AD, Banapour B, Levy JA: Characterization of the AIDS associated retroviruses reverse transcriptase and optical conditions for its detection *in vitro*. Virology 147: 326–335, 1985
8. Mosmann T: Rapid colorimetric assay for cellular growth and survival: Application to proliferation and cytotoxicity assays. J Immunol Methods 65: 55–63, 1983
9. Twentyman PR, Luscombe M: A study of some variables in a tetrazolium dye (MTT) based assay for cell growth and chemotherapy. Br J Cancer 56: 279–285, 1987
10. Hattori T, Shiozaki K, Eda Y, Tokiyoshi S, Matsushita A, Inaba H, Fujimaki M, Mguro T, Yamada K, Honda M: Characteristics of the principal neutralizing determinent of HIV-1 prevalent in Japan. AIDS Res Hum Retroviruses 7: 825–830, 1991
11. Saiki RK, Scharf S, Faloona F, Mullis KB, Horn GT, Erlich HA, Arnheim N: Enzymatic amplification of β-globin genomic sequences and restriction site analysis for diagnosis of sickle cell anemia. Science 230: 1350–1354, 1985
12. Dalgleish AG, Beverley PCL, Clapham PR, Weiss RA, Axel R: The CD4 antigen is an essential component of the receptor for the AIDS retrovirus. Nature 312: 763–766, 1984
13. Maddon PJ, Dalgleish AG, McDougal J, Clapham PR, Weiss RA, Axel R: The T4 gene encodes the AIDS virus receptors and expressed in the immune system and the brain. Cell 47: 333–348, 1986
14. Gartner S, Markovits P, Markovits DM, Kaplan MH, Gallo RC, Popovic M: The role of mononuclear phagocytes in HTLV-III/LAV infection. Science 233: 215–219, 1986
15. Gendelman HE, Orenstein JM, Baca LM, Weiser B, Burger H, Kalter DC, Meltzer MS: The macrophage in the persistence and pathogenesis of HIV infection. AIDS 3: 475–496, 1989
16. Ho DD, Rota TR, Hirsch MS: Infection of monocyte macrophages by human T lymphotropic virus type III. J Clin Invest 77: 1712–1715, 1986
17. Furman PA, Fyfe JA, St. Clair MH, Weinhold K, Rideout JL, Freeman GA, Nusinoff-Lehrman S, Bolognesi DP, Brider S, Mitsuya H, Barry DW: Phosphorylation of 3′-azido-3′-deoxythymidine and selective interaction of the 5′-triphosphate with human immunodeficiency virus reverse transcriptase. Proc Natl Acad Sci USA 83: 8333–8337, 1986
18. Dewar RI, Vasudevachari MB, Natarajan V, Salzman NP: Biosynthesis and processing of human immunodeficiency virus type 1 envelope glycoproteins: Effects of monencin on glycosylation and transport. J Virol 6: 2452–2456, 1989
19. Datema R, Olofsson S, Romero PA: Inhibitors of protein processing in viral systems. Pharmacol Ther 33: 221–286, 1987

Molecular and Cellular Biochemistry **119**: 43–50, 1993.

Function and energy metabolism of isolated hearts obtained from hyperthyroid spontaneously hypertensive rats (SHR)

A ^{31}P-nuclear magnetic resonance study

M. Heckmann[1], S. Lortet,[2] J. Aussedat,[2] A. Ray,[2], A. Rossi,[2] and H.-G. Zimmer[1]

[1] *Department of Physiology, University of Munich, Germany, and* [2] *Laboratoire de Physiologie Cellulaire Cardiaque, Université Joseph Fourier, Grenoble, France*

Abstract

It was the aim of this study to evaluate the effects of hyperthyroidism on heart function and cardiac energy metabolism of spontaneously hypertensive (SHR) rats. Hyperthyroidism was induced by daily injections of T_3 (0.2 mg/kg s.c.) for 14 days. The hearts were then isolated and perfused in the Langendorff mode. ATP, phosphocreatine (PCr), and inorganic phosphate (Pi) were measured continuously by means of ^{31}P-nuclear magnetic resonance (NMR) spectroscopy. Work load was altered by varying stepwise the Ca^{++} concentration in the perfusion fluid from 0.5 to 1.0, 1.5, and 2.0 mM, respectively. At every elevation of the Ca^{++} concentration, the increase in left ventricular developed pressure (LVDP) was higher in the hyperthyroid SHR than in the untreated SHR hearts. The ATP and PCr concentrations were lower in the hyperthyroid SHR compared to the untreated SHR hearts throughout the perfusion period. PCr decreased at every Ca^{++} elevation in both the untreated and hyperthyroid SHR hearts. The PCr/ATP ratio was not altered at any Ca^{++} concentration neither in the untreated SHR nor in the hyperthyroid SHR hearts. The Ca^{++}-induced stepwise elevation in LVDP was higher at any given PCr/Pi ratio in the hyperthyroid SHR than in the untreated SHR hearts. Thus, the Ca^{++}-inducible contractile reserve was greater in the hyperthyroid SHR heart. (Mol Cell Biochem **119:** 43–50, 1993)

Key words: ATP, adenine nucleotides, cardiac hypertrophy, nuclear magnetic resonance spectroscopy, phosphocreatine, rat heart function

Introduction

The spontaneously hypertensive rat (SHR) is an interesting animal model [1, 2] since the development of hypertension and cardiac hypertrophy is similar to that in patients with essential hypertension [3, 4]. Results of the Framingham heart study have shown that left ventricular hypertrophy is a substantial risk factor for morbidity and mortality. There was a good correlation between left ventricular mass as assessed by echocardiography and the incidence of and mortality from cardiovascular disease in subjects participating in this study [5].

We have shown recently that cardiac enlargement in

Address for offprints: H.-G. Zimmer, Physiologisches Institut, Universität München, Pettenkoferstr. 12, 8000 München 2, Germany

SHR can be further increased by triiodothyronine (T_3) treatment for 14 days. In these hyperthyroid SHR, mortality was higher than in the untreated SHR, and heart rate as well as the pressure-rate-product were elevated. However, left ventricular systolic pressure (LVSP) and the inotropic state as assessed by LV dp/dt_{max} were not altered significantly by T_3 [6]. From these results it appeared that T_3 had lost its positive inotropic effect in SHR.

It was one aim of this study to examine whether the inotropic response of the hyperthyroid SHR heart can be elicited. We have therefore evaluated the effect of Ca^{++} ions [7] in the isolated perfused Langendorff preparation [8] to eliminate the humoral and hormonal influences that are inextricably associated with the *in vivo* situation.

Another aim was to characterize the energy metabolism of the hyperthyroid SHR heart. We have used ^{31}P-nuclear magnetic resonance (NMR) spectroscopy to measure the cytosolic concentrations of free phosphorylated compounds such as adenosine triphosphate (ATP), phosphocreatine (PCr), and inorganic phosphate (Pi) in the isolated Langendorff heart. The concentration changes of the phosphorylated compounds were continuously monitored when cardiac work was increased [9] by stepwise elevation of the Ca^{++} concentration in the perfusion fluid from the control value of 0.5 mM to 1.0, 1.5 and 2.0 mM, respectively [8].

The results of this study show that the isolated hyperthyroid SHR heart responded to increased Ca^{++} concentration in the perfusion fluid with a graded elevation in left ventricular developed pressure (LVDP). Thus, there was some contractile reserve in the hyperthyroid SHR heart that could be activated by Ca^{++}. The overall ATP and PCr concentrations were lower in the hyperthyroid than in the untreated SHR hearts throughout the perfusion period. The Ca^{++}-induced stepwise elevation in LVDP was associated with a decrease in the PCr concentration and was higher at any given PCr/Pi ratio in the hyperthyroid SHR compared to the untreated SHR hearts. Thus, the contractile reserve that was recruitable by Ca^{++} was greater in the hyperthyroid SHR heart, but still lower than that reported earlier in control Wistar rat hearts [8].

Materials and methods

Female spontaneously hypertensive rats (SHR) with a body weight between 200 and 250 g were obtained from Savo GmbH, Kissleg, FRG and were of the same stock as those used for a recently published *in vivo*-investigation [6]. They were fed a control rat chow diet (Altromin C 100, Altromin GmbH, Lage, FRG) with free access to tap water. They were injected s.c. with 3,3'5-triiodo-L-thyronine (T_3) obtained from Sigma Chemie (München, FRG) in daily doses of 0.2 mg/kg body weight [10, 11]. T_3 was dissolved in 100 μl of 1 N NaOH, and this solution was made up to 1 l with 0.9% NaCl. The animals were divided into two experimental groups: 1. Daily s.c. injections of 0.9% NaCl for 14 days, 2. daily s.c. injection of T_3 for 14 days.

At the end of the 14 days treatment period, the animals received an i.p. injection of pentobarbital (50 mg/kg body weight) and an i.v. injection of heparin (1500 IU/kg body weight). The chest was opened, the heart was quickly excised and immediately placed in ice-cold physiological saline solution. A cannula was inserted into the aorta, and the perfusion was started at a constant flow of 14 ml/min/g tissue and at 37° C. The perfusion medium was a modified Krebs-Henseleit-solution containing (mM): NaCl 118, KCl 5.6, $MgCl_2$ 2.4, $NaHCO_3$ 21, $CaCl_2$ 0.5, glucose 9, pyruvate 2. The pH was adjusted to 7.4 by bubbling the perfusion fluid with a 95% O_2 – 5% CO_2 mixture.

A catheter was pierced through the left ventricular free wall at the apex to drain fluid which may accumulate in the left ventricular cavity. A fluid-filled latex balloon was inserted into the left ventricle via the left atrium for the measurement of pressure developed in the left ventricle. The balloon was connected to a P23 DB Statham pressure transducer via a polyethylene catheter (PE 190, 10 cm length). Pressure signals were monitored on a Gould 2200 recorder. The hearts were paced at a constant rate of 360 beats/min by two fine silver electrodes which were attached to the right atrium.

The heart was then placed in a sealed cylindrical glass chamber (15 mm diameter, 30 mm length) which was inserted into the NMR detection coil. To keep a constant temperature, a continuous flow of physiological salt solution at 37° C was maintained in the glass chamber containing the heart. The diastolic pressure of the left ventricle was fixed at 8 mm Hg by adjusting the volume of the balloon. The mechanical performance of the heart was assessed by measuring the left ventricular developed pressure (LVDP = systolic pressure – diastolic pressure).

^{31}P-NMR measurements were made at 81 MHz with a specially constructed five-turn horizontal solenoid placed in the bore of a 4.7 T vertical magnet (Oxford Instruments) interfaced with a CXP 200 Bruker spectrometer. The acquisition parameters based on optimal signal to noise ratio were pulse width = 17 μs, acquisition time = 0.34 sec, interpulse interval = 3 sec. In a separate set of experiments, fully relaxed spectra (interpulse interval = 15 sec) were recorded to calculate the saturation factors: 1.2 for PCr, 1.0 for ATP and Pi [8].

The areas under the peaks of PCr, Pi and βP-ATP of each spectrum were measured with a computerized planimetry system. The data were expressed as concentrations using the βP-ATP peak of the first spectrum as an internal standard with a concentration equal to the ATP concentration determined by high pressure liquid chromatography (see below). The concentrations were given in mM, assuming that all phosphorylated compounds detected by NMR spectroscopy are localized in the cytosol and that the cytosolic space is 600 μl/g wet weight [12].

The hearts were first perfused for 20 min with a 0.5 mM Ca^{++} concentration in the perfusion fluid. Thereafter the NMR spectra were collected, and LVDP was recorded continuously over the entire period of time. When the control measurements had been made after 30 min, the perfusate Ca^{++} concentration was raised to 1.0 mM, then to 1.5 mM and finally to 2.0 mM for 10 min, respectively. Each perfusion period with elevated Ca^{++} concentration was separated by 10 min of perfusion with 0.5 mM Ca^{++} concentration. The perfusion period with 2.0 mM Ca^{++} concentration in the perfusate was also followed by 10 min of perfusion with 0.5 mM Ca^{++} concentration. Thus the entire perfusion period lasted 90 min.

The concentration of adenine nucleotides was also determined using high pressure liquid chromatography (HPLC) methods. Briefly, the hearts were perfused for 20 min with 0.5 mm Ca^{++}, freeze-clamped, plunged into liquid nitrogen and extracted with perchloric acid (0.6 N). Measurements were done using a Partisil-10-ODS-2-Whatman column. The mobile phase was sodium pyrophosphate-phosphoric acid 25 mM, pH: 5.75, flow rate was 1.7 ml/min [13].

All results were expressed as mean ± SEM. For calculating significance, the Kruskal-Wallis rank sum test was used [6, 14]. A p-value smaller than or equal to 0.05 was considered to be statistically significant.

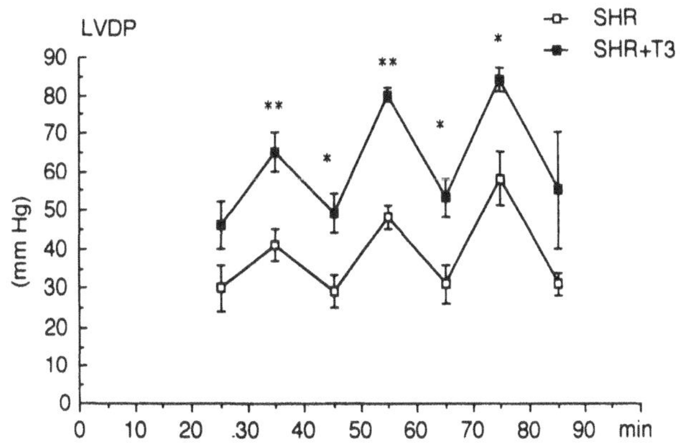

Fig. 1. Left ventricular developed pressure (LVDP) in isolated perfused Langendorff hearts obtained from untreated spontaneously hypertensive rats (SHR) and from SHR treated daily for 14 days with triiodothyronine (SHR + T_3). The perfusate contained 0.5 mM Ca^{++} during the first 30 min of perfusion time. The Ca^{++} concentration was then switched to 1.0, 1.5, and 2.0 mM for 10 min each followed by perfusion with 0.5 mM Ca^{++} for 10 min. Data are mean values ± SEM; number of SHR: 7, of SHR + T_3: 10. *$p < 0.05$, **$p < 0.01$.

Results

After the initial perfusion of the SHR hearts with a Ca^{++} concentration of 0.5 mM, the hyperthyroid group had a higher left ventricular developed pressure (LVDP) than the untreated group. The difference, however, was not statistically significant. When the Ca^{++} concentration in the perfusion fluid was stepwise elevated to 1.0, 1.5, and 2.0 mM, respectively, LVDP increased in both experimental groups, but to a greater extent in the hyperthyroid SHR hearts. The peak LVDP obtained at each perfusate Ca^{++} level was significantly higher in the hyperthyroid SHR than in the untreated SHR hearts. It returned to the control values in the untreated SHR hearts when the Ca^{++} concentration was switched back to 0.5 mM. In the hyperthyroid SHR hearts, however, there was a tendency to a progressively elevated LVDP during the control perfusion period with 0.5 mM Ca^{++} (Fig. 1).

When the hearts were perfused for 20 min with control Ca^{++} concentration and then freeze-clamped for HPLC measurements, the ATP and total adenine nucleotide concentrations were lower in the hyperthyroid SHR hearts (Table 1). Figure 2 shows the original NMR spectra of a heart isolated from a control SHR rat when it was perfused with 0.5 mM Ca^{++} (top panel) and with 2.0 mM Ca^{++} concentration (bottom panel). In Fig. 3,

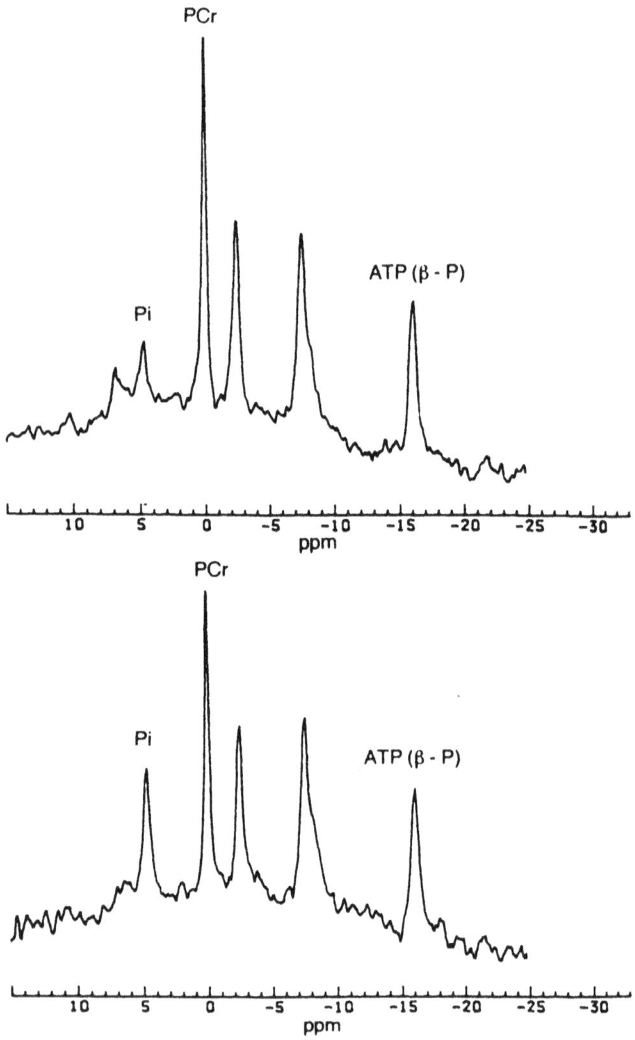

Fig. 2. ^{31}P-Nuclear magnetic resonance spectra obtained during the 3. control perfusion period with 0.5 mM Ca^{++} (see legend to Fig. 1) of a Langendorff heart isolated from a SHR (top panel) and during perfusion with 2.0 mM Ca^{++} concentration (bottom panel).

the respective NMR spectra of a hyperthyroid SHR heart are shown. In both experimental conditions, Ca^{++} induced an elevation in the intracellular concentration of inorganic phosphate (Figs 2 and 3).

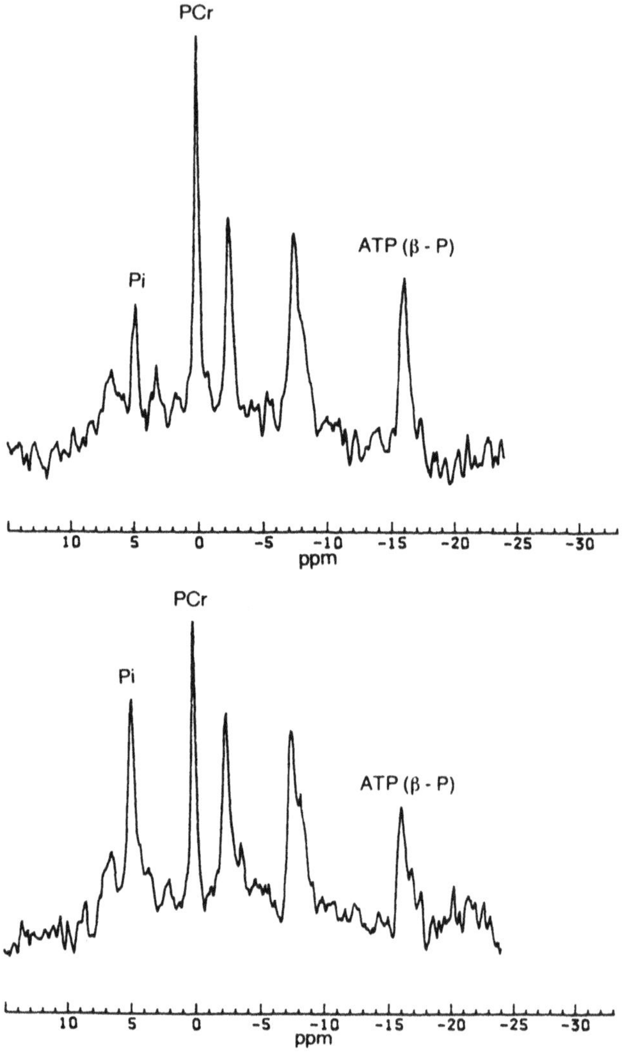

Fig. 3. ^{31}P-Nuclear magnetic resonance spectra obtained from a Langendorff heart isolated from a hyperthyroid SHR perfused with 0.5 mM Ca^{++} (top panel) and with 2.0 mM Ca^{++} concentration (bottom panel). Experimental protocol as in legend to Fig. 2.

Continuous ^{31}P-NMR monitoring throughout the entire perfusion period revealed that the ATP concentration was lower in the hyperthyroid than in the untreated SHR hearts, and that it declined slightly, but progres-

Table 1. Concentrations of adenine nucleotides (moles/g wet weight) in the hearts of spontaneously hypertensive rats (SHR) without and with T_3 treatment for 14 days. The hearts were perfused for 20 min in the Langendorff mode, and then freeze-clamped and processed for HPLC measurements

	n	ATP	ADP	AMP	ATP + ADP + AMP
SHR	7	3.9 ± 0.1	0.98 ± 0.05	0.17 ± 0.04	5.1 ± 0.1
SHR + T3	5	2.9 ± 0.1*	1.13 ± 0.12	0.38 ± 0.09	4.4 ± 0.1*

Data are mean values ± SEM; n = number of experiments. * $P < 0.05$ vs. SHR.

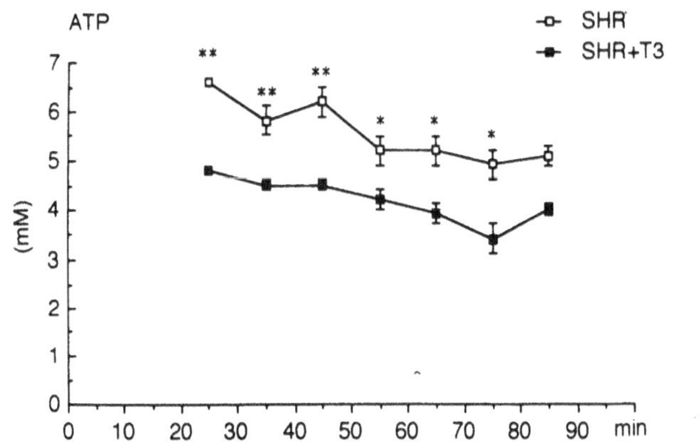

Fig. 4. ATP concentration in isolated perfused Langendorff hearts of SHR and SHR + T_3 during the entire perfusion period and with the same experimental protocol as described in the legend to Fig. 1. Mean values ± SEM; number of animals the same as in Fig. 1. * $p < 0.05$; ** $p < 0.01$.

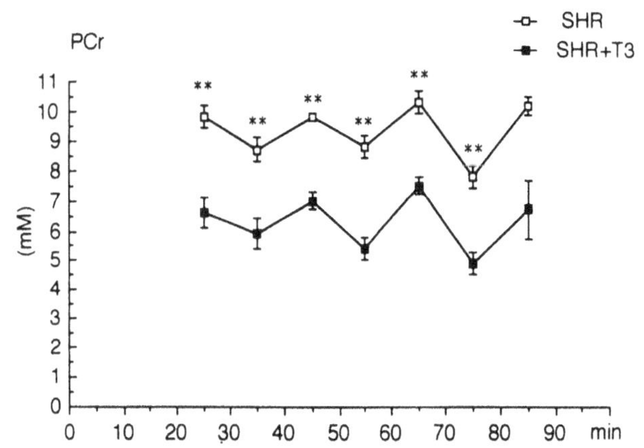

Fig. 5. Concentration of phosphocreatine (PCr) in isolated perfused SHR and SHR + T_3 hearts. Same protocol of perfusate Ca^{++} change, same number of animals as in legend to Fig. 1. * $p < 0.05$; ** $p < 0.01$.

sively in both experimental groups independently of the respective Ca^{++} concentration in the perfusate (Fig. 4).

The PCr concentration was also lower in the hyperthyroid compared to the untreated SHR hearts. It declined at every step of the Ca^{++} concentration elevation in the untreated and hyperthyroid SHR hearts (Fig. 5). The Pi concentration tended to be higher in the hyperthyroid SHR hearts. It increased stepwise with each Ca^{++} elevation in both experimental groups (Table 2). As a consequence, the PCr/Pi ratio decreased with increasing Ca^{++} concentration in both groups. It was always lower in the hyperthyroid SHR hearts, but this difference was not statistically significant (Table 2). The PCr/ATP ratio did not change throughout the experimental protocol and was only slightly, though not significantly lower in the hyperthyroid SHR hearts (Table 2).

When the PCr/Pi ratio was plotted versus the LVDP in both the untreated and hyperthyroid SHR hearts under control conditions and at every Ca^{++} concentration, two separate curves were obtained (Fig. 6). The curve of the hyperthyroid SHR hearts was shifted to the right so that the Ca^{++}-induced stepwise elevation in LVDP was higher in the hyperthyroid SHR than in the untreated SHR hearts at any given PCr/Pi ratio.

Discussion

In this study we have used the spontaneously hypertensive rat (SHR), since the development of hypertension and cardiac hypertrophy in these animals occurs in successive stages which are similar to those in patients with essential hypertension [3, 4]. It has been shown recently, that the SHR used in this study were clearly hypertensive and had already developed substantial cardiac hypertrophy when compared with sex- and body weight-mached control rats [6]. It was therefore of in-

Table 2. Intracellular concentrations of inorganic phosphate (Pi), the phosphocreatine/Pi (PCr/Pi) ratio, and the PCr/ATP ratio in isolated Langendorff perfused hearts of spontanously hypertensive rats (SHR) without and with triiodothyronine (T_3) treatment for 14 days. Data are mean values ± SEM for different Ca^{++} concentrations in the perfusion medium. Number of experiments in parentheses

$[Ca^{2+}]$	0.5 mM	1.0 mM	1.5 mM	2.0 mM
Pi (mMol)				
SHR	1.3 ± 0.3 (7)	3.2 ± 0.5 (7)	3.6 ± 0.4 (7)	4.4 ± 0.5 (6)
SHR + T_3	2.5 ± 1.0 (10)	3.5 ± 1.0 (10)	4.3 ± 0.9 (10)	5.8 ± 1.1 (8)
PCr/Pi				
SHR	10.4 ± 2.2 (7)	3.2 ± 0.5 (7)	2.7 ± 0.5 (7)	2.0 ± 0.3 (6)
SHR + T_3	8.2 ± 2.2 (10)	2.6 ± 0.5 (10)	1.7 ± 0.3 (10)	1.2 ± 0.4 (8)
PCr/ATP				
SHR	1.5 ± 0.1 (7)	1.5 ± 0.1 (7)	1.7 ± 0.2 (7)	1.7 ± 0.1 (6)
SHR + T_3	1.4 ± 0.1 (10)	1.3 ± 0.1 (10)	1.3 ± 0.1 (10)	1.5 ± 0.2 (8)

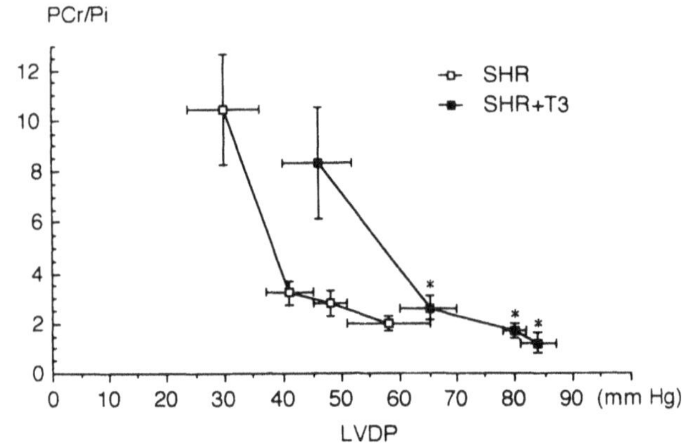

Fig. 6. PCr/Pi ratio plotted against left ventricular developed pressure (LVDP) for isolated perfused SHR and SHR + T_3 hearts at different perfusate Ca^{++} concentrations (see legend to Fig. 1). Data are mean values ± SEM; number of animals as in legend to Fig. 1. * $p < 0.05$.

terest to examine whether the Langendorff hearts obtained from the untreated SHR may respond to an increase in work load similarly to or differently from control rats [8]. The response of the SHR hearts was continuously monitored in terms of energy metabolism and functional performance.

Several previous studies have shown that the PCr, Pi and ATP contents of the isolated rat heart were affected by the work load imposed by different conditions of perfusion [8, 12, 15]. The advantages of ^{31}P-NMR spectroscopy to monitor continuously the changes in phosphorylated compound levels are obvious [16–19]. As in a recently published study, work load was changed by elevation of the perfusate Ca^{++} concentration [8]. This approach [9] was adopted since graded changes in cardiac work can easily be induced over a wide range, and since the basal cardiac work can be maintained at a low level at which substrates and oxygen supply can be neglected as limiting factors for the activity of this isolated Langendorff preparation.

Comparable to the results obtained in Langendorff experiments on Wistar rats [8] the stepwise elevation of the perfusate Ca^{++} concentration induced a graded increase in the peak LVDP of the untreated SHR hearts (Fig. 1). This was associated with a graded decline in PCr (Fig. 5) and with an elevation in Pi (Table 2). Similar to the situation in the Langendorff Wistar rat heart preparation [8], the ATP concentration in the untreated SHR heart did not change in response to the elevation of the perfusate Ca^{++} concentration. It tended to decline over time indicating a slight deterioration of the metabolic status of this heart preparation (Fig. 4).

Despite the similarities in the response to Ca^{++} ions of the Langendorff Wistar rat and SHR heart preparation, there was one remarkable difference, and this was related to the magnitude of the Ca^{++}-elicited functional performance. In the Wistar rat heart, the increase in the LVDP with the highest Ca^{++} concentration amounted to 500% [8], whereas the peak LVDP response of the SHR heart was 100% compared to the initial value (Fig. 1). Thus, the Ca^{++}-inducible contractile reserve appeared to be diminished in the SHR heart. This can be explained at the molecular level of the contractile apparatus. It has been shown that a redistribution of the myosin isoenzyme pattern towards V_3 occurs in SHR [20]. In the normal rat heart, V_1 is the predominant myosin isoenzyme. This is associated with a high myofibrillar ATPase activity and increased shortening velocity. Predominance of V_3 is reflected in a reduction of both parameters [21]. Thus, when the SHR heart is stimulated with Ca^{++}, it cannot generate as much pressure as a normal control heart.

The situation became different when the SHR were made hyperthyroid for two weeks. Experimental hyperthyroidism has been shown to lead to an increase in the number of cardiac β-adrenergic receptors [10, 11, 22–24] and to the exclusive presence of V_1 in the myocardium [25, 26]. Since swimming training prevented the myosin isoenzyme redistribution towards V_3 in SHR [20], it can safely be assumed that T_3 had a similar effect. This is supported by our findings that the Ca^{++}-induced elevation in LVDP in the hyperthyroid SHR hearts was higher than in the untreated SHR hearts (Fig. 1), and that peak LVDP of the hyperthyroid SHR hearts was shifted to the right at any given PCr/Pi ratio (Fig. 6). Thus, the Ca^{++}-recruitable contractile performance was clearly higher in hyperthyroidism.

In a recent *in vivo* -study it has been shown that the effect of T_3 in SHR is similar in some respects to that in Sprague-Dawley rats. T_3 increased heart rate in SHR to exactly the same degree as it did in normal Sprague-Dawley rats and induced an additional increase in heart weight. In contrast to the results in Sprague-Dawley rats, T_3 did not affect significantly LVSP and LV dp/dt_{max}, and cardiac output in SHR [6]. Although T_3 induced furthe enlargement of the heart in SHR, this additional increase in the heart weight/body weight ratio was less pronounced (+ 38%) than in Sprague-Dawley rats (+ 57%). Thus, the T_3-mediated additional growth response seems to be limited in the SHR.

The present studies characterize the hyperthyroid SHR in terms of cardiac energy metabolism. The ATP and PCr concentrations were lower in the Langendorff hearts of hyperthyroid SHR (Figs 4, 5). In a recent ^{31}P-NMR study it has been shown that thyroxine treatment for 7 days resulted in an increase in Pi and a decrease in PCr with no change in ATP or ADP concentration under the same conditions of cardiac work [27]. When the perfusate Ca^{++} concentration was stepwise elevated, the PCr decreased progressively, and the Ca^{++}-induced contractile force was higher than in the untreated SHR hearts (Fig. 6). Thus, although T_3 could not elicit the usual inotropic response *in vivo* [6], there was a contractile reserve that was recruitable by Ca^{++}. However, it was still lower than in control Wistar rat hearts [8].

Thc rcsults of this *in vitro*-study may also be of interest from the point of view of clinical pathophysiology. It may be concluded that chronically hypertrophied hearts of multiple causes can be forced to increased contraction by an inotropic stimulus (Figs 1, 6), and that they have the potential for further enlargement [6]. However, one has to realize that these hearts have a diminished level of high energy phosphate compounds, and that the PCr concentration becomes further reduced with any inotropic intervention. Of particular relevance is the observation that the inotropic reserve of the hypertrophied hearts is much lower than in untreated healthy control hearts. Thus, this type of combined cardiac hypertrophy may be regarded to be 'pathophysiological' [21].

Acknowledgements

This work was supported by grants from Deutsche Forschungsgemeinschaft (Zi 199/8-1,2) and from Deutscher Akademischer Austauschdienst (French/German PROCOPE exchange program 311-pro-ca). The excellent technical assistance of Mrs. Birgit Sender, Ms. Anne Bölle, Ms. Sabine Weiland, and Mr. H. Reutenauer is gratefully acknowledged.

References

1. Folkow B, Hallbäck M, Lundgren Y, Weiss L: Background of increased flow resistance and vascular reactivity in spontaneously hypertensive rats. Acta Physiol Scand 80: 93–106, 1970
2. Spech MM, Ferrario CM, Tarazi RC: Cardiac pumping ability following reversal of hypertrophy and hypertension in spontaneously hypertensive rats. Hypertension 2: 75–82, 1980
3. Frohlich ED: Is the spontaneously hypertensive rat a model for human hypertension? J Hypertension 4 (suppl 3): S15–S19, 1986
4. Trippodo NC, Frohlich ED: Similarities of genetic (spontaneous) hypertension. Man and rat. Circ Res 48: 309–319, 1981
5. Levy D, Garrison RG, Savage DD, Kannel WB, Castelli WP: Prognostic implications of echocardiographically determined left ventricular mass in the Framingham heart study. N Engl J Med 322: 1561–1566, 1990
6. Heckmann M, Zimmer H-G: Effects of triiodothyronine in spontanously hypertensive rats. Studies on cardiac metabolism, function, and heart weight. Basic Res Cardiol 87: 333–343, 1992
7. Lortet S, Zimmer H-G, Rossi A: Inotropic response of the rat heart during development and regression of triodothyronine-induced hypertrophy. J Cardiovasc Pharmacol 14: 707–712, 1989
8. Aussedat J, Ray A, Lortet S, Reutenauer H, Grably S, Rossi A: Phosphorylated compounds and function in isolated hearts: a ^{31}P-NMR study. Am J Physiol 260: H110–H117, 1991
9. Chevalier B, Mouas C, Mansier P, Aumont MC, Swynghedauw B: Screening of inotropic drugs on isolated rat and guinea pig hearts. Methods Pharmacol 17: 313–326, 1987
10. Zierhut W, Zimmer H-G: Triiodothyronine-induced hemodynamic and metabolic changes: effects of α- and β-adrenergic blockade. Basic Res Cardiol 84: 359–370, 1989
11. Zimmer H-G, Peffer H: Metabolic aspects of the development of experimental cardiac hypertrophy. Basic Res Cardiol 81 (Suppl 1): 127–137, 1986
12. Giesen J, Kammermeier H: Relationship of phosphorylation potential and oxygen consumption in isolated perfused rat hearts. J Mol Cell Cardiol 12: 891–907, 1980
13. Aussedat J: Effect of uridine supply on glycogen resynthesis after ischaemia in the isolated perfused rat heart. Cardiovasc Res 17: 145–151, 1983
14. Rosner B: Fundamentals of biostatistics. PWS-Kent Publishing Company, Boston 1990
15. Kobayashi K, Neely JR: Effects of increased cardiac work on pyruvate dehydrogenase activity in hearts from diabetic animals. J Mol Cell Cardiol 15: 347–357, 1983
16. Clarke K, Willis RJ: Energy metabolism and contractile function in rat heart during graded, isovolumic perfusion using ^{31}P nuclear magnetic resonance spectroscopy. J Mol Cell Cardiol 19: 1153–1160, 1987
17. From AHL, Petein MA, Michurski SP, Zimmer SD, Ugurbil K: ^{31}P-NMR studies of respiratory regulation in the intact myocardium. FEBS Lett 206: 257–261, 1986
18. Matthews PM, Williams SR, Seymour AM, Schwartz A, Dube G, Gadian DG, Radda GK: A ^{31}P-NMR study of some metabolic and functional effects of the inotropic agents epinephrine and ouabain, and the ionophore RO2-2985 (X537A) on the isolated, perfused rat heart. Biochim Biophys Acta 720: 163–171, 1982
19. Zweier JL, Jacobus WE: Substrate-induced alterations of high energy phosphate metabolism and contractile function in the perfused heart. J Biol Chem 262: 8015–8921, 1987
20. Rupp H, Jacob R: Response of blood pressure and cardiac myosin polymorphism to swimming training in the spontaneously hypertensive rat. Can J Physiol Pharmacol 60: 1098–1103, 1982
21. Jacob R, Ebrecht G, Kissling G, Rupp H, Takeda N: Functional consequences of cardiac myosin isoenzyme redistribution. In: H Rupp (ed). The regulation of heart function. Basic concepts and clinical applications. Thieme, Stuttgart New York 1986, pp 305–326
22. Limas C, Limas CJ: Influence of thyroid status on intracellular distribution of cardiac adrenoceptors. Circ Res 61: 824–828, 1987

23. Tsai JS, Chen A: Effect of L-triiodothyronine on (-)^{3}H-dihydroalprenolol binding and cyclic AMP response to (-) adrenaline in cultured heart cells. Nature 275: 138–140, 1978
24. Williams LT, Lefkowitz RJ, Watanabe AM, Hathaway DR, Besch HR: Thyroid hormone regulation of β-adrenergic receptor number. J Biol Chem 252: 2787–2789, 1977
25. Hoh JFY, McGrath PA, Hale PT: Electrophoretic analysis of multiple forms of rat cardiac myosin: Effects of hypophysectomy and thyroxine replacement. J Mol Cell Cardiol 10: 1053–1076, 1977
26. Hoh JFY, Egerton LJ: Action of triiodothyronine on the synthesis of rat ventricular myosin isoenzymes. FEBS Lett 101: 143–148, 1979
27. Seymour A-ML, Eldar H, Radda GK: Hyperthyroidism results in increased glycolytic capacity in the rat heart: A ^{31}P-NMR study. Biochim Biophys Acta 1055: 107–116, 1990.

Molecular and Cellular Biochemistry **119**: 51–57, 1993.

Increased calcium release from sarcoplasmic reticulum stimulated by inositol trisphosphate in spontaneously hypertensive rat heart cells

Hideaki Kawaguchi, Hitoshi Sano, Hitoshi Okada, Kenji Iizuka, Hiroshi Okamoto, Toshiyuki Kudo, Takeshi Murakami and Akira Kitabatake
Department of Cardiovascular Medicine, Hokkaido University School of Medicine, Sapporo 060 Japan

Abstract

It is known that inositol (1, 4, 5)-trisphosphate (IP_3) stimulates Ca^{2+} release from sarcoplasmic reticulum (SR) in several tissues, but in cardiac myocytes this phenomenon has not been confirmed. The purpose of the present study was to confirm the effect of (1, 4, 5)-IP_3 on Ca^{2+} release from SR in cardiac myocytes. The effect of IP_3 on Ca^{2+} release from SR in hypertrophic cardiac cells was also determined.

We examined the effects of IP_3 on Ca^{2+} release from cardiac myocyte SR by the bigital-image method in a single cell. We also determined the effect of IP_3 on calcium release from isolated SR. SR was prepared from spontaneous hypertensive rat hearts and Wistar kyoto rat hearts. The SR was prelabeled with $^{45}Ca^2+$, and then incubated with the indicated concentrations of IP^3 for 1 min at 37° C. In cardiac myocytes treated with saponin, Ca^{2+} release stimulated by 10 μM (1, 4, 5)-IP_3 was detected by fura-2. In $^{45}Ca^{2+}$ prelabeled SR, the maximal Ca^{2+} release was achieved at 10 μM IP_3 incubated for 1 min. The release of Ca^{2+} was higher in SR of SHR than in the SR of WKY. IP_3 stimulates Ca^{2+} release from cardiac SR, and this release is greater in SHR than in WKY. However, it is uncertain whether this phenomenon plays a role in cardiac hypertrophy. (Mol Cell Biochem **119:** 51–57, 1993)

Key words: Inositol-1, 4, 5-trisphosphate, sarcoplasmic reticulum, calcium release, cardiac hypertrophy, PI-response, cardiac myocyte

Introduction

The phosphatidylinositol (PI) turnover pathway generates two second messengers: IP_3 and sn-(1, 2)-diacylglycerol (DAG) [1]. DAG stimulates membrane-bound phospholipid-dependent, Ca^{2+}-dependent protein kinase C [2], while IP_3 releases Ca^{2+} from sarcoplasmic reticulum stores [3, 4]. Removal of IP_3 occurs by two different pathways: hydrolysis to inositol (1, 4)-bisphosphate (IP_2), and phosphorylation to inositol (1, 3, 4, 5)-tetrakisphosphate (IP_4) [5, 6]. IP_4 is subsequently dephosphorylated to (1, 3, 4)-IP_3, which is relatively inactive in releasing Ca^{2+} compared with the (1, 4, 5)-IP_3 [7]. While the physiological significance of this PI-turnover pathway is unknown in any mammalian cell system, IP_4 has been shown to enhance Ca^{2+} entry into the cell in sea urchin eggs [8]. However, this phenomenon has not been proven in cardiac myocytes. This

Address for offprints: H. Kawaguchi, Department of Cardiovascular Medicine, Hokkaido University School of Medicine, Sapporo 060, Japan

phosphorylation/dephosphorylation pathway is generally considered an integral part of the PI-turnover pathway. This PI-turnover pathway may have an important role in inducing cardiac myocyte hypertrophy, even in adult rat myocytes. Recently, we showed that PI turnover pathway in cardiomyopathic hamster heart cells was more active than in control cells [9]. Cardiomyopathic hamster heart cell develops hypertrophy without any pressure overload. Therefore, whether or not PI-turnover pathways increase in myocytes with a pressure overload remains to be shown on next step. Hence, we determined that phosphatidylinositol and polyphosphoinositide metabolism [10, 11] and IP_3 kinase activity [6] increased in spontaneously hypertensive rat heart. But it is still a matter of dispute whether (1, 4, 5)-IP_3 can release Ca^{2+} from sarcoplasmic reticulum in cardiac myocytes [12, 13].

The purpose of this experiment is to confirm the effect of (1, 4, 5)-IP_3 on Ca^{2+} release from sarcoplasmic reticulum in cardiac myocytes. We have also determined the effect of IP_3 on Ca^{2+} release from sarcoplasmic reticulum in hypertrophic cardiac cells.

Materials and methods

Subcellular fractions

Hearts were excised from spontaneously hypertensive rats (SHR; 10, 15 and 20 weeks old) and age matched Wistar Kyoto rats, (WKY) and washed with saline. Left ventricles were minced and homogenized for 60 seconds by a polytron homogenizer (PT-19, Kinematica, Switzerland) in 5 volumes (vol/wt) of 0.1 M Tris-HC1 (pH 7.4) [14]. The homogenate was centrifuged at 1,000 g for 10 minutes, and the resultant supernatant was centrifuged at 14,000 g for 20 minutes to remove the mitochondria. The supernatant was centrifuged at 45,000 g for 30 minutes according to previously published methods [15]. The pellet was washed twice with the same buffer and suspended in 0.6 M KCl and 10 mM tris(hydroxymethyl)methyl-2-aminomethane-sulfonic acid (TES) pH 7.4. The suspension was centrifuged at 45,000 g for 30 min. The resultant pellet was suspended in 150 mM KCl and 1 mM TES pH 7.1 and used for each assay as sarcoplasmic reticulum. The yield of sarcoplasmic reticulum protein in SHR hearts was similar to WKY hearts (0.82 ± 0.1 and 0.84 ± 0.2 mg/g, respectively). Na^+-K^+p-ATPase activity and cytochrome c oxidase activity in sarcoplasmic reticulum fraction were $<1\%$ of the total activities and were similar in two groups.

Calcium release from sarcoplasmic reticulum

$^{45}Ca^{2+}$ (4 μCi) was loaded into a freshly prepared sarcoplasmic reticulum by incubating 400 μg of sarcoplasmic reticulum in a final reaction mixture of 400 μl in cytosolic buffer (KCl 120 mM, NaCl 10 mM, KH_2PO_4 1 mM, $NaHCO_3$ 5 mM, HEPES 10 mM, pH 7.1), $MgCl_2$ 5 mM, and ATP 5 μM for 10 min [4]. The final free Ca^{2+} concentration was calculated according to the method of Fabiato and Fabiato [16] and adjusted to 1 μM. In Ca^{2+} release experiments, IP_3 was added to the 100 μl-aliquot of $^{45}Ca^{2+}$-loaded sarcoplasmic reticulum in the presence of 1 mM EGTA, 10 mM LiCl and 5 mM 2, 3-diphosphoglycerate (2, 3-DPG); this concentration of 2, 3-DPG inhibited the dephosphorylation of IP_3), and it was immediately mixed with the reaction medium. Ca^{2+} release started from this time. The experiments were carried out at 37° C. After the indicated incubation periods, the incubation mixture was filtered through a 0.45 μ Millipore filter, washed three times with 3 ml of cytosolic buffer, and counted in 5 ml of Ready Gel. The zero-time Ca^{2+} content of the sarcoplasmic reticulum was obtained by diluting the original incubate and immediately filtering it through a Millipore filter. The total sarcoplasmic reticulum calcium content was 388 ± 22 nmol/mg protein. There was no difference between two groups in calcium content of sarcoplasmic reticulum.

Cell preparation

Cardiac myocytes of SHR and WKY were prepared according to a previously reported method in phosphate buffer (PB) [17], and cultured in Hams F-10 medium with 10% fetal calf serum (FCS) until use. The freshly prepared cells were maintained at 37° C in a humidified 5% Co_2/95% air atmosphere [18]. The cells were then subcultured at 1.0×10^5 cells/dish in 0.5 ml of Krebs-Henselite (KH), HEPES buffer containing (mM) NaCl 140, KCl 5.9, NaH_2PO_4 1.2, $MgCl_2$ 1.4, $CaCl_2$ 1.8, HEPES 5, and glucose 11.5, pH 7.4 [19]. More than 90% of cells in this medium (0.3 mM $CaCl_2$) were rod-shaped, and they did not beat until NE and 1 mM $CaCl_2$ (final concentration) were added. They were lightly attached to dishes. Fura-2 AM (acetoxymethyl ester) was loaded into cardiac myocytes for 60 min. The final fura-2 AM concentration was 4 μM. At the end of enzymatic digestion, the remaining extracellular fura-2 AM was removed by washing three times with 3 Ca^{2+}-free KH, HEPES buffer [20], and the cells were made permeable with saponin (10 μg/ml). This concentration of saponin had no effect on enzyme activity [21]. The cells were preincubated with saponin for 1 min at 37° C in cytosolic buffer pH 7.1, 10 mM LiCl, 5 mM 2, 3-DPG, EGTA 0.2 mM and $CaCl_2$ 150 nM. Then the indicated concentrations of IP_3 were added to the cells, and intracellular Ca^{2+} was analyzed by a digital-image method. In these cells treated with saponin, there was little leakages of fura-2 from saponin-treated cells during the measuring of the calcium concentration.

Measurement of intracellular Ca^{2+}

Cytosolic Ca^{2+} concentrations $[Ca^{2+}]i$ were measured using furo 2-AM (acetoxymethl ester)-loaded myocytes stimulated by 1 μM NE in the presence of 1 μM metoprolol. For loading with fura-2, cell suspensions (1×10^5/ml) were incubated with 5 μM of furo 2-AM for 30 min in PB containing 0.1% bovine serum albumin. Cells were then washed with Hanks' balanced salt solution with 1.26 mM calcium and incubated in 0.7 ml of Hanks' balanced salt solution (also containing 1.26 mM Ca^{2+}). The ratio of the absorbance at 340 and 380 nm was monitored by fluorospectrometer [9]. The $[Ca^{2+}]i$ was calibrated by exposing cells to 4–10 μM ionomycin followed by the addition of 6.6 mM EGTA.

Digital image method

The experiments were performed on an Olympus OSP microscone system (Casals; Olympus, Japan), frequencies (340 and 380 nm) were selected by means of an interference filter. The excitation source was a 150 W burner (Olympus, Japan). The fluorescent image was detected with a digital-image analyzer (Hamamatsu C2741 and DVS-3000; Hamamatsu, Japan) [22].

Table 1. Left ventricular weights of SHR and WKY

Age (weeks)		5	10	20
WKY	LV/BW	150 ± 11	145 ± 3	160 ± 0
	BW	150 ± 5	170 ± 5	290 ± 10
SHR	LV/BW	175 ± 18	287 ± 12*	370 ± 22*
	BW	153 ± 7	166 ± 9	300 ± 11

* $P < 0.001$. Figures represent mean left ventricular weight (mg/100 g body weight) ± SE (LV/BW) or body weight (g) ± SE (BW).

Statistical analysis

Triplicate samples were analyzed in all experiments. Results are expressed as means ± SEM. Statistical significance was estimated using the previously described method (ANOVA) taking $p < 0.05$ as the limit of significance [23].

Results

Heart weight

Left ventricular weight was increased with age. SHR heart weights increased markedly compared to WKY heart weights (Table 1).

Intracellular Ca^{2+}

The accelerated PI-turnover may affect on intracellular free Ca^{2+} concentrations. Figures 1a and 1b show the effect of NE on intracellular free Ca^{2+} concentrations. In WKY increased Ca^{2+} signal was transient (Fig. 1b). NE markedly stimulated cytosolic Ca^{2+} concentrations in SHR. The effect of NE on peak cytosolic free Ca^{2+} concentrations were higher in SHR than in age matched WKY.

Effect of IP_3 on calcium release from SR

Figure 2 shows that the addition of IP_3 to $^{45}Ca^{2+}$-preloaded sarcoplasmic reticulum from 10 week-old SHR and WKY, in a medium containing 1 μM free Ca^{2+}, induced a transient release of Ca^{2+} and achieved a maximum effect after 1 min. The Ca^{2+} release from sarcoplasmic reticulum in SHR was significantly higher than in WKY. Figure 3 shows the concentration dependence of the IP_3-induced Ca^{2+} release. The maximum release was reached at 10 μM IP_3 for 1 min incubation at 37° C, this release was higher in SHR than in WKY. The IP_3-induced calcium increase with age was

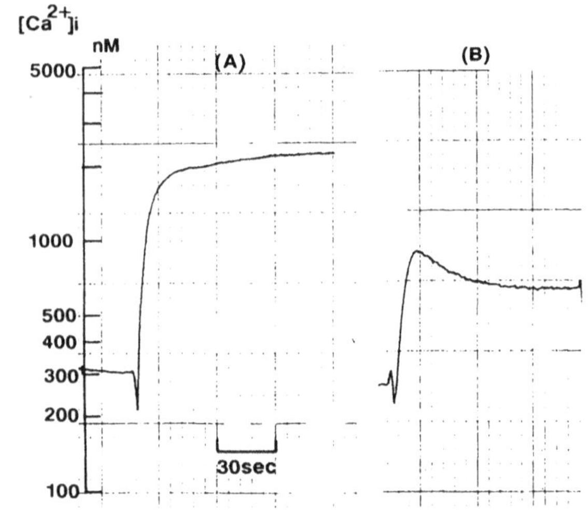

Fig. 1. Effect of norepinephrine on cytosolic free Ca^{2+} concentration. The intracellular free Ca^{2+} concentration stimulated with 1 μM NE in the presence of 1 μM metoprolol using fura-2- loaded cardiac myocytes were determined as described in Materials and methods. (a) SHRSP aged 10 weeks, (b) WKY aged 10 weeks.

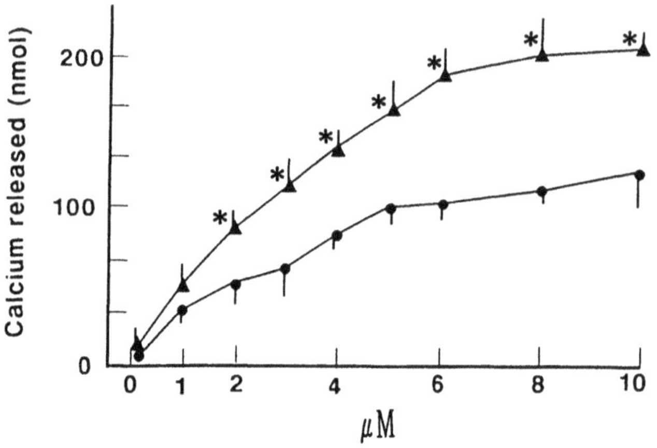

Fig. 3. Effect of concentrations of IP_3 on Ca^{2+} release from sarcoplasmic reticulum. The Ca^{2+} release was determined in the presence of the indicated concentrations of IP_3 for 1 min as described in Materials and methods. SHR; (▲), WKY (●) with IP_3, Ca^{2+} release was expressed as means ± SEM nmol/mg protein.

significant in both SHR and WKY, with SHR showing greatest increase (Fig. 4).

Effect of IP_3 on calcium release in single cell

In order to study the effect of IP_3 on calcium release from sarcoplasmic reticulum, the release of Ca^{2+} stimulated by IP_3 in permeable cells was determined. After a 1 min treatment with saponin in cytosolic buffer, 10 μM IP_3 was added to the permeable cells. The intracellular Ca^{2+} concentration was markedly increased within 1 min after the addition of IP_3 (Fig. 5). The average peak Ca^{2+} concentration was 800–1,000 nM. Cells slightly contracted in response to the IP_3 (Figs 5c and 5d compared to 5a and 5b).

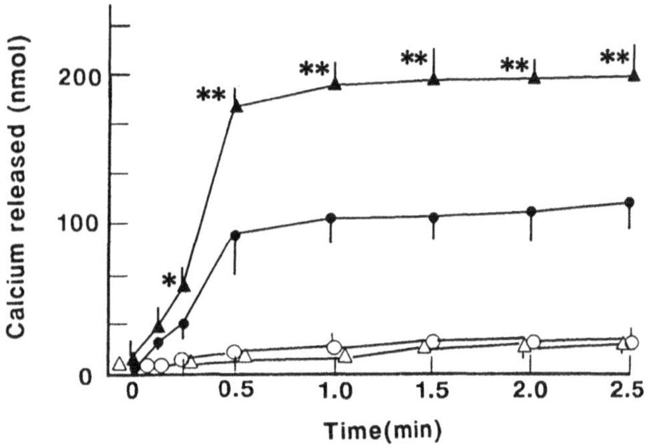

Fig. 2. The effect of IP_3 on Ca^{2+} release from sarcoplasmic reticulum. Time course of Ca^{2+} release from sarcoplasmic reticulum was determined in the presence of 10 μM IP_3 for the indicated periods as described in Materials and methods. WKY (●) with IP_3, WKY without IP_3 (○), SHR with IP_3 (▲), SHR without IP_3 (△), Ca^{2+} release was expressed as means ± SEM nmol/mg protein.

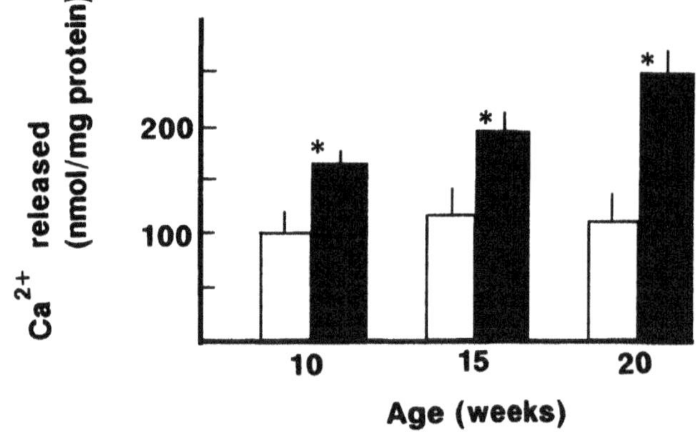

Fig. 4. Effect of age on calcium release from sarcoplasmic reticulum by IP_3. Calcium release from sarcoplasmic reticulum was determined as described in Materials and methods. Open bar; WKY, Hatched bar; SHR, Ca^{2+} release was expressed as means ± SEM nmol/mg protein.

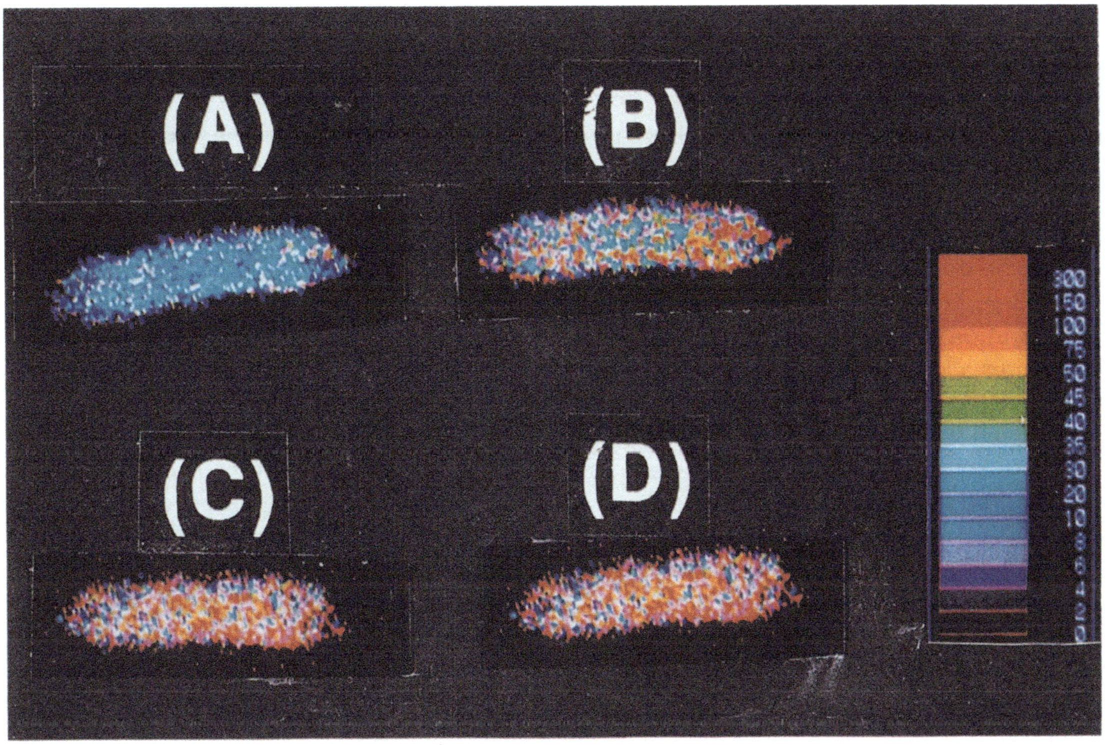

Fig. 5. Effect of IP_3 on Ca^{2+} release from SR. The release of Ca^{2+} from SR in the presence of 10 μM IP_3 was determined as described in Materials and methods. A pseudocolour image of Ca^{2+} in a single cardiac myocyte is shown. The color reads from low (purple) to high (red) Ca^{2+} concentrations. (A) control (B) 30 sec, (C) 1 min and (D) 2 min after addition of IP_3, respectively.

Discussion

The present study determined that IP_3 stimulates Ca^{2+} release from cardiac sarcoplasmic reticulum and that this release is higher in SHR than in WKY.

Previously, we reported that the metabolism of phosphatidylinositol and polyphosphoinositide are enhanced in SHR hearts and that intracellular calcium concentration is higher in SHR than in WKY [6, 10, 11]. These results suggest that phosphatidylinositol specific phospholipase C activity plays an important role in the development or maintenance or cardiac hypertrophy in SHR heart cells.

An α_1-adrenergic receptor-mediated increase in IP_3 has been reported in human cardiac tissue [24]. It was also reported that α_1-adrenergic stimulation causes IP_4 to accumulate in adult canine myocytes [25]. These reports suggest that the accumulations of IP_3 and IP_4 after hormonal stimulation play some physiological role that may involve the alteration of Ca^{2+} levels in cardiac tissues. It is further reported that 1, 3, 4, 5-IP_4 enhances Ca^{2+} entry into some cells and potentiates the Ca^{2+} mobilizing effect of 1, 4, 5-IP_3 [8]. Therefore, 1, 4, 5-IP_3 and 1, 3, 4, 5-IP_4 appear to play an important role in controlling Ca^{2+} levels in response to hormonal stimulation [25, 26]. However, it remains to be proved that this phenomenon occurs in cardiac cells. In previous studies, we clarified that PLC, as well as IP_3 kinase activity increased in isolated SHR heart cells [10, 11].

It is reported that α_1-adrenergic stimulation plays an important role in inducing hypertrophy of neonatal rat heart cells [27, 28]. The expression of protooncogene c-myc is known to increase with growth factor-induced cell division, according to studies involving cardiac myocytes [28]. It was shown that NE increased levels of c-myc encoded mRNA to 10-fold over control levels. This response was abolished by α_1-antagonists but was not affected by β-adrenergic antagonists. These findings show that α_1-adrenergic stimulation plays an important role in inducing neonatal cardiac myocyte hypertrophy. But it has not been reported that the effect of IP_3 on calcium release from SR in hypertrophic cardiac cells.

The present study determined the effect of IP_3, polyphosphoinositide metabolite, on calcium release from cardiac sarcoplasmic reticulum. While it is known that IP_3 stimulates Ca^{2+} release from sarcoplasmic reticulum in hepatocytes (3) and platelets (4), its effect on Ca^{2+} release from sarcoplasmic reticulum in cardiac cells remains controversial [12, 13]. Further, while it was not previously known whether (1, 4, 5)-IP_3 and (1, 3, 4, 5)-IP_4 exist in the heart [24], it has been shown that α_1-adrenergic stimulation enhances (1, 4, 5)-IP_3 release from PIP_2 in the heart [26, 29].

In the author's experience using cell suspension for calcium concentration, as previously described [12], fura-2 does not enter all myocytes homogeneously, such that some cells are weakly stained by fura-2. Since the measurement of cellular calcium in the cuvette decreases the sensitivity, a single-cell method was used. This allowed for cell identification under a microscope so that the most viable and clearly stained cells could be selected. IP_3 stimulation caused a response in 50–60% of myocytes. It is not clear why other cells do not respond to it, although it may be because the non-responding cells do not load of fura-2 adequately.

It has also been found that IP_3 receptors exist in cerebellar Purkinje neurons and its cDNA has been also cloned [30]. Recently, it is also suggested that these polyinositides are related to inotropic action in the heart [31], and further, that IP_3 plays an important role during excitation-contraction coupling in the heart. In the present experiment, the release of Ca^{2+} from sarcoplasmic reticulum was increased in cardiac myocytes of SHR. It is not yet clear whether this phenomenon plays a significant role in cardiac hypertrophy. Further studies will investigate this possibility.

Acknowledgment

This research was supported in part by a Research Grant for Cardiomyopathy from the Ministry of Health and Welfare of Japan and a Grant-in-Aid for Scientific Research from the Ministry of Education, Science and Culture of Japan, 02404042, 02454250, 03557038, and 04454259.

References

1. Berridge MJ: Inositol trisphosphate and diacylglycerol as second messengers. Biochem J 220: 345–360, 1984
2. Nisizuka Y: Phospholipid degradation and signal translation for protein phosphorylation. Trend Biochem Sci 8: 13–16, 1983
3. Williamson JR, Cooper RH, Joseph SK, Thomas AP: Inositol trisphosphate and diacylglycerol as intracellular second messengers in liver. Am J Phys 248: C203–216, 1985
4. Enouf J, Giraud F, Bredoux R, Bourdeau N, Toledano SL: Possible role of cAMP-dependent phosphorylation in the calcium release mediated by inositol 1, 4, 5-trisphosphate in human platelet membrane vesicles. Biochem Biophys Acta 928: 76–82, 1987
5. Irvine RF, Letcher AJ, Heslop JP, Berridge MJ: The inositol tris/tetrakisphosphate pathway-demonstration of Ins(1, 4, 5)P_3 3-kinase activity in animal tissues. Nature 320: 631–634, 1986
6. Kawaguchi H, Iizuka K, Takahashi H, Yasuda H: Inositol trisphosphate kinase activity in hypertrophied rat heart. Biochem Med Metabol Biol 44: 42–50, 1990
7. Batty IR, Nahorski SR, Irvine RF: Rapid formation of inositol 1, 3, 4, 5-tetrakisphosphate following muscarinic receptor stimulation of rat cerebral cortical slices. Biochem J 232: 211–215, 1985
8. Irvine RF, Moor RM: Micro-injection of inositol 1, 3, 4, 5-tetrakisphosphate activities sea urchin eggs by a mechanism dependent on external Ca^{2+}. Biochem J 240: 917–920, 1986
9. Kawaguchi H, Shoki H, Sano H, Kudo T, Sawa F, Okamoto H, Sakata Y, Yasuda H: Phospholipid metabolism in cardiomyopathic hamster heart cells 69: 1015–1021, 1991
10. Shoki M, Kawaguchi H, Okamoto H, Sano H, Sawa H, Kudo T, Hirao H, Sakata Y, Yasuda H: Phosphatidylinositol and inositolphosphatide metabolism in hypertrophied rat heart. Jpn Circ J 56: 142–147, 1992
11. Kawaguchi H, Shoki M, Sano H, Kudo T, Sawa H, Mochizuki N, Okamoto H, Endo Y, Kitabatake A: Polyphosphoinositide metabolism in hypertrophic rat heart. J Mol Cell Cardiol, in press
12. Movsesian MA, Thomas AP, Selak M, Williamson JR: Inosi tol trisphosphate does not release Ca^{2+} from permeabilized cardiac myocytes and sarcoplasmic reticulum. FEBS Lett 185: 328–332, 1985
13. Nosek TM, Williams MF, Zeigler ST: Inositol trisphosphate enhances calcium release in skinned cardiac and skeletal muscle. Am J Physiol 250: C807–811, 1986
14. Fujii S, Kawaguchi H, Okamoto H, Saito H, Togashi H, Yasuda H: Fatty acid binding protein of cardiac muscle in spontaneously hypertensive rats: Effect of hypertrophy and its regression. J Moll Cell Cardiol 20: 779–787, 1988
15. Nakanishi T, Jarmakini JM: Developmental changes in myocardial mechanical function and subcellular organelles. Am J Physiol 246: H615–625, 1988
16. Fabiato A, Fabiato F: Contractions induced by a calcium-triggered release of calcium from the sarcoplasmic reticulum of single skinned cardiac cells. J Physiol (London) 249: 469–495, 1975
17. Glick MR, Burns AH, Reddy WJ: Dispersion and isolation of beating cells from adult rat heart. Anal Biochem 61: 32–42, 1974
18. Kawaguchi H, Yasuda H: Effect of elastase on prostacyclin synthesis in aortic smooth muscle cells. Biochim Biophys Acta 878: 42–48, 1986
19. Kawaguchi H, Yasuda H: Prostacyclin biosynthesis and phospholipase activity in hypoxic rat myocardium. Circ Res 62: 1175–1181, 1988
20. Kawaguchi H, Sawa F, Yasuda H: Mechanism of increased angiotensin-converting enzyme activity stimulated by platelet-activating factor. Biochim Biophys Acta 1052: 503–508, 1990
21. Tennes KA, McKinney JS, Putney JW: Metabolism of inositol 1,

4, 5-trisphosphate in guinea pig hepatocytes. Biochem J 242: 797–802, 1987

22. Geerts H, Nuydens R, Nuyens R, Donck LV: The effect of flunarizine on intracellular calcium in isolated rat cardiomyocytes. A digital image processing study. Cardiovasc Res 23: 797–806, 1989
23. Wallenstein S, Zucker CL, Fleiss J: Some statistical methods useful in circulation research. Circ Res 47: 1–9, 1980
24. Kohl C, Schmitz W, Scholz H, Scholz J, Toth M, Doring V, Kalmar P: Evidence for α_1-adrenoceptor-mediated increase of inositol triphsophate in the human heart. J Cardiovasc Pharmacol 13: 324–327.22, 1989
25. Renard D, Poggioli J: Does the inositol tris/tetrakisphosphate pathway exist in rat heart. FEBS lett 217: 117–123, 1987
26. Heathers G, Corr PB, Rubin LJ: Transient accumulation of inositol (1, 3, 4, 5)-tetrakisphosphate in response to α_1-adrenergic stimulation in adult cardiac myocytes. Biochem Biophys Res Commun 156: 485–492, 1988
27. Simpson P: Stimulation of hypertrophy of cultured neonatal rat heart cells through an α_1-adrenergic receptor and induction of beating through an α_1- and β_1-adrenergic receptor interaction. Circ Res 56: 884–894, 1985
28. Starksen NF, Simpson PC, Bishopric N, Coughlin SR, Lee WF, Escobedo JA, Williams LT: Cardiac myocyte hypertrophy is associated with c-myc protooncogene expression. Proc Natl Acad Sci USA 83: 8348–8350, 1986
29. Steinberg SF, Kaplan LM, Inoue T, Zhang JF, Robinson RB: Alpha-$_1$ adrenergic stimulation of 1, 4, 5-inositol tris phosphate formation in ventricular myocytes. J Pharm Exp Ther 250: 1141–1148, 1989
30. Furuichi T, Yoshikawa S, Miyawaki A, Wada K, Maeda N, Mikoshiba K: Primary structure and functional expression of the inositol 1, 4, 5-trisphosphate-binding protein P_{400}. Nature 342: 32–48, 1989
31. Kohl C, Schmitz W, Scholz H, Scholz J: Evidence for the existence of inositol tetrakisphosphate in mammalian heart. Effect of α_1-adrenoceptor stimulation. Circ Res 66: 580–583, 1990

Molecular and Cellular Biochemistry **119**: 59–66, 1993.

Radial spread of aequorin Ca^{2+} signal in single frog skeletal muscle fibers

Masato Konishi and Satoshi Kurihara
Department of Physiology, The Jikei University School of Medicine, 3-25-8 Nishi-shinbashi, Minato-ku, Tokyo 105, Japan

Abstract

The Ca^{2+}-sensitive photoprotein aequorin was injected into single frog skeletal muscle fibers, and the intracellular aequorin light intensity during muscle activation with different maneuvers was mapped with digital imaging microscopy. During 50 Hz electrical activation (tetanus), the aequorin light intensity from different locations in the muscle fiber rose with very similar time course. Caffeine (10 mM) application, on the other hand, caused aequorin light signals to show significantly different time courses, with an earlier increase in Ca^{2+} concentration near the surface of the fiber than near the core. The non-uniform rise of intracellular Ca^{2+} concentration with caffeine treatment is consistent with the slow inward diffusion of caffeine and subsequent Ca^{2+} release from sarcoplasmic reticulum. (Mol Cell Biochem **119:** 59–66, 1993)

Key words: skeletal muscle, imaging, aequorin, intracellular calcium, caffeine

Introduction

In skeletal and cardiac muscle cells, caffeine has been widely used to elicit contraction without membrane excitation, an effect considered to be due to direct effect of the drug to induce Ca^{2+} release from sarcoplasmic reticulum. In frog skeletal muscle, high concentration of caffeine causes massive Ca^{2+} release from the sarcoplasmic reticulum [1], and as a consequence, causes large tension development (caffeine contracture) [2, 3]. Although the peak tension of caffeine contracture is comparable to that of tetanus, the rate of tension development is considerably slower than that of tetanus [2, 3]. The increase of intracellular Ca^{2+} concentration ($[Ca^{2+}]_i$) was also found to be very slow [4]. Morphologically, it has been reported that significant inhomogeneity of myofilament structure develops during caffeine contracture, as revealed by over-stretched and over-contracted sarcomeres in electron micrographs [5]. From the experimental evidence obtained so far, it has been speculated that the slow diffusion of caffeine from the cell surface causes a slow and inhomogeneous increase of $[Ca^{2+}]_i$ (and therefore activation of the myofilaments). However, there is no direct experimental evidence of a non-uniform distribution of $[Ca^{2+}]_i$ during caffeine application.

In this study, we have used the Ca^{2+}-sensitive bioluminescent protein aequorin [6] in combination with digital imaging microscopy to map the distribution of $[Ca^{2+}]_i$ in the muscle fiber activated by rapid electrical stimulation and by caffeine. The results with these two different protocols were compared. Preliminary results of this work have been published in an abstract form [7].

Address for offprints: M. Konishi, Department of Physiology, The Jikei University School of Medicine, 3-25-8 Nishi-shinbashi, Minato-ku, Tokyo 105, Japan

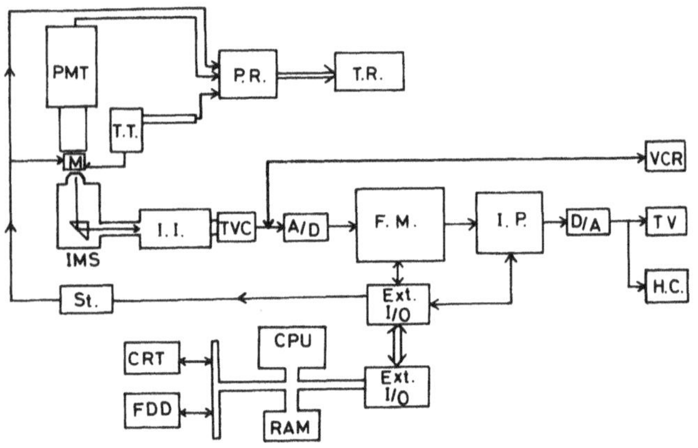

Fig. 1. Schematic drawing of an experimental set-up. Abbreviations: M, muscle fiber; IMS, inverted microscope; I.I., image intensifier tube; TVC, TV camera (vidicon); A/D, analog to digital converter; FM, frame memory; I.P., image processor; D/A, digital to analog converter; VCR, video cassette recorder; H.C., hard copy printer; ST, stimulator; PMT, photomultiplier tube; TT, tension transducer; P.R., pen recorder; T.R., tape recorder. Other abbreviations have their usual meanings. Arrows indicate the flow of signals.

Material and methods

Single fast twitch muscle fibers were dissected from the tibialis anterior muscle of frogs (Rana temporaria), and were mounted horizontally in the narrow trough of the muscle chamber on the stage of an inverted microscope. One end of the muscle fiber was attached to a hook and the other end was attached to a tension transducer. Single fibers were slightly stretched to a sarcomere length of 2.4 μm by observation of laser diffraction lines, and were continuously superfused by the Ringer's solution at 17–18° C. Aequorin was pressure-injected into the muscle fiber from the micropipette containing concentrated aequorin solution (see below), and the aequorin light signal, a reflection of myoplasmic free Ca^{2+} concentration was measured [8]. The muscle fibers were stimulated by either 1) electrical stimulation or 2) rapid application of caffeine (10 mM). For electrical stimulation, rectangular pulses (amplitude, 1.5 × threshold; duration, 500 μs) were applied through a pair of platinum-black electrodes running parallel to the muscle fiber. Caffeine was applied by rapid exchange of the solution in the trough of the muscle chamber (3.5 mm width, 3 mm depth) within several hundred millisecond. The tension was measured by a semiconductor transducer (Akers AE801, Horten, Norway) attached to one end of the muscle fiber. The condition of each fiber was checked before and after aequorin injection by observation of tetanic tension induced by 1.5 sec of 50 Hz stimuli. Fibers which showed sustained plateau tension were used in this study. In order to obtain the aequorin light signal with S/N ratio high enough for the analysis, it was required to inject a large amount of aequorin into the fiber in the optical field of a TV camera (≃250 μm length). This was technically difficult, since the injection of a large amount of solution into the restricted region of the fiber often caused injection damage. Successful injection were obtained from 6 muscle fibers which were heavily loaded with aequorin without obvious indication of the injection damage. Each data shown in this article is a representative of at least three experiments conducted on different muscle fibers.

Solutions and chemicals

The Ringer's solution contained (mM): 115 NaCl, 2.5 KCl, 1.8 $CaCl_2$, 0.85 NaH_2PO_4, and 2.15 Na_2HPO_4, pH 7.05. Caffeine (Nakarai Inc., Kyoto, Japan) was dissolved in the Ringer's solution to set the desired final concentration. It was found *in vitro* that caffeine, at a concentration of 10 mM, had no detectable effect on aequorin light emission [4]. For injection, aequorin was dissolved in a buffer solution containing 150 mM KCl and 5 mM HEPES (N-2-hydroxyethylpiperazine-N′-2-ethanesulphonic acid; pH 7.0) (at a final concentration of 2 mg/ml).

Experimental set-up

The aequorin light signal emitted from the muscle fiber lying horizontally on the stage of an inverted microscope (Diaphot TMD, Nikon, Tokyo, Japan) was simultaneously measured with two different detectors (Fig. 1): 1) a photomultiplier tube (PMT; 9789A, Thorn EMI, U.K.) with a quartz cylinder light guide (10 mm diameter) placed above the preparation, and 2) a vidicon TV camera (TVC) equipped with an image intensifier tube (I.I) (C2400-20; Hamamatsu Photonics, Shizuoka, Japan) in combination with a microscope objective (40 ×, Nikon, Tokyo, Japan). Since the frog skeletal muscle fibers have a roughly cylindrical shape with a diameter of 100–200 μm, the entire thickness of the muscle fiber was not in focus. The objective was focussed on the lateral edge of the muscle fiber, so the detected light signal was thought to primarily come from an image plane across the middle of the muscle fiber. Aequorin light and tension signals were displayed on a pen recorder (P.R.; RECTI-HORIZ 8K21, NEC San-ei, Tokyo, Japan) and recorded in a FM tape recorder (T.R.; NFR 3515, SONY, Tokyo, Japan). Output signals of the TV camera (voltage outputs of 512 × 512 pixels) of the standard video rate (30 frames/sec) were analog-to-digital converted (8 bit), and were stored in a frame memory (F.M.; VM01B1, Shibasoku Co. Ltd., Tokyo, Japan). The image data in the frame memory were displayed on a TV monitor (TV in Fig. 1) and a hard copy printer (H.C.), if necessary. The TVC output signals were also recorded in a video cassette recorder (VCR). The real time control of image-data-collection and off-line analysis were done by a personal computer (CPU in Fig. 1) (VM07A1, Shibasoku Co. Ltd.). The data taken by the TVC were corrected, pixel by pixel, for the slight non-linearity of the camera, assuming the output signal of the camera was proportional to the 0.95th power of the input light intensity (as indicated by the manufacturer). The stimulation for the muscle fiber was also timed by the computer, which enabled us to compare the time course of the light signal taken by the PMT and the TVC. We did not convert aequorin light intensity to $[Ca^{2+}]_i$, because of the lack of any reliable method for calibration of image data. Since it seems very difficult to get Lmax (maximum light intensity of aequorin) from each pixel, it is probably not appropriate to rely on a conventional calibration method for aequorin light signal taken by PMT [9].

Results

Aequorin light signal during tetanus

Figure 2 shows the video images of aequorin light signal during the tetanus (50 Hz, 1.5 sec) taken sequentially at 67 ms interval. The aequorin light intensity clearly increased after the initiation of stimulation (the first stimulation pulse was applied between panel a2 and a3), and decreased towards the resting level after the cessation of stimulation (between panel g1 and g2), as expected for the activated increase in $[Ca^{2+}]_i$. In most panels taken during activation (e.g. panel f3), aequorin light intensity in the left third appears to be slightly lower than that in the rest. This apparent non-uniformity of the light intensity is not likely due to the optical 'shading' of the apparatus, because this non-uniformity was not consistently observed in 3 other fibers. Instead, the

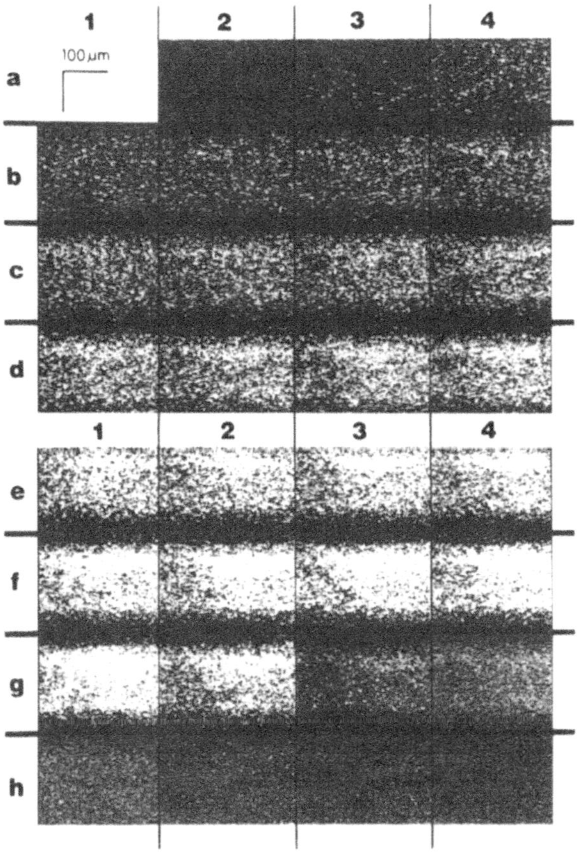

Fig. 2. A series of video images recorded (simultaneously with light (PMT) and tension signals) during tetanus elicited by 50 Hz stimuli for 1.5 sec. The image data were signal averaged during 5 successive tetani. The field of images (265 × 235 µm) covered the entire diameter and a part of the length of a muscle fiber which was oriented horizontally in the field. Each panel (separated by black lines) shows image data of 33 ms (time of 1 video frame). Panels were sequentially taken every 2 video frames (67 ms interval) in the order of row a to row h (top to bottom) and, in each row, 1 to 4 (2 to 4 in row a). In each of 5 tetanic stimulations, the first stimulus to a muscle fiber was synchronized to a starting pulse of the second panel (a3), and the image data of the same timing were accumulated pixel by pixel in 5 tetani. The first panel (a2) was, therefore, obtained in the resting state, and the second panel was obtained at 0 to 33 ms after the first stimulus. Fiber diameter 200 µm, sarcomere length 2.4 µm, temperature 17° C.

non-uniformity is probably due to non-homogeneous distribution of aequorin molecules in the fiber, since aequorin is known to diffuse only very slowly away from the site of injection in myoplasm [10]. Because of this inhomogeneous distribution of aequorin molecules which is likely to occur in large cells like skeletal muscle fibers, it was not meaningful to directly compare the size of aequorin light signals in different portions of muscle fibers. We, therefore, decided to primarily ana-

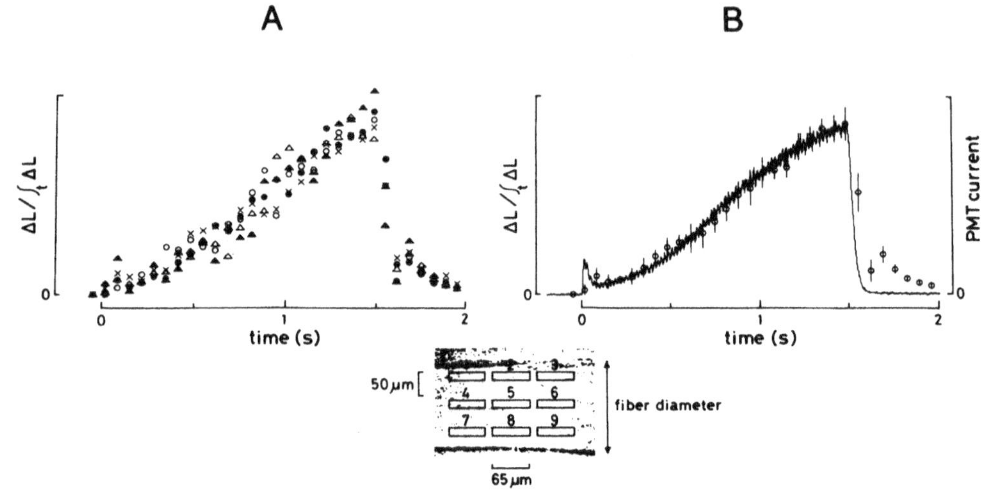

Fig. 3. Temporal comparison of aequorin light signals during tetanus recorded from different areas in the image plane of a muscle fiber (image data shown in Fig. 2) and that simultaneously recorded from the entire lenght of the fiber by PMT. The inset at the bottom of the figure (taken by white light illumination during tetanus) indicates the position of a muscle fiber in the image plane, and also indicates small areas (numbered 1 to 9) in which the data were analyzed. The dimension of each area was (horizontal × vertical direction) 31 × 10 pixels which corresponds to 64 × 20 μm. Spatially averaged light intensity in the resting state was taken as a background, and the change in spatially averaged light intensity (△L) in each area was normalized to the time integral (for 2 seconds after the first stimulus) of △L ($\int_t$△L) in the area. In part A and B, the stimulation to a muscle fiber was started at time 0, and normalized △L values in the areas (△L/$\int_t$△L) were plotted as a function of time. Although the timing of image data within a video frame have, at maximum, a 33 ms difference, normalized △L data were plotted at the midpoint of 33 ms. In part A, normalized △L's from areas 1 (○), 3 (●), 5 (△), 7 (▲) and 9 (×) were plotted. Vertical scale is arbitrary. In part B, mean ± s.d. values (open circle and vertical bar) of normalized △L's from 9 areas (area 1 to 9) were superposed on PMT current data form the entire length of the fiber (continuous trace). The PMT current trace was scaled to give least-squares-fitting to mean values of normalized △L's in the time range of 0.1 to 1.5 s after the first stimulus. Vertical scale for normalized △L (left ordinate) is arbitrary, but is the same as that in part A. Vertical scale for PMT current (right ordinate) corresponds to 37 μA. For fiber information and experimental conditions, see the legend of Fig. 2.

lyze the time course of the light signals in different locations. Figure 3 shows the aequorin light signals during tetanus from five different areas in the image plane of a muscle fiber shown in Fig. 2. The change in average counts (△L) in the small areas (numbered 1, 3, 5, 7, 9 in the inset picture) were normalized to the time integral of the △L in each area for the temporal comparison. The normalized light signals in the 5 different areas of the muscle fiber showed very similar time courses within the noise of the data (Fig. 3A). In Fig. 3B, the mean (± s.d.) values of normalized light signals from all of nine areas (numbered 1 to 9 in the inset picture) were shown (circles), and they were compared with the light signal measured with the PMT from the entire part of the muscle fiber (solid line). The solid trace in Fig. 3B shows a typical time course of the aequorin light signal during tetanus: i.e., initial peak, slow rising phase, rapid relaxation after the stimulation [10] (It is still unclear if the slow rising phase truly reflects a slow rise of $[Ca^{2+}]_i$. This may be, at least partly, due to the slow kinetics of aequorin light emission [9]). It is clear in Fig. 3B that the data from 9 areas (circles) generally follows the PMT data (solid trace) well. The agreement of both types of data (circles and solid line) was particularly good in slow rising phase, but the circles somewhat lag behind the solid line in the rapidly changing phase (initial peak and rapid relaxation), which could be attributed to the limited time resolution of the standard video rate plus the slow response of the image intensifier. The image data (circles) also show the hump after the PMT data (solid line) has returned very close to the base line (around 1.7 sec on the time axis in Fig. 3B). This hump, which was not consistently observed in three other muscle fibers, may be due to the artifact of the fiber movement during muscle relaxation, because the timing of the hump matches the falling phase of the tension simultaneously measured with the aequorin light signal (not shown). The generally close similarity of the time course of aequorin light signal taken in different parts of the muscle image plane suggests that the entire part of the muscle fiber is rapidly activated during tetanus by conduction of action potentials into the transverse-tubular system.

Aequorin light signal induced by caffeine

The image data taken during electrical stimulation (Figs 2 and 3) showed little or no indication of inhomogeneous activation of the muscle fiber, at least with the time and spatial resolutions of our system. The lack of detectable inhomogeneity in $[Ca^{2+}]_i$ could be attributed to the very rapid spread of excitation through the entire cell. It was then of interest to know if any non-uniformity of $[Ca^{2+}]_i$ existed during slow activation. For this purpose, we utilized caffeine, the drug known to directly induce Ca^{2+} release from sarcoplasmic reticulum [1]. It has been shown that the time course of caffeine-induced contraction (caffeine contracture) is much slower than that induced by electrical stimulation [2, 3]. The video images of aequorin light signal taken every 200 ms during 10 mM caffeine application were shown in Fig. 4 (row a to row h, and in each row, 1 to 4). In contrast to data taken during electrical stimulation (Fig. 2), aequorin light intensity appeared to rise inhomogeneously. Aequorin light intensity in the image plane first increased at the upper and lower boundaries of the muscle fiber, and slowly spread to the center (d3–f2). Finally, the light intensity started to fall at the boundaries (f3–g4). This non-uniform increase and decrease of aequorin light intensity is clearly shown in Fig. 5A which illustrates the temporal comparison in the light intensities in different parts of the muscle fiber (part A, B and C shown in the inset picture). The histogram of the light intensity shown in Fig. 5B also points out the differential increase in the aequorin light intensity across the image plane during activation by caffeine. Since we focussed on the edge of the muscle fiber, the light signals were thought to primarily come from the image plane across the middle of the fiber. In this case, the rapid rise of the light signal at the edge (and slow rise in the center) probably reflects radial spread of activation, i.e., Ca^{2+} release from sarcoplasmic reticulum. To estimate the consumption of aequorin molecules during caffeine treatment, the muscle fibers were lysed by triton-X (1%) in some experiments after the caffeine treatment to discharge all residual aequorin molecules in the muscle fiber. From the comparison of the integral of the light signal during caffeine treatment with that obtained from lysed fibers, consumption of aequorin during caffeine treatment was estimated to be 5–21% (in 3 fibers). The caffeine induced light signal shown in Fig. 5 consumed 21% of aequorin molecules.

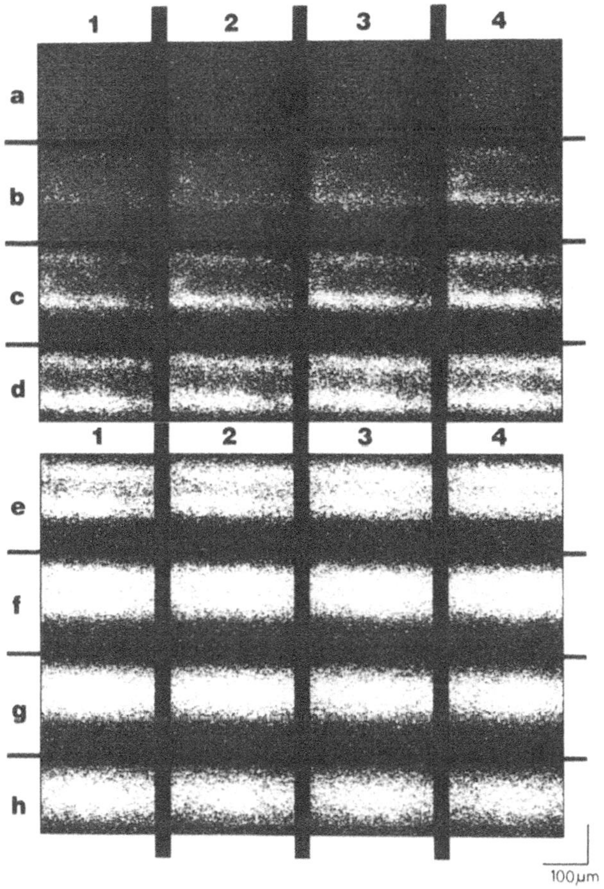

Fig. 4. A series of video images taken on application of 10 mM caffeine. The field of images covered the entire diameter and a part of the length of the muscle fiber lying horizontally. Each panel (separated by black lines) shows image data of one video frame. Panels were sequentially taken every 6 video frames (200 ms interval) in order of row a to row h and, in each row, 1 to 4. The first panel (a1) was taken at approximately 1.1 sec after perfusion of caffeine, and caffeine was continuously perfused during these measurements. Fiber diameter 150 μm, sarcomere length 2.4 μm, temperature 17° C.

The consumption during tetanus was not checked in the present study, but was considered to be negligible [10].

Discussion

Since aequorin gives a signal that is proportional to $[Ca^{2+}]^n$, where n is 2.5 to 3, aequorin light intensity changes very steeply with the change in free Ca^{2+} concentration [11]. This non-linearity makes it difficult to quantitatively interpret aequorin signals (in terms of spatially averaged $[Ca^{2+}]_i$ in the presence of significant $[Ca^{2+}]_i$ gradients within a preparation when the signal is

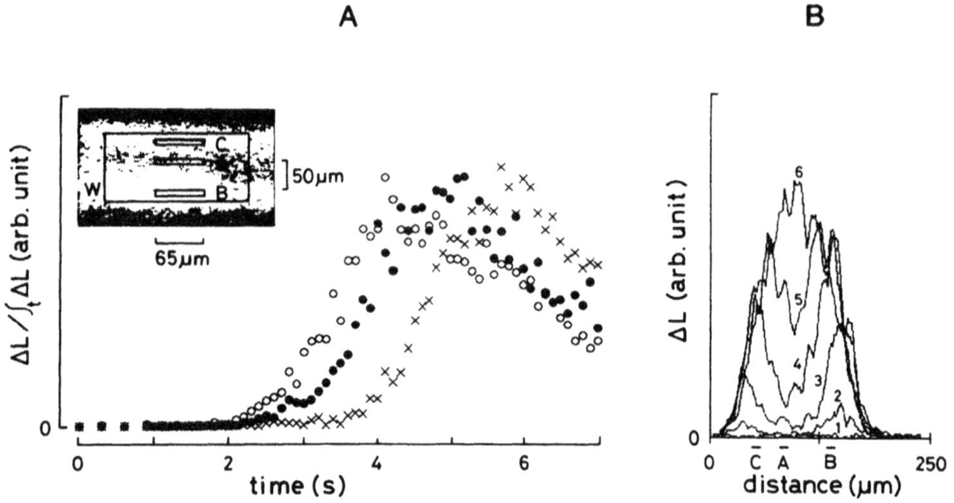

Fig. 5. Temporal comparison of aequorin light signals during caffeine contracture obtained from different areas in the image plane of the muscle fiber shown in Fig. 4. The inset of part A shows an example of an image of aequorin light from a part of the length of the muscle fiber oriented horizontally in the field. Areas of analysis are labeled A, B and C. The dimensions of areas A, B and C were 32 × 5 pixels (horizontal × vertical direction), which corresponds to 66 × 10 μm. The change in spatially averaged light intensity ($\triangle$L) in each of area was normalized to the time integral (for 14 seconds after caffeine application) of $\triangle$L ($\int_t \triangle$L) in each area. The normalized date ($\triangle L/\int_t \triangle L$) from area A (×), area B (○) and area C (●) are plotted as a function of time in part A (data plotted at the midpoint of 33 ms). Zero time is a moment of valve opening and rapid perfusion of 10 mM caffeine. Part B shows the histograms of $\triangle$L across the fiber diameter ('vertical histogram'). Each data point of the histogram is the change in the average light intensity of 32 pixels horizontally aligned between 100 μm and 166 μm from the left edge of the image field (see inset of part A). Histograms were digitally smoothed with 6 adjacent data points (3 points on each side). The abscissa of the histograms is the distance from the upper edge of the image field, and the locations corresponding to the areas A, B and C (see inset of part A) are indicated by small bars marked A, B and C, respectively. Histograms numbered 1 to 6 were obtained from image data taken at (for histogram # 1 to # 6) 0.7, 1.5, 2.3, 3.1, 3.9, 4.7 and 5.6 seconds, respectively, after caffeine application. Ordinate is the change in average light intensity in arbitrary units. For fiber information and experimental conditions, see the legend of Fig. 4.

detected by a photomultiplier tube from the entire part of the preparation [9]. The non-linearity, however, may be even advantageous in detection of a localized change in $[Ca^{2+}]_i$ using an imaging microscopy technique. Imaging of aequorin light signals was first described by Rose and Loewenstein [12] who measured the light intensity image from two adjacent cells of Chironomus salivary gland. Several additional applications have also been reported in a variety of cells (either single or multiple cells) injected with aequorin [12–16]. Our results in the present study show the usefulness of aequorin for the mapping of $[Ca^{2+}]_i$ (with the limited time and spatial resolution) in single frog skeletal muscle fibers. The spatial resolution in the present experiment is partly limited by the resolution of the microscope (40 × objective), but also limited by the small number of photons emitted by aequorin molecules. The time resolution is, on the other hand, limited by the video rate of the camera. With this spatial and time resolution, it was obviously impossible to obtain information on the Ca^{2+} concentration gradient which has been predicted to exist within a sarcomere during the first 10–20 ms after the initiation of Ca^{2+} release from sarcoplasmic reticulum [17].

During the electrical stimulation (tetanus), aequorin light signals from different parts of the muscle fiber showed a similar time course (Fig. 3). These data suggest the uniform rise of $[Ca^{2+}]_i$ in the fiber. The results of the present study are consistent with those of Westerblad *et al.* [18], who did not find the grossly non-uniform distribution of $[Ca^{2+}]_i$ throughout the skeletal muscle fiber during tetanus using fura-2, a fluorescent Ca^{2+} indicator.

In contrast to the aequorin light distribution during tetanus, the aequorin light signal revealed an apparent $[Ca^{2+}]_i$ gradient during slow activation by caffeine (Figs 4, 5). The spread of aequorin light signal from the edge to the center of the muscle fiber seems to reflect the radial spread of the region of high Ca^{2+} concentration from the surface to the core of the muscle fiber. The slow spread of the region of high Ca^{2+} concentration is consistent with the slow diffusion of caffeine applied to the surface of the muscle fiber and resultant Ca^{2+} release from sarcoplasmic reticulum induced by caffeine.

Based on the measurements of Kushmerick and Podolsky [19] in frog skinned muscle fibers, a molecule with the size of caffeine (m.w. 194) is expected to have an apparent diffusion constant of approximately $2.8 \times 10^{-6} cm^2 s^{-1}$ in myoplasm (20° C). With this diffusion constant, it is expected to take many seconds to raise caffeine concentration at the core of the fiber to the level close to that in the surface, which was predicted by a simple computer simulation of the radial diffusion of molecules from the surface to the core of a cylinder (data not shown). When Ca^{2+} is released from sarcoplasmic reticulum by caffeine, released Ca^{2+} itself might regeneratively cause Ca^{2+} release from adjacent sarcoplasmic reticulum (Ca^{2+}-induced Ca^{2+} release [1]). Although the radial spread of activation observed in the present study is roughly consistent with the diffusion of caffeine, it is also possible that the spread of activation is caused by the propagation of regenerative Ca^{2+} release from sarcoplasmic reticulum [20]. Further study is required to clarify the contribution of these two mechanisms (caffeine diffusion and propagation of regenerative Ca^{2+} release). The simple diffusion of Ca^{2+} from the surface to the core of the fiber is probably too slow to explain the spread of high Ca^{2+} concentration region, because the diffusion of Ca^{2+} itself is considered to be extremely slow in myoplasm (apparent diffusion constant of Ca^{2+} is approximately $1.4 \times 10^{-7} cm^2 s^{-1}$ in myoplasm, according to Kushmerick and Podolsky [19]). After the aequorin light signal reached a peak, it declined to the basal level even in the continuous presence of caffeine (Fig. 5). This apparent 'inactivation' could be attributed to either depletion of Ca^{2+} in sarcoplasmic reticulum or desensitization of sarcoplasmic reticulum to caffeine, since the aequorin light signal was generally much smaller in the second application of caffeine, when tested in other muscle fibers. (However, it is also possible that the small light signal on the second application of caffeine is due to some irreversible change of the muscle fiber caused by the inhomogeneous activation during the first application of caffeine.) Consumption of aequorin molecules during caffeine treatment (21% for the data in Fig. 5) was likely to somewhat affect the light signal, but probably only in a minor way.

In conclusion, the results suggest that the applications of drugs in the perfusate do cause slow activation of the muscle fiber, which spreads from the surface to the core of the fiber, and the significant non-uniformity of $[Ca^{2+}]_i$ is formed within the muscle fibers.

Acknowledgements

We thank Dr. J.R. Berlin for helpful comments on the manuscript. We also thank Dr. J.R. Blinks for supplying aequorin. This work was supported by the Grant-in-Aid for Scientific Research from the Ministry of Education, Science and Culture.

References

1. Endo M: Calcium release from the sarcoplasmic reticulum. Physiol Rev 57: 71–108, 1977
2. Axelsson J, Thesleff S: Activation of the contractile mechanism in striated muscle. Acta Physiol Scand 44: 55–56, 1958
3. Sakai T, Geffner ES, Sandow A: Caffeine contracture in muscle with disrupted transverse tubules. Am J Physiol 220: 712–717, 1971
4. Konishi M, Kurihara S, Sakai T: Change in intracellular calcium ion concentration induced by caffeine and rapid cooling in frog skeletal muscle fibers. J Physiol (London) 365: 131–146, 1985
5. Yoshioka T, Somlyo AP: Calcium and magnesium contents and volume of the terminal cisternae in caffeine-treated skeletal muscle. J Cell Biol 99: 558–568, 1984
6. Shimomura O, Johnson FH: Properties of the bioluminescence protein aequorin. Biochem 8: 3991–3997, 1969
7. Kurihara S, Konishi M, Okazaki O, Suda N: Imaging of intracellular Ca ion concentration change in aequorin injected skeletal muscle fibers. J Physiol Soc Japan 50: 534, 1988
8. Konishi M, Kurihara S: Effects of caffeine on intracellular calcium concentration in frog skeletal muscle fibres. J Physiol (London) 383: 269–283, 1987
9. Blinks JR, Wier WG, Hess P, Prendergast FG: Measurement of Ca^{2+} concentrations in living cells. Prog Biophys Molec Biol 40: 1–114, 1982
10. Blinks JR, Rudel R, Taylor SR: Calcium transients in isolated amphibian skeletal muscle fibres: Detection with aequorin. J Physiol (London) 277: 291–323, 1978
11. Allen DG, Blinks JR, Prendergast FG: Aequorin luminescence: Relation of light emission to calcium concentration – A calcium-independent component. Science 196: 996–998, 1977
12. Rose B, Loewenstein WR: Calcium ion distribution in cytoplasm visualized by aequorin: Diffusion in cytosol restricted by energized sequestering. Science 190: 1204–1206, 1975
13. Rose B, Loewenstein WR: Permeability of a cell junction and the local cytoplasmic free ionized calcium concentration: A study with aequorin. J Memb Biol 28: 87–119, 1976
14. Ridgway EB, Gilkey JC, Jaffe LF: Free calcium increases explosively in activating medaka eggs. Proc Natl Acad Sci USA 74: 623–627
15. Steinhardt RA, Zucker RS, Schatten G: Intracellular calcium release at fertilization in the sea urchin eggs. Dev Biol 58: 185–196, 1977
16. Miyazaki S, Hashimoto N, Yoshimoto Y, Kishimoto T, Igusa Y, Hiramoto Y: Temporal and spatial dynamics of the periodic

increase in intracellular free calcium at fertilization of golden hamster egg. Dev Biol 118: 259–267, 1986

17. Channell MB, Allen DG: Model of calcium movements during activation in the sarcomere of frog skeletal muscle. Biophys J 45: 913–925, 1984
18. Westerblad H, Lee JA, Lamb AG, Bolsover SR, Allen DG: Spatial gradients of intracellular calcium in skeletal muscle during fatigue. Pflügers Arch 415: 734–740, 1990
19. Kushmerick MJ, Podolsky RJ: Ionic mobility in muscle cells. Science 166: 1297–1298, 1969
20. Wier WG, Cannell MB, Berlin JR, Marban E, Lederer WJ: Fura-2 fluorescence imaging reveals cellular and subcellular heterogeneity of $[Ca^{2+}]_i$ in single heart cells. Science 235: 325–328, 1987

Molecular and Cellular Biochemistry **119**: 67–71, 1993.

Cardiac contractile proteins and autoimmune myocarditis

Tohru Izumi, Haruo Hanawa, Makihiko Saeki and Makoto Kodama
First Department of Internal Medicine, Niigata University School of Medicine, Niigata, Japan

Abstract

Concerning cardiac contractile proteins, antigenicity and myocarditogenicity were discussed. In normal states, these proteins are immunologically tolerant, and can not provoke any heart-specific disease. Why the proteins can provoke such lethal autoimmune myocarditis has not been completely elucidated. Shortly after cardiac infection or myocardial ischemia, these proteins may work as the antigen for the autoimmune myocardites. First of all, the role of cardiac myosin has been strongly emphasized. But, the antigen determinants: epitope proteins remain unclear. Either cross-activity to the streptococcal M protein and/or the α-helical coiled-coil protein may be an important factor to determine antigenicity. In this autoimmune myocarditis, the roles of T-lymphocyte and cardiac dendritic cell are noticeable. Through further study on the relation between antigen epitope and the infectious agents in the heart; on cardio-cytotoxity of the T-lymphocyte and on the precise contribution of cardiac dendritic cells, this autoimmune myocarditis will be more clarified. (Mol Cell Biochem **119:** 67–71, 1993)

Key words: myosin, actin, autoimmune myocarditis, T-lymphocyte, dendritic cell

Introduction

Myocarditis, cardiomyopathy [1–13], the post stage of acute myocardial infarction, the state after pericardiotomy [14] and rheumatic heart disease [15–20] have been noticed to be closely related to autoimmune heart damage [21]. Employing cardiac muscle samples biopsied from patients with myocarditis and cardiomyopathy, enteroviral genoms are able to be detected in the hearts nowadays [22]. To our best understanding, these viruses are important not only as agents which invade into the heart and damage the cardiocyte directly, but also as agents which may provoke secondary autoimmune damage to the myocardium after a viral infection [1]. For this reason, we would like to raise and discuss the theme: the relationship between cardiac proteins and autoimmune myocardites. To make clear the cardiac damage, it is required to investigate in several ways whether the cardiac protein can produce its auto-antibody in the blood circulation, and then whether it can provoke autoimmune myocarditis in experimental animals. Up to this date, out of forty thousand cardiac proteins, a myosin [1, 3, 14–16, 18–20, 23–28], an actin [3, 14], troponin complex [21], a tropomyosin [21, 29], sarcoplasmic proteins, β adrenergic receptor proteins [6–8], Ca++ channel proteins [10], muscarine-cholinergic receptor proteins [30], ADP-ATP carrier proteins [5], laminin, fibronectin, collagen [9, 21], and vimentin [16] have been nominated as the possible related candidates. Using these proteins, provocative autoimmune myocarditis has been repeatedly experimented [5, 25, 27]. To achieve further advances in this

Address for offprints: T. Izumi, First Department of Internal Medicine, Niigata University School of Medicine, Niigata 951, Japan

theme, we need to briefly review the clinical and experimental studies closely related to cardiac contractile proteins.

Myosin

The cardiac myosin is a large protein, which is composed of two heavy chains, approximately 200 kDa in molecular weight, and four light chains ranging from 17 to 27 kDa in weight. This protein is the most abundant of the contents of the cardiac muscle. The amount of myosin alone occupies more than 60% of the whole heart. Structurally, this protein is divided into three portions, two spherical ones and an α-helical coiled-coil rod. In the trypsin, it is enzymatically fragmented into two parts: a heavy meromyosin and a light meromyosin. The heavy meromyosin further consists of two regions, S-1 and S-2. The S-1 region contains the spherical portion, and the S-2 region is a part of the rod. The light meromyosin corresponds with the other rod portion.

In literature, cardiac myosin has been the most frequently investigated among the cardiac contractile proteins whether it can provoke cardiac inflammatory lesions with immunization [25, 27]. Theoretically, to provoke an organ-specific autoimmune disease, the protein must be specific to the organ. From this point of view, this cardiac protein has dominancy as an antigen candidate, because the primary structure of cardiac myosin differs in many amino acid sequences from the ones of skeletal muscle or smooth muscle. Recent advances in analysis of the primary amino acid structure of this myosin has been very remarkable. From these, the whole amino acid sequence of the heavy chain has already been completed in the human cardiac β chain, the rat cardiac α and β chains, and the chicken skeletal chain [31, 32]. Compared with those amino acid sequences, homologies between the human and rat cardiac β heavy chains are detectable in 96% of the sequence. However, between the human cardiac heavy chain and the chicken skeletal chain, only 80% of the sequence has these homologies. That is, 20% of the whole structure is different from one another. Unfortunately, the primary structure of the human skeletal myosin has not yet been completed. For this reason, human cardiac myosin is not comparable with the skeletal muscle. Probably, the ratio of the sequence difference is similar to the above mentioned percent. The difference seems to be sufficient to work as an immunological specificity in organ diseases.

Among patients with viral myocarditis, dilated cardiomyopathy, the post-cardiac injury syndrome, rheumatic heart disease and infective endocarditis, anti-myosin autoantibodies are frequently detectable in the sera. Their appearance may explain why myosin is able to provoke autoimmune disease in the heart. In addition, we now know that antibodies against the infectious agents causing viral myocarditis or rheumatic heart disease have a cross-reactivity to cardiac myosin. Cunningham *et al.* emphasized the presence of cytotoxic and enteroviral neutralizing antibodies, which can react not only to streptococcus pyogenes, but also to human cardiac myosin [15]. Streptococcus pyogenes are known to have various kinds of M proteins as a virulence factor. Out of these, types 1, 5, 6 and 19 can also cross-react with human cardiac myosin. Cunningham's research group has insisted that the viral neutralizing antibodies can be blocked by synthetic peptides which contain the amino acid sequence of the M5 proteins and overlap their homologies. In case of the M5 protein, a peptide (Q K S K Q in the primary structure of the sequence) can react with cardiac myosin [28]. However, this attractive sequence is limited to the antibodies in the human and mouse. In the rabbit, the cross-reactive sequence localizes on the residues 84-116 in the M5 protein. Concerning the M6 protein, a primary sequence, L T E K E K A E L Q A K L E shows this cross-reactivity [15]. Interestingly, these cross-reactive sequences reveal that 40% of the homologies in the amino acid overlap with the M6 protein, the human cardiac myosin and the coxsackie virus B3 capsid protein VP1. Furthermore, the cross-reactive antibodies indicate a cytotoxity to either the cardiocyte and/or the fibroblast in the presence of complement. Besides, the homological sequence with the M protein was found in the α-helical rod region in the cardiac light meromyosin. Through those data, an interesting hypothesis has been developed: that the α-helical structural antigen of the streptococcus pyogenes, and the cardiac myosin may induce heart-specific autoimmune disease.

Experimentally, we have proven that cardiac myosin can provoke a lethal autoimmune myocarditis [25]. With myosin extracts obtained from human hearts, young healthy Lewis rats were immunized. On the 18th day after the first immunization, all the animals documented a remarkable inflammatory cell infiltrate and myocardial necrosis in the heart as shown in Fig. 1, but no specific changes were seen in any organs. Neu *et al.* also immunized A/J and A.SW/SnJ mice with cardiac myosin, and provoked a serious but not lethal myocarditis [27]. In both experiments, the skeletal myosin failed to provoke the lesion. From these data, an epitope producible to the autoimmune myocarditis seems to localize only on the cardiac myosin. This lethal myocarditis is also inducible by both rat and porcine cardiac myosins. Certainly, the leading antigen has a common epitope in any cardiac myosins. It is very interesting that the novel autoimmune myocarditis is mediated by the T-lymphocyte, but not by immuno-globulin. This myocarditis is transferable into healthy syngeneic rats only with the lymphocytes stimulated by Concanavalin A, but impossible with high titer anti-myosin antibodies [23]. Conclusively, this heart-specific autoimmune disease belongs to the type of delayed hyper-sensitivity. In the mouse model, myocarditis was also transferable into SCID mouse with the T-lymphocyte [24]. Through these two different experimental animal models, we stress on the cellular effector pathomechanism in autoimmune myocarditis provoked by cardiac myosin. The humoral effector is not important for cardio-cytotoxity. Probably, a myosin auto-reactive T-lymphocyte clone is provoked in the local lymphnode near the immunized site by an antigen epitope on the cardiac myosin. The T-lymphocyte stimulates the interleukins and enhances activity of the cardio-cytotoxic T-lymphocyte. The T-lymphocytes attack the cardiocyte. The cardiac myosin which has spread out from the damaged heart expands the T-lymphocyte clone furthermore and enhances the activity much more. Finally, the cardio-cytotoxic T-lymphocytes concentrate on the target. To our present understanding, the lethal autoimmune myocarditis seems to be built up in such an accelerated vicious cycle. Also, the role of cardiac dendritic cells (Fig. 2) and the genetic differences among animal species must be taken into account. In the novel autoimmune myocarditis, apparently the macrophages increase in number as the cell-infiltrate. Of these large nuclear cells, cardiac dendritic cells, which are ultrastructurally characterized by an irregularly shaped nucleus, a relatively clear cytoplasma and an invaginating cell membrane in the figure of a glove, occupy a relatively large number. This cell is quite rare in the normal state. In autoimmune myocarditis, it plays the role of the antigen presenting cell. In addition, it works as a mediator, because the cell is in close contact to the cardiocyte as can be seen in some areas of the inflammation.

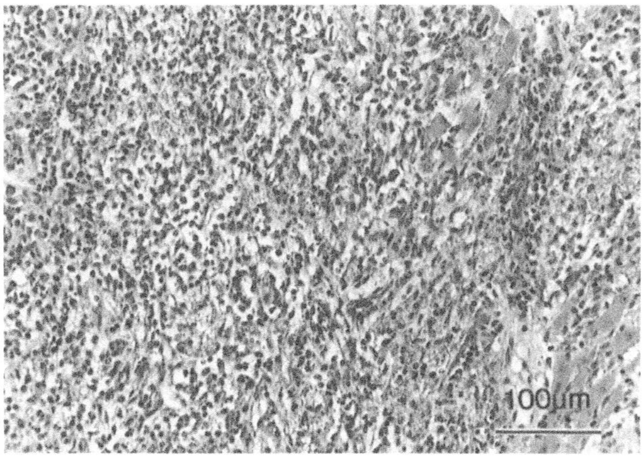

Fig. 1. Experimentally, with human cardiac myosin, young healthy Lewis rats were immunized. On the 18th day after the first immunization, the rats documented a remarkable inflammatory cell infiltrate and myocardial necrosis in the heart.

Actin

Actin is a 43 kDa protein in molecular weight. Structurally, it presents an α-helical coiled-coil. This protein is seen in all organs of the human body. Its isoform depends on the organ. The isoforms have a very high homology except 18 amino-terminal residues. For example, four differences out of 375 amino acids can be detected between the cardiac and the skeletal actins. Considering theoretical background, a heart-specific inflammation can not easily be provoked by this protein.

A circulating anti-actin autoantibody has been documented in some types of cardiac diseases. But, the autoantibody occurs fewer in number than the myosin ones. In a study on cardiac injury syndrome, 46% positive sera was found in the anti-heart autoantibody, 38% in the anti-myosin one and 26% in the actin one [14]. Also in the case of acute myocardial infarction, the occurrence ratio was 20%, 16% and 9%, respectively. Neumann *et al.* analyzed the antigens reactive to the circulating antibodies with Western immunoblot staining [3]. They emphasized that, in cardiomyopathy with a high antibody titer, 40 to 49 kDa and 100 to 109 kDa cardiac proteins were the most reactive among the antigen candidates. On the other hand, in the case of the myocarditis with a high titer, 40 to 49 kDa proteins were significantly immuno-stained. In contrast, low titer myocardities was reactive to 190 to 199 kDa proteins. The actin belongs to between 40 to 49 kDa in molecular

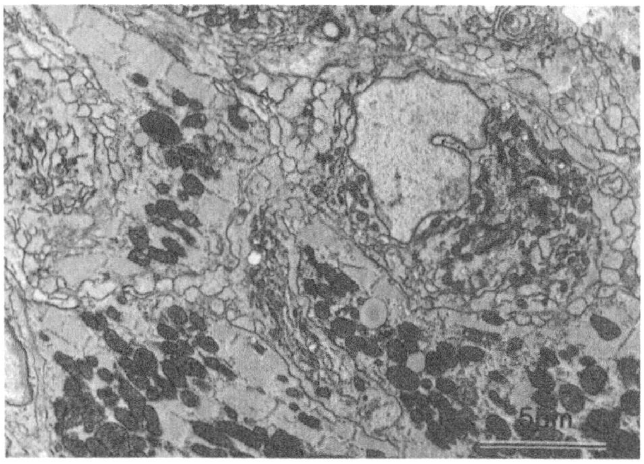

Fig. 2. The role of cardiac dendritic cell must be taken into account in novel autoimmune myocarditis. The dendritic cells, which are ultrastructurally characterized by an irregularly shaped nucleus, a relatively clear cytoplasma and an invaginating cell membrane in the shape of a glove, occupy a relatively large number in the cell infiltrate. In some place, the cellular contact to the cardiocyte was seen in the lesion.

weight. The 190 to 199 kDa proteins seem to be the heavy chain of cardiac myosin.

Until now, this provocative autoimmune myocarditis has not been experimented on any animals.

The other contractile proteins

The tropomyosin is an α-helical coiled-coil cardiac protein. Recently, this protein has been also revealed to show a cross-reactivity with the streptococcal M protein. Especially, the M6 protein has 31% homologies of the tropomyosin sequence [29]. However, the pathogenic role of this cross-reactivity remains unresolved.

Very little is known concerning the roles of the troponin complex in the autoimmune heart disease. The troponin complex is comprised of three subunits, troponin-I, troponin-C and troponin-T. Not all troponin isoforms have yet been identified, though these proteins are less likely to be heart-specific autoantigens.

Acknowledgements

We sincerely thank Dr. Y. Matsumoto (Niigata University) for his immunological advice, and Drs. A. Fujiwara, S. Motoda (Denkaseiken Co.), E. Okazaki (Niigata Shimin Hospital) and H. Tsunoda (Niigata Cancer Center) for the preparation of human cardiac myosin. This investigation was in part supported by a grant to the Research Committee for Epidemiology and Etiology of Idiopathic Cardiomyopathy from the Ministry of Health and Welfare of Japan.

References

1. Beisel KW, Srinivasappa J, Prabhakar BS: Identification of a putative shared epitope between Coxsackie virus B4 and alpha cardiac myosin heavy chain. Clin Exp Immunol 86: 49–55, 1991
2. Traystman MD, Beisel KW: Genetic control of Coxsackievirus B3-induced heart-specific autoantibodies associated with chronic myocarditis. Clin Exp Immunol 86: 291–298, 1991
3. Neumann DA, Burek CL, Baughman KL, Rose NR, Herskowitz A: Circulating heart-reactive antibodies in patients with myocarditis or cardiomyophaty. J Am Coll Cardiol 16: 839–846, 1990
4. Caforio ALP, Bonifacio E, Stewart JT, Neglia D, Parodi O, Bottazzo GF, Mckenna WJ: Novel organ-specific circulating cardiac autoantibodies in dilated cardiomyopathy. J Am Coll Cardiol 15: 1527–1534, 1990
5. Schulze K, Becker BF, Schauer R, Schultheiss HP: Antibodies to ADP-ATP carrier – an autoantigen in myocarditis and dilated cardiomyopathy – impair cardiac function. Circulation 81: 959–969, 1990
6. Limas CJ, Goldenberg IF, Limas C: Influence of anti-beta-receptor antibodies on cardiac adenylate cyclase in patients with idiopathic dilated cardiomyopathy. Am Heart J 119: 1322–1328, 1990
7. Magnusson Y, Marullo S, Hoyer S, Waagstein F, Anderson B, Vahlne A, Guillet JG, Strosberg AD, Hjalmarson A, Hoebeke J: Mapping of functional autoimmune epitope on the β_1-adrenergic receptor in patients with idiopathic dilated cardiomyopathy. J Clin Invest 86: 1658–1663, 1990
8. Limas CJ, Goldenberg F, Limas C: Autoantibodies against β-adrenoreceptors in human idiopathic dilated cardiomyopathy. Circ Res 64: 97–103, 1989
9. Wolff PG, Kuhl U, Schultheiss HP: Laminin distribution and autoantibodies to laminin in dilated cardiomyopathy and myocarditis. Am Heart J 117: 1303–1309, 1989
10. Schultheiss HP, Kuhl U, Janda I, Melzner B, Ulrich G, Morad M: Antibody-mediated enhancement of calcium permeability in cardiac myocytes. J Exp Med 168: 2105–2119, 1988
11. Klein R, Maisch B, Kochsiek K, Berg PA: Demonstration of organ specific antibodies against heart mitochondria (anti-M7) in sera from patients with some form of heart diseases. Clin Exp Immunol 58: 283–292, 1984
12. Maisch B, Deeg P, Liebau G, Kochsiek K: Diagnostic relevance of humoral and cytotoxic immune reactions in primary and secondary dilated cardiomyopathy. Am J Cardiol 52: 1072–1078, 1983
13. Maisch B, Trostel-Soeder R, Stechemesser E, Berg PA, Kochsiek K: Diagnostic relevance of humoral and cell-mediated immune reactions in patients with acute viral myocarditis. Clin Exp Immunol 48: 533–545, 1982
14. Scheerder ID, Vandekerckhove J, Robbrecht J, Algoed L, Buyzere MD, Langhe JD, Schrijver GD, Clement D: Post-cardiac injury syndrome and an increased humoral immune response against the major contractile proteins (actin and myosin). Am J Cardiol 56: 631–633, 1985

15. Cunningham MW, Antone SM, Gulizia JM, Mcmanus BM, Fischetti VA, Gaunt CJ: Cytotoxic and viral neutralizing antibodies cross react with streptococcal M protein, enteroviruses, and human cardiac myosin. Proc Natl Acad Sci USA 89: 1320–1324, 1992
16. Gulizia JM, Cunningham MW, McManus BM: Immunoreactivity of anti-streptococcal monoclonal antibodies to human heart valves. Am J Pathol 138: 285–301, 1991
17. Cunningham MW, McCormack JM, Talaber LR, Harley JB, Ayoub EM, Muneer RS, Chun LT, Reddy DV: Human monoclonal antibodies reactive with antigens of the group A streptococcus and human heart. J Immunol 141: 2760–2766, 1988
18. Dale JB, Beachey EH: Sequence of myosin-crossreactive epitopes of streptococcal M protein. J Exp Med 164: 1785–1790, 1986
19. Dale JB, Beachey EH: Epitopes of streptococcal M proteins shared with cardiac myosin. J Exp Med 162: 583–591, 1985
20. Krisher K, Cunningham MW: Myosin: A link between streptococci and heart. Science 227: 413–415, 1985
21. Beisel KW: Immunogenetic basis of myocarditis: Role of fibrillary antigens. Springer Semin Immunopathol 11: 31–42, 1989
22. Jin O, Sole MJ, Butany JW, Chia WK, McLaughlin PR, Liu P, Liew CC: Detection of enterovirus RNA in myocardial biopsies from patients with myocarditis and cardiomyopathy using gene amplification by polymerase chain reaction. Circulation 82: 8–16, 1990
23. Kodama M, Matsumoto Y, Fujiwara M: *In vivo* lymphocyte-mediated myocardial injuries demonstrated by adoptive transfer of experimental autoimmune myocarditis. Circulation 85: 1918–1926, 1992
24. Smith S, Allen PM: Myosin-induced acute myocarditis is a T cell-mediated disease. J Immunol 147: 2141–2147, 1991
25. Kodama M, Matsumoto Y, Fujiwara M, Masani F, Izumi T, Shibata A: A novel experimental model of giant cell myocarditis induced in rats by immunization with cardiac myosin fraction. Clin Immunol Immunopathol 57: 250–262, 1990
26. Neu N, Beisel KW, Traystman MD, Rose NR, Craig SW: Autoantibodies specific for the cardiac myosin isoform are found in mice susceptible to coxsackievirus B_3-induced myocarditis. J Immunol 138: 2488–2492, 1987
27. Neu N, Rose NR, Beisel KW, Herskowitz A, Gurriglass G, Craig SW: Cardiac myosin induces myocarditis in genetically predisposed mice. J Immunol 139: 3630–3636, 1987
28. Cunningham MW, McCormack JM, Fenderson PG, Ho MK, Beachey EH, Dale JB: Human and murine antibodies cross-reactive with streptococcal M protein and myosin recognize the sequence gly-lys-ser-lys-gln in M protein. J Immunol 143: 2677–2683, 1989
29. Fenderson PG, Fischetti VA, Cunningham MW: Tropomyosin shares immunologic epitopes with group A streptococcal M proteins. J Immunol 142: 2475–2481, 1989
30. Leiros CP, Sterin Borda Leonor, Cossio P, Borda ES: Muscarinic cholinergic antibody in experimental autoimmune myocarditis regulates cardiac function. P.S.E.B.M. 195: 356–363, 1990
31. Mcnally EM, Kraft R, Bravo-Zehnder M, Taylor DA, Leinwand LA: Full-length rat alpha and beta cardiac myosin heavy chain sequences. Comparisons suggest a molecular basis for functional differences. J Mol Biol 210: 665–671, 1989
32. Maita T, Yajima E, Nagata S, Miyanishi T, Nakayama S, Matsuda G: The primary structure of skeletal muscle myosin heavy chain. J Biochem 110: 75–87, 1991

Molecular and Cellular Biochemistry **119**: 73–78, 1993.

Force-frequency-relation in human atrial and ventricular myocardium

Robert H.G. Schwinger,[1] Michael Böhm,[1] Andrea Koch,[1] Rainer Uhlmann,[1] Peter Überfuhr,[2] Eckhart Kreuzer,[2] Bruno Reichart[2] and Erland Erdmann[1]
Universität München, [1] Medizinische Klinik I; [2] Herzchirurgische Klinik, Klinikum Großhadern, Marchioninistr. 15, D-8000 München 70, Germany

Abstract

In human heart failure, an increase in frequency of stimulation is followed by a reduced force of contraction *in vivo* and *in vitro*. The present study aimed to investigate whether a different origin of the myocardial sample or pretreatment with the cardioprotective agent 2,3-butanedione-monoxime (BDM) influences the force-frequency-relationship in electrically driven muscle strips taken from failing and nonfailing human myocardium. With as well as without pretreatment with BDM, the altered force-frequency-relationship in failing compared to nonfailing human ventricular myocardium can be observed. The effectiveness and the potency to increase force of contraction following an increase in frequency of stimulation was significantly higher in atrial than in ventricular myocardium in nonfailing and failing tissue. The different observations in atrial and ventricular myocardium provide evidence for functionally relevant differences in the electromechanical coupling between the human atrial and ventricular myocardium. (Mol Cell Biochem **119:** 73–78, 1993)

Key words: human myocardium, heart failure, force-frequency-relationship, inotropy, cardiac function

Introduction

In patients without clinical signs of heart failure, an increase in frequency of stimulation is followed by an increase in contractility *in vitro* [1–4]. In patients with normal left ventricular function, an increased contractility in response to rapid atrial pacing was demonstrated also *in vivo* [5]. In contrast, in patients with dilated cardiomyopathy only little or no enhancement in systolic and diastolic function during atrial pacing tachycardia was seen [5], thus, suggesting an important depression of inotropic and lusitropic function. *In vitro* studies using isolated electrically driven left ventricular papillary muscle strips taken from patients with severe heart failure have demonstrated that an increase in frequency of stimulation is accompanied by a reduced force of contraction [3]. Force of contraction increased in nonfailing myocardial tissue but it did not change significantly or it even decreased in failing human myocardium [3]. These studies, however, were performed after addition of 30 mmol/l 2,3-butanedione-monoxime (BDM) to the storage-solution to protect myocardial

Address for offprints: R.H.G. Schwinger, Universität München, Medizinische Klinik I, Klinikum Großhadern, Marchioninistr. 15, D-8000 München 70, Germany

tissue from dissection injury. BDM, suggested to be a cardioprotective agent [6], exerts a variety of effects on myocytes, e.g. BDM decreases the sensitivity towards Ca^{2+} [7], BDM affects the slow inward Ca^{2+}-current [8, 9], and BDM influences cross-bridge-kinetics [10]. Altogether, BDM has been suggested to affect both Ca^{2+} availability and responsiveness of the myofilaments to Ca^{2+} [11]. Therefore, it is not unreasonable to speculate that pretreatment with BDM might influence myocyte properties and function. In order to isolate undamaged human left ventricular myocardium to study the influence of BDM-containing-solution on the force-frequency-relationship, we examined left ventricular human trabeculae. Left ventricular trabeculae can be used without causing dissection injury. To explore whether or not the different origin of the myocardial sample or the hemodynamic work load imposed on atria compared to ventricles influences the force-frequency-relationship, we studied the force-frequency-relationship in left ventricular and right auricular trabeculae from nonfailing myocardium.

Materials and methods

1. Myocardial tissue

Experiments were performed on isolated, electrically stimulated, human left ventricular and right atrial trabeculae. Tissue was obtained during cardiac transplantation (n = 12, 12 male; age: 47 years, range 19–62 yr; dilated cardiomyopathy). Patients suffered from heart failure clinically classified as NYHA IV on the basis of clinical symptoms and signs as judged by the attending cardiologist shortly before operation. Human atrial myocardium was also obtained from 9 nonfailing patients who underwent aortocoronary bypass operations. All patients gave written informed consent before surgery. Medical therapy consisted of diuretics, nitrates, ACE-inhibitors and cardiac glycosides. Patients receiving catecholamines, β-adrenoceptor- or Ca^{2+}-antagonists were withdrawn from the study. Drugs used for general anesthesia were flunitrazepam and pancuroniumbromide with isoflurane. Cardiac surgery was performed on cardiopulmonary bypass with cardioplegic arrest during hypothermia. Nonfailing human left ventricular myocardium was obtained from 4 donors who were brain dead as a result of traumatic injury. The cardioplegic solution (a modified Bretschneider solution) contained (in mmol/l): NaCl 15, KCl 10, $MgCl_2$ 4, histidine 180, tryptophan 2, mannitol 30 and potassium dihydrogen oxoglutarate 1.

2. Contraction experiments

Immediately after excision, the left ventricular muscles as well as the atrial tissue were placed in ice-cold prep-arated 2,3-butanedione-monoxime-containing Tyrode's solution (composition in mM: NaCl 119,8; KCl 5.4; $MgCl_2$ 1.05; $CaCl_2$ 1.8; Na HCO_3 22.6; NaH_2PO_4 0.42; glucose 5.05; ascorbic acid 0.28; Na_2EDTA 0.05, BDM 30) at room temperature or in Tyrode's solution without BDM and oxygenated by bubbling with 95% $O_2$5% CO_2 and delivered to the laboratory within 10 min. The experiments were performed on isolated, electrically driven muscle preparations. The preparations were attached to a bipolar platinum stimulating electrode and suspended individually in 75 ml glass tissue chambers for recording of isometric contractions. The bathing solution used was a modified Tyrode's solution, maintained at 37° C; its pH was 7.4. Isometric force of contraction was measured with an inductive force transducer (W. Fleck, Mainz, FRG) attached to a Hellige Helco Scriptor (Hellige, Freiburg, FRG) or Gould recorder (Gould Inc, Cleveland, Ohio, USA). Each muscle was stretched and resting load was kept constant throughout the experiments. This was imperative, as both, relaxation and force generation are load dependent. The preparations were electrically paced at 1 Hz with rectangular pulses of 5 ms duration (Grass stimulator SD 9), the voltage was 20% above threshold. All preparations were allowed to equilibrate at least 90 min in a drug-free bathing solution until complete mechanical stabilization. After 45 min the solution was changed. The force-frequency-relationship was studied starting with a rate of 0.5 Hz. Control strips performed in Tyrode's solution with identical composition as original experiments revealed maximally 20% reduction of baseline isometric tension over the period necessary to complete testing. The experiments were performed as described previously in detail [4].

Materials

2,3-Butanedione-monoxime was from SERVA (Heidelberg, FRG). All other chemicals were of analytical grade or the best grade commercially available. Applied agents did not change the pH of the medium.

Statistics

The data shown are means ± SEM. The drug concentration producing 50% of the maximum effect (EC_{50}) was graphically determined in each individual experiment. The EC_{50}-values are given with 95% confidence limits. Statistical significance was analysed using the Student's t-test for unpaired or paired observations (SPSS PC plus); $p < 0.05$ was considered significant [12].

Results

Figure 1 shows the frequency-dependent change in force of contraction in electrically driven left ventricular trabeculae from terminally failing myocardium after pretreatment with or without BDM-containing Tyrode's solution. Following an increase of frequency of stimulation, force of contraction was reduced in left ventricular trabeculae of terminally failing human hearts (dilated cardiomyopathy). BDM-pretreatment did not influence the force-frequency-relationship in the examined failing (see Fig. 1) and nonfailing (data not shown) human myocardium. In myocardium from patients without heart failure a positive force-frequency-relationship was observed ($p < 0.05$). Data are given in Table 1.

Figure 2 demonstrates the effect of increasing fre-

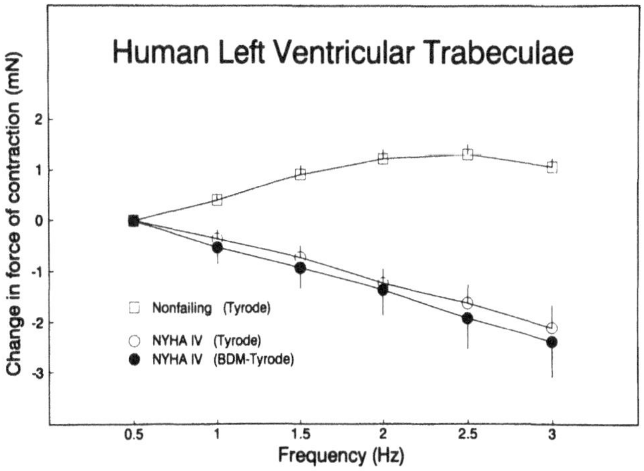

Fig. 1. Change in isometric force of contraction (ordinate, mN) plotted as a function of frequency of stimulation (0.5 Hz to 3 Hz) (abscissa) in electrically driven left ventricular trabeculae from terminally failing myocardium due to dilated cardiomyopathy and nonfailing control tissue (n = 5; basal force of contraction: 2.5 ± 0.3 mN). Muscle strips of terminally failing myocardium were either stored in BDM-containing Tyrode's solution (basal force of contraction: 3.9 ± 0.7 mN, n = 8) or in Tyrode's solution (basal force of contraction: 4.3 ± 0.7 mN, n = 6). Results are presented as mean ± SEM.

quencies of stimulation in nonfailing and terminally failing ventricular and atrial trabeculae. In muscle strips from nonfailing human hearts, an increase in frequency of stimulation was followed by an increase in force of contraction. This holds true for atrial and ventricular myocardium. However, the effectiveness (Δ mN atrial 2.4 ± 0.3 mN, ventricular 1.4 ± 0.12 mN) and the potency (mean EC_{50}: atrial 0.7 Hz, ventricular 1.3 Hz) to

Table 1. Force-frequency-relationship in ventricular and atrial trabeculae from nonfailing and failing myocardium

	n	basal FOC (mN)	1 Hz: delta mN (% basal)	2 Hz: delta mN (% basal)
NONFAILING				
AUT	13	4.1 ± 0.5*+	2.4 ± 0.3*+	2.2 ± 0.3*+
EC_{50}		0.7 (0.5–0.8)*+	(163 ± 8)*+	(159 ± 9)*
PAP	13	1.8 ± 0.3+	0.4 ± 0.1+	1.4 ± 0.1+
EC_{50}		1.3 (1.2–1.5)	(123 ± 4)+	(190 ± 12)+
FAILING				
AUT	4	1.2 ± 0.2*	0.2 ± 0.1*	1.0 ± 0.2*
EC_{50}		1.6 (1.0–2.4)	(117 ± 6)*	(180 ± 10)*
PAP	48	2.5 ± 0.2	− 0.3 ± 0.1	− 0.6 ± 0.1
EC_{50}		1.7 (1.5–1.9)	(92 ± 2)	(82 ± 4)

* $p < 0.05$ vs PAP
+ $p < 0.05$ vs Failing
AUT = Right auricular trabeculae
PAP = Left ventricular trabeculae
FOC = Force of contraction
(% basal) = Change in force of contraction 1 Hz (2 Hz) vs. 0.5 Hz stimulation frequency in mN.

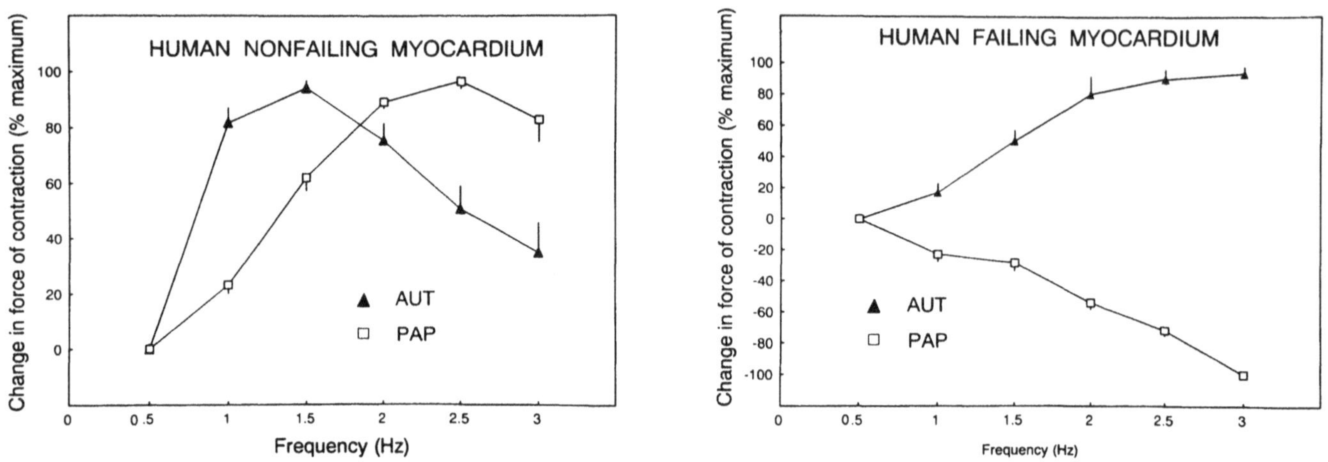

Fig. 2. Change in isometric force of contraction (ordinate, %) plotted as a function of frequency of stimulation (0.5 Hz to 3 Hz) (abscissa) in electrically driven left ventricular trabeculae from nonfailing myocardium (A) (PAP, n = 13; basal force of contraction 1.8 ± 0.2 mN; AUT, n = 9; basal force of contraction: 4.1 ± 0.5 mN) and from terminally failing myocardium (B) (PAP, n = 48; basal force of contraction 2.4 ± 0.2 mN; AUT, n = 4; basal force of contraction: 1.2 ± 0.2 mN). Results are presented as mean ± SEM.

increase force of contraction following an increase in frequency of stimulation was significantly lower in atrial than in ventricular nonfailing human myocardium. In human failing myocardium the force-frequency-relationship was found positive in right atrial tissue, but negative in ventricular trabeculae.

Discussion

The different force-frequency-relationship in nonfailing and failing human myocardium can be demonstrated in freshly isolated human myocardium with [3] as well as without the use of BDM [this study]. Consistently, we report that the use of BDM as 'cardioprotective solution' does not artifactually influence the force-frequency-relationship in human myocardium. However, the force-frequency-relationship is different in human atrial and ventricular myocardium.

In vivo, adrenergic reflex mechanisms may be activated to compensate reduced myocardial contractility in heart failure. These mechanisms may mask changes in contractility in response to different frequencies of stimulation. Therefore, isolated, electrically driven cardiac preparations may provide a useful tool to examine isometrically the effect of different stimulation frequencies on force of contraction. Under this experimental approach, additional influencing factors like changes of preload or afterload do not affect the developed myocardial force of contraction. In previous studies, it has been demonstrated that BDM protects the human myocardium from cutting injury [3], thus facilitating dissection and preservation of muscle strip preparations [3]. Addition of BDM resulted in a significantly higher force of contraction in electrically driven human papillary muscle strips after longterm storage (4° C, 10 h, 24 h) in BDM-containing tyrode's solution than in tyrode's solution [13]. BDM affects via a variety of mechanisms myocardial contractility [7–11] and exerts concentration-dependently negative inotropic activity in isolated muscle strip preparations from animals [10, 14] and humans [3, 15]. Experiments performed without the use of BDM showed a significant difference between the force-frequency-relationship (up to 3 Hz) in nonfailing and terminally failing human tissue [14]. Consistently in patients with normal left ventricular function, rapid atrial pacing initiated an increase in contractility, i.e. increase in peak + dP/dt and an increase in the peak-systolic pressure to end-diastolic volume rate [5]. In patients with dilated cardiomyopathy, however, these increases in contractility indices were absent or diminished [5]. *In vitro*, stimulation frequencies from 0.5 Hz up to 3 Hz were followed by a progressive decrease in force of contraction in muscle preparations from the failing myocardium [4]. Human papillary muscle strip preparations and left ventricular trabeculae from nonfailing hearts increased, demonstrating that the positive force-frequency-relationship can be detected in preparations with [3] as well as without pretreatment with BDM [4, this study]. Therefore, the demonstrated reversibility of BDM-mediated effects [15, this study] opens the possibility to use BDM as a cardioprotective agent for isolating preparations of animal and human myocardium. In addition, BDM might be a beneficial supplement for cardioplegic solution for open heart surgery as well.

An important aspect of these studies is that the origin of the myocardial tissue greatly influences the force-frequency-relationship. In isolated trabeculae from nonfailing and failing myocardium, the frequency-dependent increase in force of contraction was more pronounced in atrial than in ventricular tissue. In atrial and ventricular myocardium the density of 1,4-dehydropyridine-binding sites is unchanged [16]. Thus, differences in the number of sarcolemmal Ca^{2+}-channels do not influence the different force-frequency-relationship. In addition, in human failing and nonfailing myocardium the binding to 1,4-dihydropyridine-binding sites was also unchanged [17]. The basic characteristics of L-type Ca^{2+}-currents of ventricular myocytes are qualitatively similar to those described in human atrial cells and other mammalia species [18, 19]. Thus, the significantly altered force-frequency-relationship in failing compared to nonfailing tissue and the differences observed in atrial and ventricular myocardium most likely are not due to differences in transsarcolemmal Ca^{2+}-currents, but may be due to differences in intracellular Ca^{2+}-handling. This is in accordance with measurements in isolated human myocytes. Beuckelmann et al. [20] in this respect observed significant differences in isolated human cells from myopathic and control hearts; e.g. resting Ca^{2+}-levels were higher, and the rate of diastolic Ca^{2+}-decay was slowed. Possible mechanisms may relate to a reduced Ca^{2+}-efflux through the Na^+/Ca^{2+}-exchange system, or may be due to a reduced re-uptake of Ca^{2+} into the sarcoplasmatic reticulum.

In simultaneous measurements with the Ca^{2+}-indicator protein aequorin, the peak amplitude of the Ca^{2+}-transient and the resting intracellular Ca^{2+}-concentration increased along with increasing stimulation frequency [21]. Therefore, the mechanical deterioration in the failing heart, e.g. during a frequency increase, is not due to a reduced availability of cystolic Ca^{2+} rather than due to a diminished Ca^{2+}-re-uptake from the cytosol. Hence, pronounced elevation of Ca^{2+} intracellularly will worsen the force-frequency-relationship [21]. Consistently, interventions that increase intracellular Ca^{2+} excessively such as high Ca^{2+} decreased augmentation of tension following an increase in stimulation frequencies in human muscle strip preparations [21].

The sensitivity of the myofibrils to Ca^{2+} might also influence the force-frequency-relationship in atrial and ventricular myocardium. Using skinned fiber preparations from human myocardial tissue, the sensitivity to Ca^{2+} was observed to be different in atrial and ventricular myocardium [22]. The different sensitivity to Ca^{2+} in atrial and ventricular myocardium may be related to differences in myosin isoenzyme expression [23–25]. For instance, the change in Ca^{2+}-sensitivity in atrial myocardium of patients with heart failure has been reported to be related to an elevation of right atrial pressure [22]. Differences in contraction coupling might also explain the differences observed in the frequency-dependent increase in force of contraction in atrial and ventricular tissue. The present study clearly demonstrates that experiments performed with atrial tissue may be not necessarily comparable to the situation in ventricular myocardium. This holds true for receptor-dependent inotropic mechanisms, e.g. β-adrenoreceptor-mediated increase in force of contraction [24], as well as for receptor-independent inotropic mechanisms, e.g. force-frequency-relationship (this study). On the other hand, the missing increase in $+\ dp/dt_{max}$ or in end systolic pressure-volume ratio after atrial pacing tachycardia in patients with dilated cardiomyopathy might have their origin in the changes demonstrated in these experiments in isolated trabeculae.

Acknowledgement

We thank Mrs. Heidrun Villena Hermoza for her excellent technical help. Experimental work was supported by the Deutsche Forschungsgemeinschaft (DFG) and by a grant of the MMW.

References

1. Bodwitch HP: Über die Eigentümlichkeiten der Reizbarkeit, welche die Muskelfasern des Herzens zeigen. Berl Sächs Ges (Akad) Wiss 652–689, 1871
2. Buckley NM, Penefsky ZJ, Litwak RS: Comparative force-frequency relationship in human and other mammalian ventricular myocardium. Pflügers Arch 332: 259–270, 1972
3. Mulieri LA, Hasenfuss G, Leavitt B, Allen PD, Alpert NR: Altered myocardial force-frequency relation in human heart failure. Circulation 85: 1743–1750, 1992
4. Schwinger RHG, Böhm M, Erdmann E: Inotropic and lusitropic dysfunction in myocardium from patients with dilated cardiomyopathy. Am Heart J 123: 116–128, 1992
5. Feldman MD, Alderman JD, Aroesty JM, Royal HD, Ferguson JJ, Owen RM, Grossman W, McKay RG: Depression of systolic and diastolic myocardial reserve during atrial pacing tachycardia in patients with dilated cardiomyopathy. J Clin Invest 82: 1661–1669, 1988
6. Mulieri LA, Hasenfuss G, Ittleman F, Blanchard EM, Alpert

NR: Protection of human left ventricular myocardium from cutting injury with 2,3-butanedione monoxime. Circ Res 65: 1441–1444, 1989
7. Li T, Sperelakis N, Teneick RE, Solaro RJ: Effects of diacetyl monoxime on cardiac excitation-contraction coupling. J Pharmacol Exp Ther 232: 688–695, 1992
8. Coulombe A, Lefevre IA, Deroubaix E, Thuringer D, Coraboeuf E: Effect of 2,3-butanedione 2 monoxime on slow inward and transient outward currents in rat ventricular myocytes. J Mol Cell Cardiol 22: 921–932, 1990
9. Sada H, Sada S, Sperelakis N: The calcium channel agonist, Bay K-8644, antagonizes effects of diacetyl monoxime on cardiac tissue. Can J Physiol Pharmacol 63: 1267–1270, 1985
10. Mörner SEJN, Wohlfart B: The action of 2,3-butanedionemonoxime on the inotropic state in guinea-pig myocardium. Acta Physiol Scand 142: 211–219, 1991
11. Gwathmey JK, Warren SE, Briggs GM, Copelas L, Feldman MD, Phillips PJ, Callaham M, Schoen FJ, Grossman W, Morgan JP: Diastolic dysfunction in hypertrophic cardiomyopathy. J Clin Invest 87: 1023–1031, 1991
12. Wallenstein S, Zucker CL, Fleiss JL: Some statistical methods useful in circulation research. Circ Res 47: 1–9, 1980
13. Schwinger RHG, Böhm M, Koch A, Erdmann E: Beneficial cardioprotective actions of BDM on human myocardium. Eur Heart J 13: P762, 1992
14. Blanchard EM, Smith GL, Allen DG, Alpert NR: The effects of 2,3-butanedione monoxime on initial heat, tension, and aequorin light output of ferret papillary muscles. Pflügers Arch 416: 219–221, 1990
15. Schwinger RHG, Koch A, Erdmann E: Einfluß von Butandionmonoxim auf die Kontraktionskraft des menschlichen Herzens. Z Kardiol 81: P286, 1992
16. Finkel MS, Patterson RE, Roberts WC, Smith TD, Keise HR: Calcium channel binding characteristics in the human heart. Am J Cardiol 62: 1281–1284, 1988
17. Rasmussen RP, Minobe W, Bristow MR: Calcium antagonist binding sites in failing and nonfailing human ventricular myocardium. Biochem Pharmac 39: 691–696, 1990
18. Beuckelmann DJ, Näbauer M, Erdmann E: Characteristics of calcium-current in isolated human ventricular myocytes from patients with terminal heart failure. J Mol Cell Cardiol 23: 929–937, 1991
19. Escade D, Coulombe A, Faivre JF, Coraboeuf E: Characteristics of the time-dependent slow inward current in adult human atrial single myocytes. Mol Cell Cardiol 18: 547–551, 1986
20. Beuckelmann DJ, Näbauer M, Erdmann E: Intracellular calcium handling in isolated ventricular myocytes from patients with terminal heart failure. Circulation 85: 1046–1055, 1992
21. Gwathmey JK, Hajjar RJ, Solaro RJ: Contractile deactivation and uncoupling of crossbridges. Circ Res 69: 1280–1292, 1991
22. Wankerl M, Böhm M, Morano I, Rüegg JC, Eichhorn M, Erdmann E: Calcium sensitivity and myosin light chain pattern of atrial and ventricular skinned cardiac fibers from patients with various kinds of cardiac disease. J Mol Cell Cardiol 22: 1425–1438, 1990
23. Morano I, Beltz C, Wojciechowski, Rüegg JC: Modulation of crossbridge kinetics by myosin isoenzymes in skinned human heart fibers. Circ Res 68: 614–618, 1991
24. Schwinger RHG, Böhm M, Pieske B, Erdmann E: Different β-adrenoceptor-effector coupling in human ventricular and atrial myocardium. Eur J Clin Invest 21: 443–451, 1991
25. Solaro RJ: Myosin and why hearts fail. Circulation 85: 1945–1946, 1992.

Molecular and Cellular Biochemistry **119**: 79–88, 1993.

Depressed sliding velocity of isolated cardiac myosin from cardiomyopathic hamsters: Evidence for an alteration in mechanical interaction of actomyosin

Hiroshi Yamashita, Seiryo Sugiura, Masataka Sata, Takashi Serizawa, Masahiko Iizuka,[1] Teruo Shimmen[2] and Shin-ichi Momomura

The Second Department of Internal Medicine, Faculty of Medicine, University of Tokyo, 7-3-1, Hongo, Bunkyo-ku, Tokyo, 113, Japan

[1] *The First Department of Internal Medicine, Dokkyo University, School of Medicine, 880, Kitakobayashi, Mibu-machi, Shimotsuga-gun, Tochigi-ken, 321, Japan*

[2] *The Department of Life Science, School of Science Himeji Institute of Technology, 1479-1, Kanaji, Kamigori-cho, Ako-gun, Hyogo-ken, 678-12, Japan*

Abstract

We measured the relative sliding velocity of cardiomyopathic hamster cardiac myosin on actin cables by using an *in vitro* motility assay system. We also investigated the relationship between the velocity and both myosin isozyme content and ATPase activity. Cardiac myosin was obtained from cardiomyopathic hamsters (BIO 14.6; B) aged 3, 6, 9, and 18 months and age-matched controls (F1B; F). Long well-organized actin cables of an alga, *Nitellopsis*, were used for the motility assay. Small latex beads ($2\,\mu m$ in diameter) were coated with purified cardiac myosin. When myosin-coated beads were introduced into an algal cell in the presence of Mg-ATP, myosin interacted with actin and dragged the beads. Active movement of the beads along the actin cables was observed under a photomicroscope and the velocity was measured. The velocity was significantly lower in B than in F for each age group (0.47 vs. 0.71 $\mu m/s$ at the age of 3 months, $p < 0.05$; 0.44 vs. 0.88 $\mu m/s$ at 6 months, $p < 0.01$; 0.44 vs. 0.67 $\mu m/s$ at 9 months, $p < 0.01$; 0.35 vs. 0.52 $\mu m/s$ at 18 months, $p < 0.05$). Both Ca^{2+}-activated ATPase activity and the percentage of α-myosin heavy chain were also lower in B than in F for each age group. When examined for individual specimens, there was a positive correlation between the velocity and both myosin Ca^{2+}-activated ATPase activity ($r = 0.84$) and percentage of α-myosin heavy chain ($r = 0.83$). These data points of both control and cardiomyopathic hamsters were distributed near the regression line obtained from control and thyroxine-treated rabbits reported previously. The present results indicate that the difference in mechanical properties between control and cardiomyopathic cardiac myosin is attributed to isozyme redistribution and not to a qualitative change in each myosin molecule. (Mol Cell Biochem **119:** 79–88, 1993)

Key words: cardiomyopathic hamster, *in vitro* motility assay, myosin isozyme, ATPase activity, sliding velocity

Address for offprints: H. Yamashita, The Second Department of Internal Medicine, Faculty of Medicine, University of Tokyo, 7-3-1, Hongo, Bunkyo-ku, Tokyo, 113, Japan

Introduction

The Syrian cardiomyopathic hamster (BIO 14.6) is known to display various hereditary abnormalities in both cardiac and skeletal muscles [1]. Cardiac involvement is characterized by focal myocardial necrosis and myocardial hypertrophy followed by progressively depressed contractile performance resulting in premature death from congestive heart failure [1–3]. Since these findings mimic those observed in human cardiomyopathy in many respects, this strain has been used as an experimental model of human cardiomyopathy, and morphological, physiological, and biochemical researches have been undertaken to understand this disease [1–13]. As for the physiological abnormalities, reduction in both isometric force and isotonic shortening velocity has been identified [2, 4, 5]. Among numerous biochemical abnormalities proposed for the underlying causes of this hereditary disease, it is still controversial whether the alteration in contractile proteins is involved in the depressed mechanical performance [5, 13]. The main problem in answering this question is that the mechanical parameters have been evaluated in either isolated papillary muscles or skinned cardiac fibers, which are under the influence of complex excitation contraction coupling or other regulatory mechanisms. To resolve this problem, we adopted experimental techniques, called *in vitro* motility assay systems, that were developed recently [14, 15]. In this system the relative sliding movement of myosin and actin is reconstituted with no inhibitory effects exerted by regulatory mechanisms. Previous studies have already shown that these systems are reliable in quantitative assessment of the mechanical properties of myosin molecules [14, 15].

The present study was designed to elucidate whether the alteration in myosin isoform is involved in depressed contractile functions of cardiomyopathic hamster hearts. We used an *in vitro* motility assay system developed independently by Sheetz *et al.* [14] and Shimmen *et al.* [15], in which small latex beads coated with myosin were made to slide on algal actin cables. Biochemical assays performed on the identical myosin specimen revealed that the depressed mechanical parameters (sliding velocity) were closely correlated with myosin ATPase activity and isozyme distribution.

Materials and methods

Preparation of myosin

Cardiac myosin was obtained from BIO 14.6 hamsters aged 3, 6, 9, and 18 months. Age-matched F1B strains were used as controls. Animals were obtained from Bio Breeders (Fitchburg, MA). They were anesthetized by intraperitoneal injections of 40–50 mg/kg body weight of sodium pentobarbital, and the hearts were quickly excised. All procedures described below were carried out at 4° C. The hearts were washed in buffered saline (10 mM sodium phosphate buffer, 0.9% NaCl, pH 7.2), and both atria and large vessels were removed. Five to eight hearts of the same strain at each age were pooled for the extraction of myosin. Both left and right ventricles were homogenized in tris-maleic acid buffer (20 mM Tris (hydroxymethyl) aminomethane-maleic acid, 1 mM ethylenediaminetetraacetic acid (EDTA), pH 7.0), and centrifuged (20PR Hitachi, Japan) at 1,000 g for 15 min. After supernatant was discarded, the pellet was used for myosin extraction by two different methods: (1) For the determination of myosin ATPase activity and the *in vitro* motility assay, purified myosin was prepared as follows. Myosin was extracted in 3 volumes of Guba-Straub solution (0.3 M KCl, 0.1 M KH_2PO_4, 50 mM K_2HPO_4, 1 mM ATP, 5 μg/ml Leupeptin, 5 mM dithiothreitol (DTT), 1 mM EDTA, pH 6.5) for 10 min. After the extract was centrifuged at 11,000 g for 15 min., the supernatant was collected and 14 volumes of ice-cold distilled water were added to precipitate the myosin. After being left for 1 h, the myosin was collected by centrifugation at 11,000 g for 15 min. Myosin was again dissolved in a high ionic strength solution (0.6 M KCl, 10 mM tris-chloride, 5 mM DTT, pH 7.5), and the trace amount of actin was removed by centrifugation (65 P, Hitachi, Japan) at 120,000 g for 4 h. The purity of the final myosin solution was confirmed by the sodium dodecyl sulfate(SDS)-polyacrylamide gel electrophoresis method [16]. This myosin solution was stored in an ice-cooled container and used for both the determination of ATPase activity and the *in vitro* motility assay within 48 h after purification. (ii) For the isozyme analysis, myosin was extracted by the method of Martin *et al.* [17].

Analysis of myosin isozyme

Myosin isozyme analysis was performed by pyrophosphate polyacrylamide gel electrophoresis described by Martin *et al.* [17]. Electrophoresis was carried out at 3° C, using a Pharmacia GE4 electrophoresis chamber. The gels were fixed and stained with Coomassie brilliant blue and scanned by a laser densitometer (LKB 2202 LKB Produkter AB, Sweden) at 600 nm. The area under peak was measured to determine the relative amounts of V_1, V_2, and V_3 isomyosins. The percentage of α myosin heavy chain (%α-MHC) was determined by the following formula.

$$\%\alpha\text{-MHC} = \%V_1 + \%V_2/2$$

Determination of ATPase activity

Calcium(Ca^{2+})-activated ATPase activity of myosin was determined according to the method of Yazaki *et al.* [18] with some modifications. The reaction was run for 5 min. at pH 7.5 and 25° C. The reaction solution contained 0.1 mg/ml myosin, 60 mM KCl, 10 mM $CaCl_2$, 1 mM disodium ATP, and 50 mM tris-HCl (pH 7.5). Inorganic phosphate liberated in ATP hydrolysis was measured by the method of Youngburg *et al.* [19]. Specific activity was calculated by using protein concentrations determined by the method of Lowry [20].

In vitro *motility assay*

We used an *in vitro* motility assay system developed by Shimmen *et al.* [15]. This method is composed of three procedures, each of which is described below.

Myosin-coated bead preparation

To visualize the movement of myosin on actin cables, small beads were used as markers. Latex beads (Dow Chemical Company, Uniform Latex Particles, 2 μm in diameter) were suspended in a poly-L-lysin solution (5 M poly-L-lysin (Sigma m.w. 70,000–150,000) and 400 mM KOH) at 0° C for 3 h and washed four times by centrifugation in a buffer solution containing 400 mM KCl and 5 mM N-2-hydroxyethylpiperazine-N′-2-ethanesulfonic acid (HEPES)-KOH (pH 7.5). These poly-L-lysin coated beads were mixed with purified myosin in a high ionic strength buffer (400 mM KCl, 10 mM imidazole-HCl, pH 7.0). By reducing the ionic strength of the buffer solution, myosin formed filaments and attached to the bead surface. Unbound myosin was removed by centrifugation, and the myosin-coated beads were suspended in a Mg-ATP solution (1 mM ATP, 5 mM ethyleneglycol-bis(β-aminoethyl ether) N,N,N′,N′-tetraacetic acid (EGTA), 6 mM $MgCl_2$, 70 mM KOH, 30 mM piperazine-N,N′-bis(2-ethanesulfonic acid) (PIPES), and 200 mM sorbitol, pH 7.0) for the *in vitro* motility assay.

Preparation of actin donor cells

The internodal cell of a green alga, *Nitellopsis obtusa*, was used as an actin donor (Fig. 1). *Nitellopsis* was cultured in a plastic bucket filled with tap water and mold in the bottom. The bucket was illuminated from above with two fluorescent lamps (20 W, Toshiba, Japan) in a 14 h light, 10 h dark cycle at 25° C. This plant consists of long internodal cells connected in series, and each cell has well-organized rows of chloroplasts inside the plasmalemma. Since long straight actin cables are anchored on the inner surface of chloroplast rows, the analysis of the movement is greatly facilitated. Besides the cell wall and plasmalemma that form the outer shell, there is tonoplast, an inner membrane system. Tonoplast forms a vacuole in the core of the cell, and cytoplasmic streaming takes place in the space between the tonoplast and the actin cables. To expose the actin cables, disintegration of tonoplast is necessary (tonoplast-free cell).

Before the experiment, an internodal cell was isolated, trimmed free of branches, and stored in artificial pond water (0.1 mM each of KCl, NaCl, and $CaCl_2$, pH approximately 5.6). The tonoplast-free cell, which would be used as an actin donor cell, was prepared as follows. Both ends of an internodal cell were cut open and a vacuole was perfused with the Mg-ATP solution (vacuolar perfusion). Since the solution contained 5 mM of EGTA, a drastic decrease in the calcium concentration in the vacuole resulted in a disintegration of the tonoplast. Although most of the endoplasm dispersed after this procedure, some still remained attached to the actin cables and continued to stream. Then the cell was left on a perfusion bench placed in a moist chamber to keep from drying. After 20 min., the cell was perfused again with an EDTA solution (5 mM EDTA, 1 mM ATP, 71 mM KOH, 30 mM PIPES, and

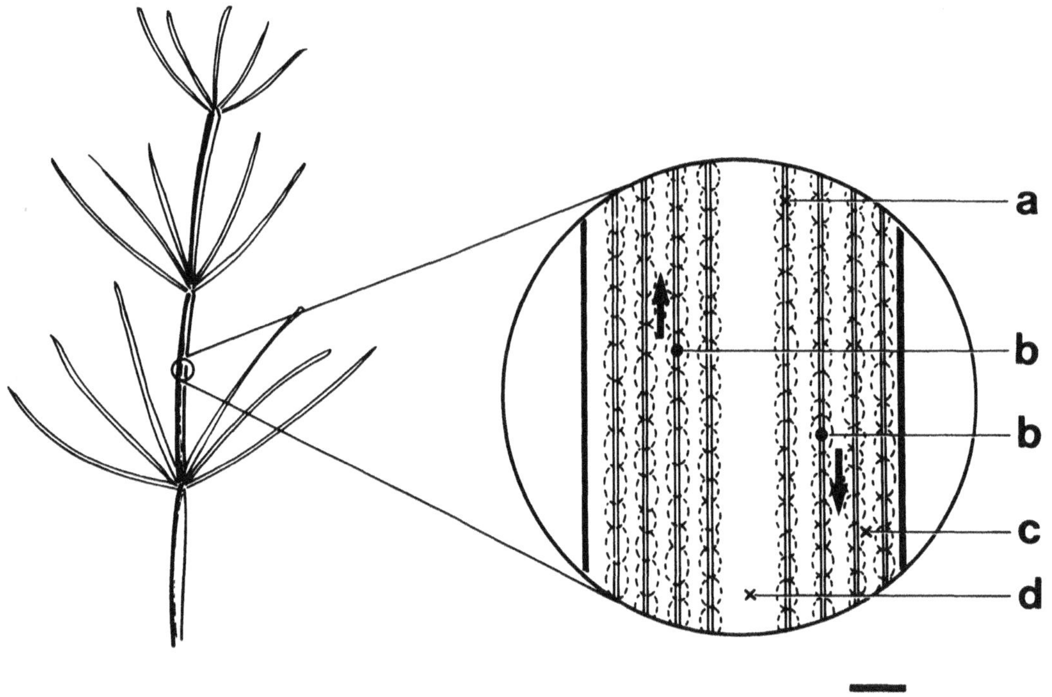

Fig. 1. Schematic drawing of the *in vitro* motility assay. Left: *Nitellopsis obtusa.* The long segment between branches is a single internodal cell. Right: A magnified view showing the intracellular structure of the internodal cell and the bead movement, a, actin cables; b, myosin-coated beads; c, chloroplasts; d, indifferent line. The solid arrows indicate the directions of the bead movement, which are reversed on each side of the indifferent line where chloroplast rows are absent. Bar: 20 μm.

200 mM sorbitol, pH 7.0). When magnesium was depleted with EDTA treatment, the endoplasm containing the alga myosin was inactivated, but the actin cables remained intact. The cell was placed in the moist chamber for a minute and again perfused with the Mg-ATP solution to effuse out the inactivated endoplasm. The tonoplast-free cell thus prepared was used for the motility assay.

Introduction of myosin-coated beads

Myosin-coated beads suspended in the Mg-ATP solution were introduced into the tonoplast-free cell by intracellular perfusion. Then both ends of the cell were ligated with polyester threads to avoid the passive movement of intracellular fluid, and the cell was placed on the slide glass in a solution containing 0.1 mM K_2SO_4, 1 mM NaCl, 0.1 mM $CaCl_2$, and 150 mM sorbitol. The movement of the beads was observed under a photomicroscope (Labophoto-2, Nikon, Japan) and recorded on videotape with a video camera (KV-24, Hitachi Denshi, Japan) and a video tape recorder (HR D-23, JVC, Japan). The movement was observed with a 40 × objective, and the final magnification on the monitor screen was 900 ×. The measurement of the velocity was performed on replays of the videotapes. Positions of the beads or bead aggregates in the frozen frame were marked on a clear plastic film placed on the monitor screen. By advancing the tape, the velocity was calculated from the distance the beads moved and the elapsed time recorded by a video timer (VTG-33, FOR·A Company, Japan). The velocity measurement was performed on the smooth and continuous movement over a distance of 5 μm or more. The mean velocity for each myosin obtained from either strain at different age was calculated from those of 20 to 30 different beads or bead aggregates. All experiments were carried out at 21° to 23° C.

Statistical analysis

Data were indicated as means ± standard deviations (S.D.). The statistical comparison of two groups was done by student t-test. The correlation between the parameters was studied by linear regression analysis. A p-value of less than 0.05 was considered significant.

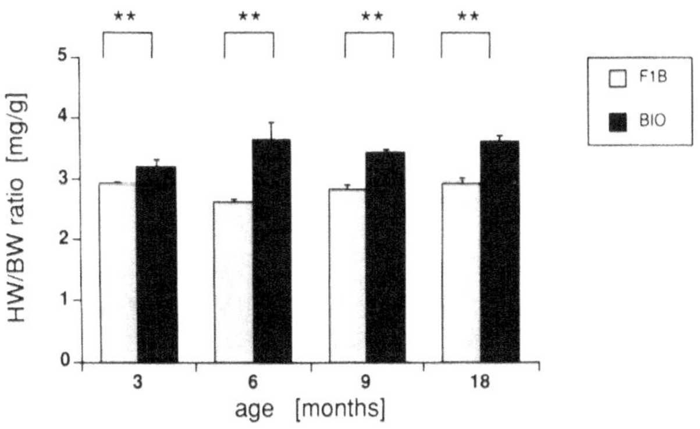

Fig. 2. The mean HW/BW ratios in control (F1B) and cardiomyopathic (BIO) hamsters for each age group. Values represent the mean ± SD of five to eight animals. Statistical comparisons were made between age-matched groups. ** p < 0.01.

Results

Cardiac hypertrophy

The mean HW/BW ratio of cardiomyopathic animals at each age was significantly greater than that of age-matched controls (Fig. 2). In cardiomyopathic animals, there was a tendency for the HW/BW ratio to increase with age.

Myosin isozyme distribution

In both control and cardiomyopathic hamsters, there was a tendency for the relative amount of V_1 to decrease with age. Typical densitometric scan patterns of pyrophosphate gel electrophoresis for native myosin obtained from 9-month-old control and cardiomyopathic hamsters are shown in Fig. 3. In cardiomyopathic animals, the relative content of V_3 increased at the expense of that of V_1 in comparison with controls. Accordingly, the percentage of α-MHC in cardiomyopathic animals was lower than in controls at every age studied.

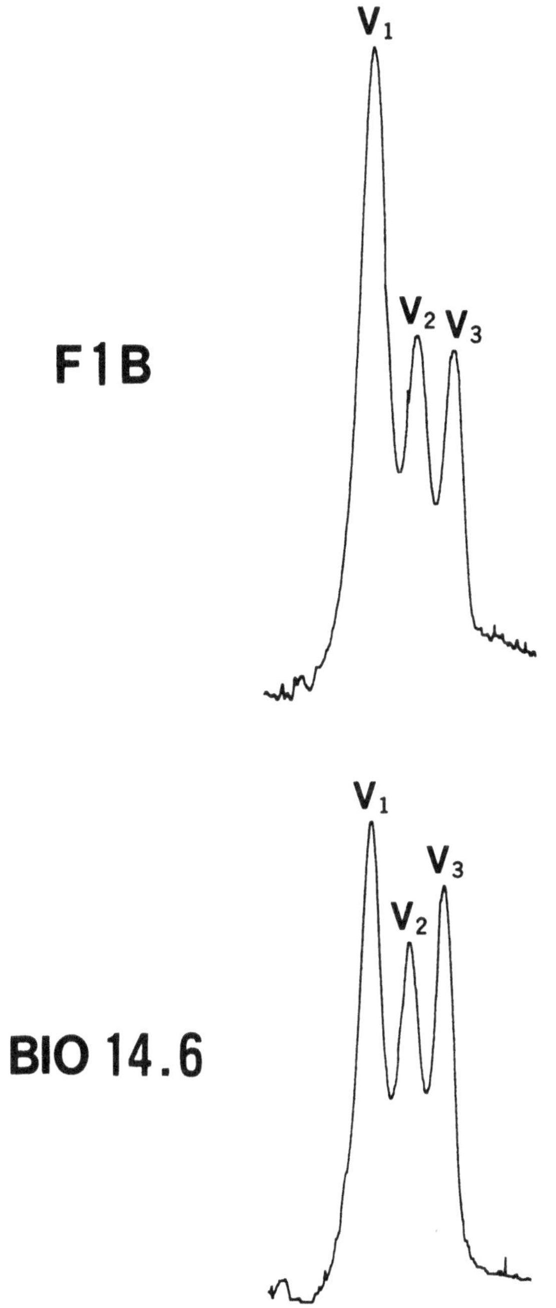

Fig. 3. Densitometric scan patterns of pyrophosphate gels of myosins from 9-month-old control (F1B) and cardiomyopathic (BIO14.6) hamster hearts. The relative amount of V_1 decreased and V_3 increased in cardiomyopathic animals.

Myosin ATPase activities

The SDS-polyacrylamide gels of purified myosin from control and cardiomyopathic hamsters are shown in Fig. 4. There was no loss or breakdown of myosins from both control and cardiomyopathic animals. The Ca^{2+}-activated ATPase activity was lower in cardiomyopathic animals than in controls at every age studied with a significant reduction in the average of all age groups (0.396 ± 0.038 vs. 0.536 ± 0.069 μMPi/mg min, $p < 0.05$).

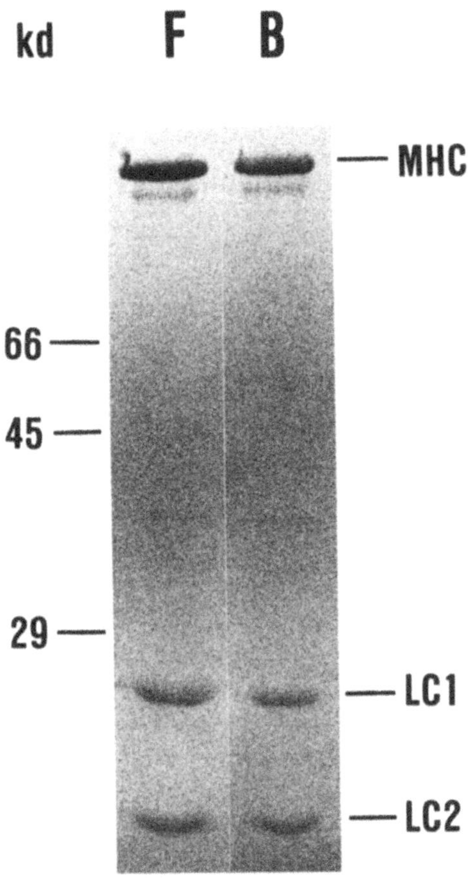

Fig. 4. Sodium dodecyl sulfate-polyacrylamide gels of purified myosin from control (F) and cardiomyopathic (B) hamster hearts. The molecular weight markers are indicated in the figure. MHC, myosin heavy chain; LC1, myosin light chain 1; LC2, myosin light chain 2.

In vitro *motility assay*

The characteristics of the bead movement observed in the previous reports on rabbit skeletal [14, 15] and cardiac myosin [21, 22] was also confirmed in hamster cardiac myosin. Furthermore, myosin from both control and cardiomyopathic animals showed the same characteristics except for their velocities: (i) The direction of the movement was reversed on each side of the indifferent line as that of cytoplasmic streaming (Fig. 1). This indicated that the movement was *active* and that the direction was determined by the polarity of the actin cables. (ii) The velocity was independent of the size of the bead aggregates, which consist of 2 to 10 beads, and fairly constant over a long distance. It was reproducible within each internodal cell preparation and also among different ones, as long as the beads were coated with the same myosin sample.

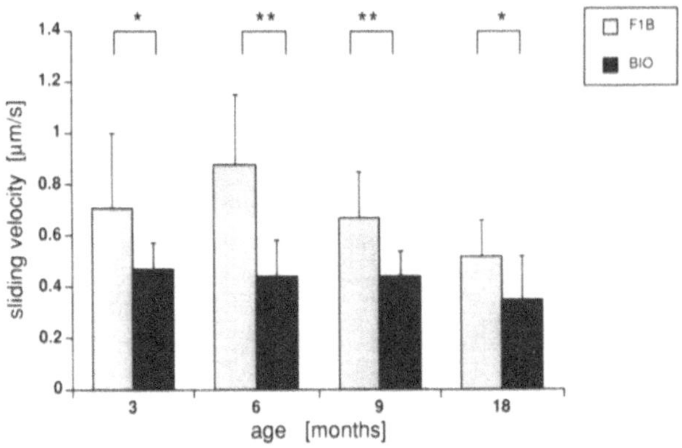

Fig. 5. The mean sliding velocities of the beads coated with control (F1B) and cardiomyopathic (BIO) hamsters for each age group. Values represent the mean ± SD of 20 to 30 beads. Statistical comparisons were made between age-matched groups. $^{**}p < 0.01$, $^{*}p < 0.05$.

The sliding velocity

The mean sliding velocity of beads coated with myosin from cardiomyopathic animals at each age was significantly lower than that of age-matched controls (Fig. 5). There was a tendency for the velocity to decrease with age in both control and cardiomyopathic animals. To relate this change in the sliding velocity to biochemical parameters, we plotted the relationship between the velocity and myosin Ca^{2+}-activated ATPase activity determined for each group in Fig. 6. A positive correlation was found between these parameters ($r = 0.84$). Also plotted in Fig. 7 was the relationship between the velocity and the relative amount of α-MHC. As expected, we observed a significant correlation between them ($r = 0.83$).

Discussion

In the present study, we measured the sliding velocity of isolated cardiac myosin from both control and cardiomyopathic hamsters on actin cables using an *in vitro* motility assay system. The sliding velocity of cardiomyopathic animals was lower than that of agematched controls, and there was a close correlation between the sliding velocity and biochemical parameters. Since both mechanical (sliding velocity) and biochemical parameters (myosin ATPase activity and isozyme content) were studied for identical pure myosin, the obtained relationship was a direct one free from the influence of any other factors.

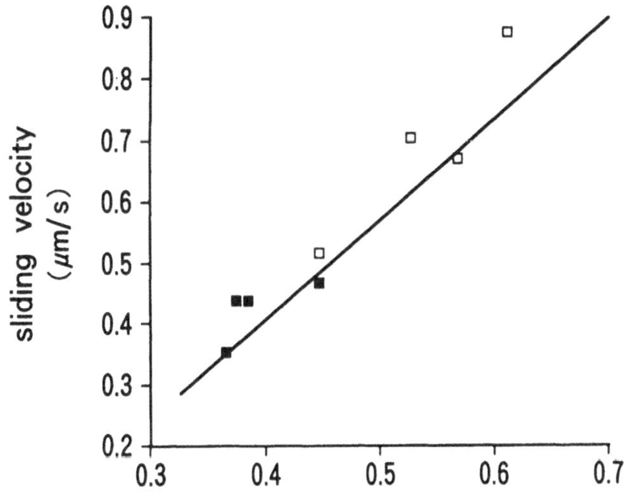

Fig. 6. Relationship between the velocity of myosin-coated beads and Ca^{2+}-activated ATPase activity of myosin in control (open squares) and cardiomyopathic (closed squares) animals. Each point represents the velocity (mean value of 20 to 30 beads) and ATPase activity for a given myosin sample. The solid line indicates the linear regression applied to them (r = 0.84).

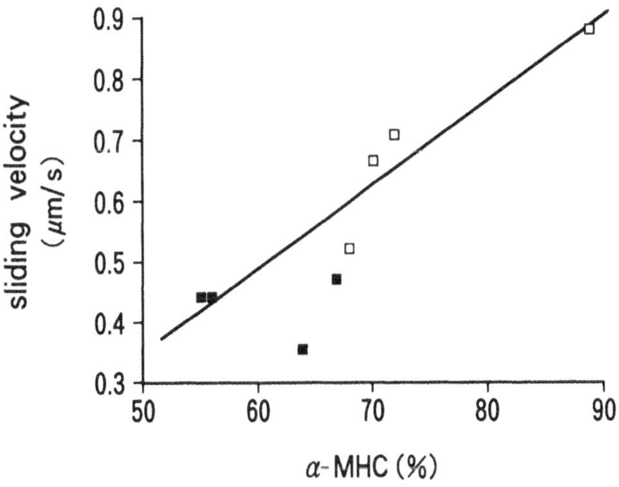

Fig. 7. Relationship between the velocity of myosin-coated beads and the relative amount of α-myosin heavy chain of control (open squares) and cardiomyopathic (closed squares) animals. Each point represents the value determined for a given animal. A linear regression analysis yields a significant correlation between these parameters (r = 0.83).

In vitro *motility assay system*

In this study, we used an *in vitro* motility assay technique developed independently by Sheetz *et al.* [14] and Shimmen *et al.* [15]. Earlier studies on various types of myosin have shown that this system is a reliable and quantitative assay for the mechanical function of contractile proteins [14, 15, 21, 22]. Furthermore, according to Sheetz *et al.* [14], the velocity measured in this system is a good index of unloaded shortening velocity of muscle (V_{max}). According to Rouleau *et al.* [4], V_{max} of isolated left ventricular papillary muscles obtained from 3-month-old control and cardiomyopathic hamsters (UM-X7.1 strain, a derivative of the BIO 14.6 strain) were 2.6 and 1.7 muscle length/second, respectively, at 29° C. Considering the temperature dependence of the shortening velocity [23, 24] and the similar pathological course in both strains [2, 4, 12], the velocity for 3-months-old animals measured in this study (0.90 and 0.65 μm/s for control and BIO 14.6 strain, respectively) was comparable to the V_{max} determined in isolated papillary muscle preparations.

BIO 14.6 as a model of human cardiomyopathy

The Syrian cardiomyopathic hamster (BIO 14.6) is known to display hereditary abnormalities in both cardiac and skeletal muscles [1]. Especially, cardiac involvement is similar to what is observed in human cardiomyopathy and often results in premature death [1–3]. Numerous abnormalities have been identified in the myocardium of these animals, such as increased density of voltage-sensitive Ca^{2+} channels [1] and α_1-adrenergic receptors [3], diminished activity of the guanine nucleotide-binding proteins stimulating adenylate cyclase [8], decreased Na^+,K^+-ATPase activity [9], decreased Ca^{2+} uptake by sarcoplasmic reticulum [10], an alteration in the Na^+/Ca^{2+} exchanger [1], a defect in mitochondrial oxidative activity [11], and a decrease in the intracellular concentration of high energy phosphates [12]. Although we know that all these factors are more or less responsible for the pathogenesis of this disease, we focused on the alterations in contractile proteins and their influence on the mechanical function in the present study.

Mechanical and biochemical disorders in BIO 14.6

The mechanical functions of cardiomyopathic hamsters have been studied by using isolated papillary muscles [2, 4, 5] or skinned cardiac fibers [13]. According to these studies, time-to-peak isometric tension was prolonged, peak rate of isometric tension rise depressed, and the velocity of isotonic shortening decreased in cardiomyopathic papillary muscles [2, 4, 5]. The de-

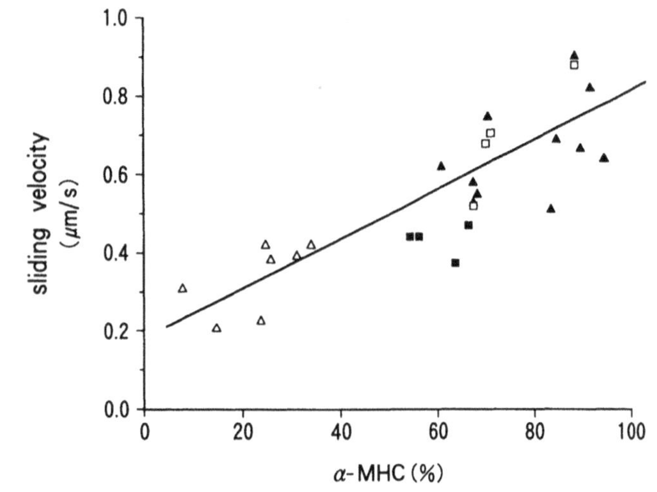

Fig. 8. Relationship between the velocity of myosin-coated beads and the relative amount of α-myosin heavy chain in hamsters superimposed on that obtained in rabbits. Open squares, control hamsters; closed squares, cardiomyopathic hamsters; open triangles, normal rabbits; closed triangles, hyperthyroid rabbits. The solid line indicates the linear regression applied to rabbit data. Note that data points of hamster myosin were distributed near the regression line.

pressed shortening velocity was also observed in isolated cardiomyopathic myocytes [25]. Bárány [23] revealed that the maximal shortening velocity (V_{max}) correlated with both Ca^{2+}-activated and actin-activated ATPase activities of myosin in various types of muscles. In mammalian cardiac muscles, a linear correlation was proposed between V_{max} and Ca^{2+}-activated ATPase activity of myosin [24, 26]. Rouleau *et al.* [2] reported that V_{max} correlated with actin-activated ATPase activity of cardiac myosin obtained from cardiomyopathic hamsters. The present results were in accordance with these findings.

Electrophoretic analysis on pyrophosphate gels shows that three distinct myosin isozymes (V_1, V_2, and V_3) exist also in hamster myocardium. V_1 has the highest Ca^{2+}-activated ATPase activity and V_3 the lowest [27]. In cardiomyopathic hamsters, both actin-activated and Ca^{2+}-activated ATPase activities of cardiac myosin were depressed as early as when the subjects were 4 months old [6]. Malhotra *et al.* [6] reported that actin-activated ATPase activity of cardiac myosin from both control and cardiomyopathic hamsters closely correlated with the proportions of V_1 isozyme content. On the other hand, Ca^{2+}, Mg^{2+}-activated ATPase activity of myopathic cardiac myofibrils *increased* after the subjects were 4 months old [6]. The discrepancy between decreased myosin ATPase activity and increased myofibrillar ATPase activity suggested abnormalities in multiple components of contractile apparatuses in myopathic hearts. Malhotra *et al.* [7] suggested that the abnormalities of myofibrillar ATPase activities in cardiomyopathic hamsters resided in their regulatory proteins. These facts make it difficult to relate the depressed contractile function observed in muscles or fibers directly to the myosin isozyme shift. However, in the *in vitro* motility assay system applied in the present study, we can investigate the mechanical property of pure myosin free from influences of any other factors [14, 15].

In Fig. 8, the data points of hamsters were superimposed on the relationship obtained from control and thyroxine-treated rabbits in our previous study [22]. Thyroxine treatment increased the sliding velocity with a proportional increase in the relative amount of α-MHC. The data points of both control and cardiomyopathic hamsters were distributed near the regression line obtained from the rabbits. These results indicate that the linear correlation is valid within and across species and that myosin isoform content is the major determinant of sliding velocity. Furthermore, the difference between control and myopathic cardiac myosin is attributed to isozyme redistribution and not to a qualitative change in each myosin molecule.

Is isoform distribution the sole determinant of sliding velocity?

In the present study, we applied a linear regression analysis [24, 26, 28–31] for the relationship between the sliding velocity and the biochemical parameters because the range of isozyme distribution was small. However, several investigators have suggested a nonlinear regression for this relationship [23, 32]. Sellers *et al.* [32] studied the sliding velocity of various mixtures of skeletal (fast-type) and smooth muscle (slow-type) myosins in a similar assay system and revealed a nonlinear relationship between the velocity and isomyosin content. They suggested that two distinct populations of crossbridges with different properties would mechanically interact, i.e., slow-type crossbridges could impose a load on fast-type ones to reduce the velocity. This type of mechanical interaction, if it exists, may be functionally important in an *in vivo* situation.

Physiological significance of isozyme shift

Hasenfuss *et al.* [33] reported that alteration in crossbridge behavior was associated with myosin isozyme composition: the average force-time integral of the individual crossbridge cycle of V_3 was greater than that of

V_1 in rat and rabbit myocardium. Assuming that each isomyosin molecule generates the same and constant amount of force during attachment time in each crossbridge cycle, they suggested that V_3 has prolonged attachment time, thus developing force with higher economy than what V_1 develops. If the sliding velocity is determined by the crossbridge detachment rate [34], an increase in slowly detaching V_3 isomyosin should prolong the attachment time with increased economy of force production at the expense of reduced sliding velocity. In this standpoint, the myosin redistribution may be an adaptive compensatory event in the evolution of the disease process.

In summary, we applied an *in vitro* motility assay technique and studied the mechanical properties of cardiac myosin in cardiomyopathic hamsters. The sliding velocity of cardiac myosin on actin cables was lower in cardiomyopathic hamsters than in age-matched controls. The velocity closely correlated with both Ca^{2+}-activated ATPase activity and isozyme content of myosin.

Acknowledgments

Supported in part by Mitsukoshi Medical Award, Tokyo, Japan.

References

1. Wagner JA, Weisman HF, Snowman AM, Reynolds IJ, Weisfeldt ML, Snyder SH: Alterations in calcium antagonist receptors and sodium-calcium exchange in cardiomyopathic hamster tissues. Circ Res 65: 205–214, 1989
2. Rouleau JL, Chuck LHS, Hollosi G, Kidd P, Sievers RE, Wikman-Coffelt J, Parmley WW: Verapamil preserves myocardial contractility in the hereditary cardiomyopathy of the Syrian hamster. Circ Res 50: 405–412, 1982
3. Kagiya T, Hori M, Iwakura K, Iwai K, Watanabe Y, Uchida S, Yoshida H, Kitabatake A, Inoue M, Kamada T: Role of increased α_1-adrenergic activity in cardiomyopathic Syrian hamster. Am J Physiol 260: H80–88, 1991
4. Rouleau JL, Juneau C, Stephens H, Shenasa H, Parmley WW, Brutsaert DL: Mechanical properties of papillary muscle in cardiac failure: Importance of pathogenesis and of ventricle of origin. J Mol Cell Cardiol 21: 817–828, 1989
5. Capasso JM, Olivetti G, Anversa P: Mechanical and electrical properties of cardiomyopathic hearts of Syrian hamsters. Am J Physiol 257: H1836–1842, 1989
6. Malhotra A, Karell M, Scheuer J: Multiple cardiac contractile protein abnormalities in myopathic Syrian hamsters (BIO 53:58). J Mol Cell Cardiol 17: 95–107, 1985
7. Malhotra A, Scheuer J: Troponin-Tropomyosin abnormalities in hamster cardiomyopathy. J Clin Invest 86: 286–292, 1990
8. Kessler PD, Cates AE, van Dop C, Feldman AM: Decreased bioactivity of the guanine nucleotide-binding protein that stimulates adenylate cyclase in hearts from cardiomyopathic Syrian hamsters. J Clin Invest 84: 244–252, 1989
9. Nørgarrd A, Baandrup U, Larsen JS, Kjeldsen K: Heart Na, K-ATPase activity in cardiomyopathic hamsters as estimated from K-dependent 3-O-MFPase activity in crude homogenates. J Mol Cell Cardiol 19: 589–594, 1987
10. Whitmer JT, Kumar P, Solaro RJ: Calcium transport properties of cardiac sarcoplasmic reticulum from cardiomyopathic Syrian hamsters (BIO53.58 and 14.6): Evidence for a quantitative defect in dilated myopathic hearts not evident in hypertrophic hearts. Circ Res 62: 81–85, 1988
11. Hoppel CL, Tandler B, Parland W, Turkaly JS, Albers LD: Hamster cardiomyopathy – A defect in oxidative phosphorylation in the cardiac interfibrillar mitochondria. J Biol Chem 257: 1540–1548, 1982
12. Sievers R, Parmley WW, James T, Wikman-Coffelt J: Energy metabolism at systole vs. diastole in normal hamster hearts vs. myopathic hamster hearts. Circ Res 53: 759–766, 1983
13. Veksler VI, Ventura-Clapier R, Lechene P, Vassort G: Functional state of myofibrils, mitochondria and bound creatine kinase in skinned ventricular fibers of cardiomyopathic hamsters. J Mol Cell Cardiol 20: 329–342, 1988
14. Sheetz MP, Chasan R, Spudich JA: ATP-dependent movement of myosin *in vitro*: Characterization of a quantitative assay. J Cell Biol 99: 1867–1871, 1984
15. Shimmen T, Yano M: Active sliding movement of latex beads coated with skeletal muscle myosin on *Chara* actin bundles. Protoplasma 121: 132–137, 1984
16. Weber K, Osborn M: The reliability of molecular weight determinations by dodecyl sulfate-polyacrylamide gel electrophoresis. J Biol Chem 244: 4406–4412, 1969
17. Martin AF, Pagani ED, Solaro RJ: Thyroxine-induced redistribution of isozymes of rabbit ventricular myosin. Circ Res 50: 117–124, 1982
18. Yazaki Y, Raben MS: Cardiac myosin adenosinetriphosphatase of rat and mouse – Distinctive enzymatic properties compared with rabbit and dog cardiac myosin. Circ Res 35: 15–23, 1974
19. Youngburg GE, Youngburg MV: Phosphorus metabolism: A system of blood phosphorus analysis. J Lab Clin Med 16: 158–166, 1930
20. Lowry OH, Rosebrough NJ, Farr AL, Randall RJ: Protein measurement with the folin phenol reagent. J Biol Chem 193: 265–275, 1951
21. Sugiura S, Yamashita H, Serizawa T, Iizuka M, Shimmen T, Sugimoto T: Active movement of cardiac myosin on Characeae actin cables. Pflügers Arch 421: 32–36, 1992
22. Yamashita H, Sugiura S, Serizawa T, Sugimoto T, Iizuka M, Katayama E, Shimmen T: Sliding velocity of isolated rabbit cardiac myosin correlates with isozyme distribution. Am J Physiol 1992 (in press)
23. Bárány, M: ATPase activity of myosin correlated with speed of muscle shortening. J Gen Physiol 50: 197–218, 1967
24. Hamrell BB, Low RB: The relationship of mechanical V_{max} to myosin ATPase activity in rabbit and marmot ventricular muscle. Pflügers Arch 377: 119–124, 1978
25. Sen L, O'Neill M, Marsh JD, Smith TW: Myocyte structure,

function, and calcium kinetics in the cardiomyopathic hamster heart. Am J Physiol 259: H1533–1543, 1990
26. Delcayre C, Swynghedauw B: A comparative study of heart myosin – ATPase and light subunits from different species. Pflügers Arch 355: 39–47, 1975
27. Wiegand V, Stroh E, Henniges A, Lossnitzer K, Kreuzer H: Altered distribution of myosin isoenzymes in the cardiomyopathic Syrian hamster (BIO 8.262). Basic Res Cardiol 78: 665–670, 1983
28. Cappelli V, Bottinelli R, Poggesi C, Moggio R, Reggiani C: Shortening velocity and myosin and myofibrillar ATPase activity related to myosin isoenzyme composition during postnatal development in rat myocardium. Circ Res 65: 446–457, 1989
29. Ebrecht G, Rupp H, Jacob R: Alterations of mechanical parameters in chemically skinned preparations of rat myocardium as a function of isoenzyme pattern of myosin. Basic Res Cardiol 77: 220–234, 1982
30. Pagani ED, Julian FJ: Rabbit papillary muscle myosin isozymes and the velocity of muscle shortening. Circ Res 54: 586–594, 1984
31. Schwartz K, Lecarpentier Y, Martin JL, Lompre AM, Mercadier JJ, Swynghedauw B: Myosin isoenzymic distribution correlates with speed of myocardial contraction. J Mol Cell Cardiol 13: 1071–1075, 1981
32. Sellers JR, Spudich JA, Sheetz MP: Light chain phosphorylation regulates the movement of smooth muscle myosin on actin filaments. J Cell Biol 101: 1897–1902, 1985
33. Hasenfuss G, Mulieri LA, Blanchard E, Holubarsch C, Leavitt BJ, Ittleman F, Alpert NR: Energetics of isometric force development in control and volume-overload human myocardium – Comparison with animal species. Circ Res 68: 836–846, 1991
34. Siemankowski RF, Wiseman MO, White HD: ADP dissociation from actomyosin subfragment 1 is sufficiently slow to limit the unloaded shortening velocity in vertebrate muscle. Proc Natl Acad Sci USA 82: 658–662, 1985

Molecular and Cellular Biochemistry **119**: 89–94, 1993.

Abnormalities of ADP/ATP carrier protein in J-2-N cardiomyopathic hamsters

Mitsutoshi Kato, Jie Yang, Takaaki Iwai, Akira Tanamura, Toru Arino, Osamu Kawashima and Nobuakira Takeda
Department of Internal Medicine, Aoto Hospital, Jikei University School of Medicine, Tokyo, Japan

Abstract

ADP/ATP carrier protein (AAC) is located in the mitochondrial inner membrane and has an important function in mitochondrial energy supply. This protein transports ATP to the cytoplasm and counter transports ADP into the mitochondria. J-2-N cardiomyopathic hamsters were investigated to determine the AAC content in cardiac mitochondria. After recording an electrocardiogram and collecting blood, the cardiac mitochondria were isolated. The mitochondrial membranes were labelled with eosin-5-maleimide (EMA) and separated on SDS polyacrylamide gels. The position of the AAC component was identified by exposing the gel under UV light, and the AAC content was determined by densitometry after staining with Coomassie blue. The AAC content ratio was significantly decreased in both 10-week-old and 1-year survived J-2-N hamsters when compared to control Golden hamster. Among 10-week-old J-2-N hamsters, the decrease in the AAC content ratio was more marked for the animals with more severe myocardial damage. The H^+-ATPase activities of mitochondrial membrane were higher in 10-week-old J-2-N hamsters than in control hamsters. These results suggest that the decrease of AAC in J-2-N hamster plays an important role in the pathogenesis of cardiomyopathy in J-2-N hamsters. (Mol Cell Biochem **119**: 89–94, 1993)

Key words: ADP/ATP carrier protein, adenine nucleotide translocase, cardiomyopathic hamster J-2-N, H^+-ATPase, cariomyopathy

Introduction

The ADP/ATP carrier protein (AAC) located in the mitochondrial inner membrane has a very important role in mitochondrial energy supply [1]. This protein transports ATP to the cytoplasm and counter transports ADP into the mitochondria. AAC is an essential protein for myocardial energy metabolism. Recently, Schultheiss *et al.* have reported on abnormalities of AAC in patients with dilated cardiomyopathy [2]. Such abnormalities appear to be closely involved with the development of idiopathic cardiomyopathy. In addition, abnormalities of the sarcolemmal membrane and impaired cardiac contractility have been observed in recent studies of J-2-N cardiomyopathic hamsters. In the present study, we investigated the relationship between quantitative abnormalities of the AAC protein and cardiomyopathy by studying mitochondrial AAC in the myocardium of J-2-N hamsters.

Address for offprints: M. Kato, Department of Internal Medicine, Aoto Hospital, Jikei University School of Medicine, 6-41-2 Aoto, Katsushika-ku, Tokyo, 125 Japan

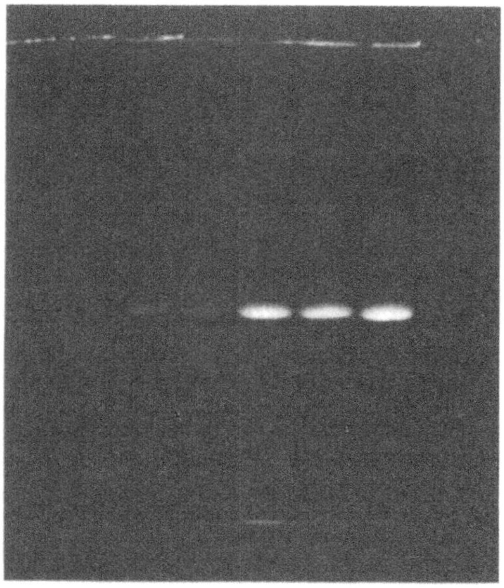

Fig. 1. The AAC bands visualized under UV light: Isolated mitochondrial membrane were labeled with eosin-5-maleimide (EMA) for 30 min at room temperature. A single spot of fluorescence under UV light was seen indicating the presence of AAC. The number on the figure is 30 kDa.

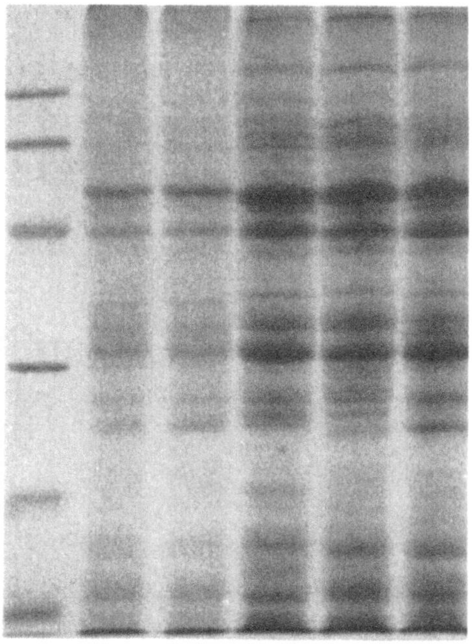

Fig. 2. SDS-polyacrylamide gel after being stained with Coomassie blue: The gel of this figure is the same one as Fig. 1. From left to right, the lanes separately represent: marker of molecular weight (dalton); mitochondrial membrane of hamster before labeling with EMA; the mitochondrial membrane pretreated with carboxyatractylate before labeling with EMA; and the mitochondrial membrane of rabbit, beef and human subject in lane 4, 5, 6 respectively. The molecular weight of AAC was slightly over 30 kDa.

Materials and methods

Animal models

We used J-2-N cardiomyopathic hamsters bred at our institute [3, 4]. As reported previously, the serum creatine kinase (CK) level correlates well with the degree of myocardial damage in this animal model, so CK was used as an indicator of myocardial injury. Ten-week-old J-2-N hamsters underwent ECG recording including both limb and chest leads, and then were laparotomized under pentobarbital anesthesia. Blood was collected from the abdominal vena cava and serum was separated for biochemical analysis. The heart was also extracted for determination of the mitochondrial AAC content.

The 10-week-old hamsters were classified into the following groups according to the serum CK level at the time of death: group A) $>$ 10,000 mU/ml, group B) 10,000–3,000 mU/ml, group C) 3,000–1,000 mU/ml, and group D) $<$ 1,000 mU/ml. The CK levels in group D were almost normal. Golden hamsters were used as the controls (group G).

Also the AAC contents of 1-year survived J-2-N hamsters were compared with the contents of age matched Golden hamsters.

Detection of AAC by eosin-5-maleimide (EMA) and determination of the AAC content

The heart was minced, homogenized and centrifuged and obtained mitochondrial membrane according to the method of Tzagdoff *et al.* [5]. N-ethylmaleimide was added to the mitochondrial fraction. After incubation at 0° C for 10 min, EMA was added and the reaction proceeded for 30 min at room temperature. Then, the reaction of EMA with the mitochondrial membrane proteins was stopped by adding dithiothreitol, and the mixture was let standing at 0° C for 30 min. After subsequent washing and centrifugation with buffer for 3 times, the membrane proteins were isolated.

The mitochondrial membrane proteins thus obtained were further processed by SDS-polyacrylamide gel electrophoresis (SDS-PAGE). Then the gels were stained with Coomassie blue and AAC was detected under UV illumination. The AAC content and AAC content ratio (the ratio of AAC to the total mitochondrial membrane proteins) were determined using a densitometer [6].

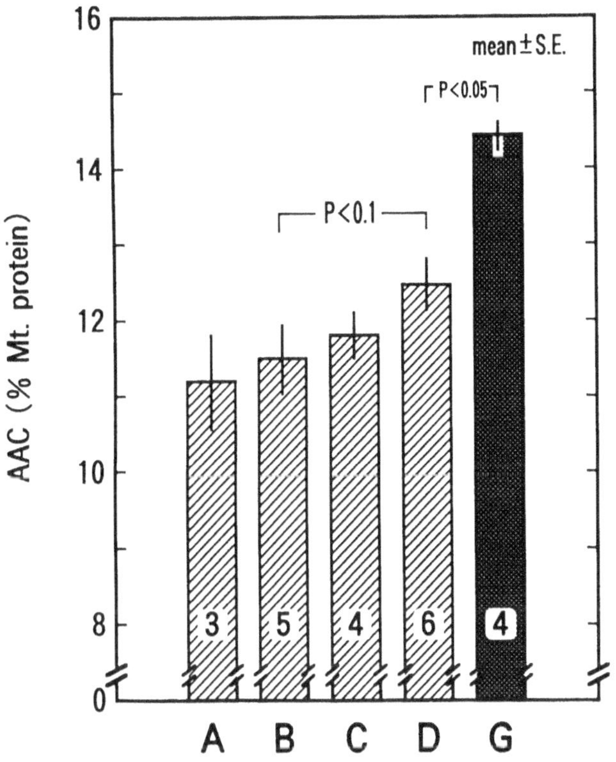

Fig. 3. AAC content ratios in 10-week-old J-2-N hamsters and control Golden hamster: According to the serum CK levels, the hamsters were divided into group A (CK > 10,000), group B (10,000–3,000), group C (3,000–1,000), and group D (1,000 mU/ml >). Disease-free Golden hamsters were used as controls. The AAC content ratio was significantly lower in all groups of J-2-N hamsters than in the control Golden hamsters, and was the lowest in group A with severe myocardial damage. In contrast, the decrease of the AAC content was the least noticeable in group D, which included hamsters with minimal myocardial damage. The values are mean ± S.E.

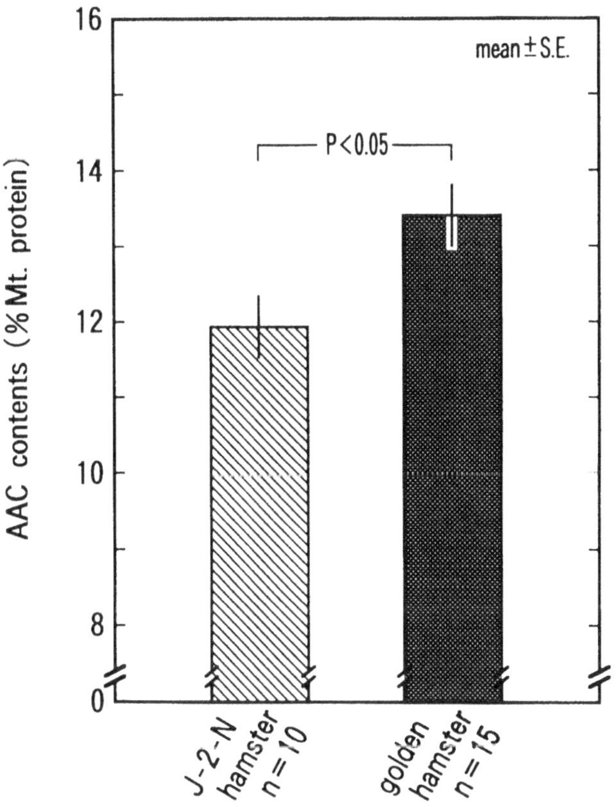

Fig. 4. The AAC content ratio of J-2-N hamsters which survived for 1-year: The ratio was significantly lower in the 1-year old J-2-N hamsters, similar to that in the young hamsters, but the extent of the decrease was milder in 1 year survived hamsters. The values are mean ± S.E.

In addition to the role of AAC, mitochondrial membrane H^+-ATPase also synthesizes ATP with energy obtained through the H^+ ionic gradient in addition to the role of AAC. Therefore, H^+-ATPase activity was also measured using the adjusted mitochondrial membrane fraction and the H^+-ATPase activities were measured by the method of Pande *et al.* [7]. Protein concentrations were determined by the method of Lowry [8].

Results

Figure 1 shows the AAC bands visualized under UV light. A single spot of fluorescence was seen indicating the presence of AAC.

Figure 2 shows SDS-polyacrylamide gel after being stained with Coomassie blue. This gel is the same one as Fig. 1. From left to right, the lanes separately represent: the first lane is a marker of molecular weight (dalton), the second lane is a mitochondrial membrane of hamster before labeling with EMA. The third is the mitochondrial membrane pretreated with carboxyatractylate before labeling with EMA, and the fourth, fifth and sixth lane is the mitochondrial membrane of rabbit, beef and human subject respectively. The molecular weight of AAC is slightly over 30 kDa.

Figure 3 shows the AAC content ratios of the 10-week-old J-2-N hamsters and the control group. The AAC content ratio was significantly lower in all J-2-N hamsters than in the control Golden hamsters, and was the lowest in group A with severe myocardial damage. In contrast, the decrease of the AAC content was the least noticeable in group D, which included hamsters with minimal myocardial damage.

The AAC content ratio of J-2-N hamsters which survived for 1 year due to relatively mild myocardial dam-

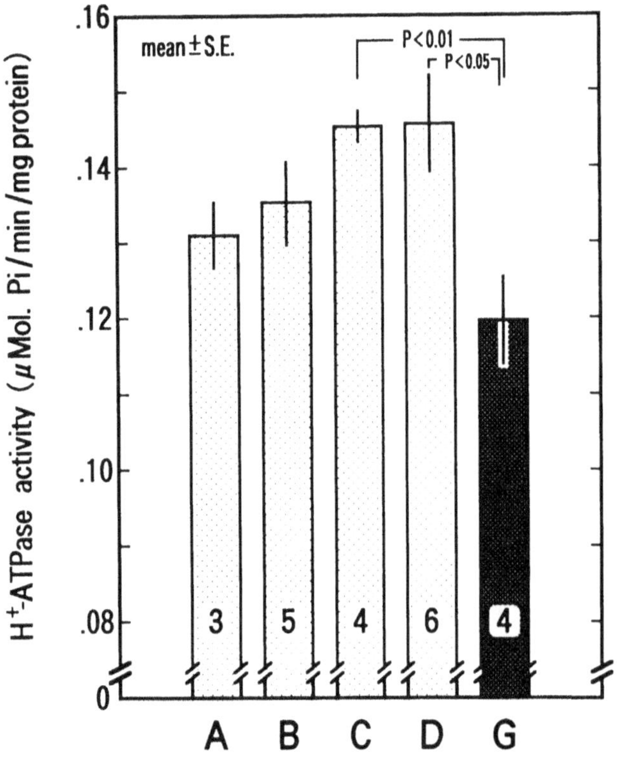

Fig. 5. Activities of total H^+-ATPase activities in J-2-N cardiomyopathic hamsters: Total H^+-ATPase activity was increased significantly in J-2-N hamsters when compared with Golden hamsters. H^+-ATPase activity tended to be lower in the J-2-N hamsters with more severe myocardial injury in each group. The values are mean ± S.E.

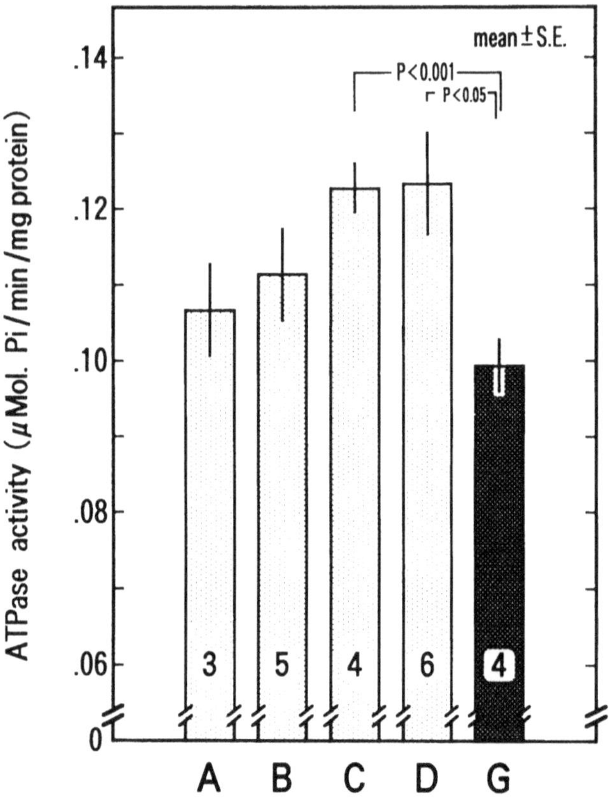

Fig. 6. Activities of oligomysin-sensitive H^+-ATPase in J-2-N cardiomyopathic hamsters: Oligomysin sensitive H^+-ATPase showed just the same tendency as total H^+-ATPase activity. The values are mean ± S.E.

age was compared with that of age-matched Golden hamsters (Fig. 4). The ratio was significantly lower in the 1-year old J-2-N hamsters, similar to that in the young hamsters, but the extent of the decrease was milder at 1 year.

Figure 5 shows total H^+-ATPase activity of each group. ATPase activity was increased significantly in J-2-N hamsters when compared with Golden hamsters, H^+-ATPase activity tended to be lower in the J-2-N hamsters with more severe myocardial injury in each group.

As shown in Fig. 6, oligomysin-sensitive H^+-ATPase showed just the same tendency as Fig. 5.

Discussion

ADP is produced in myocardium by the rapid consumption of ATP associated with myocyte contraction, is transported into the mitochondria by AAC, and then is recycled rapidly.

The formation of creatine phosphates from ATP produced in the mitochondria by oxidative phosphorylation is coupled to AAC formation [9]. It has been suggested that autoantibodies to AAC may induce cardiac dysfunction associated with abnormal energy metabolism in cardiomyopathy [10]. It has also been reported that antibodies produced by immunizing animals can decrease the cytosolic ATP concentration and increase the mitochondrial ATP concentration [2].

There is no method yet available for the direct determination of AAC activity in order to assess whether cardiac dysfunction in cardiomyopathy is related to abnormalities of AAC, but indirect determination is possible by various methods. Quench-flow method [11], maleimide spin-label [12], labeling with ATP or ADP is one method [13, 14]. However, ATP and ADP bind not only to AAC, but also to other enzymes, so such labeling cannot be used as a direct indicator of adenine nucleotide binding. Accordingly, AAC-specific agents such as carboxyatractyloside (CAT) and bongkrekate (BKA) have been used in some studies [15].

AAC is known to have two different conformational

states which orient the binding site of the protein either towards the cytosolic side of the mitochondrial membrane (c-state) or towards the mitochondrial matrix (m-state). Antibodies to the c-state bond will only detect c-state AAC, and those to the m-state bond will only detect m-state AAC. Moreover, it has been demonstrated that the specific high-affinity ligand CAT only binds to c-state AAC, while BKA only binds to m-state AAC [16, 17]. The third possible choice for such studies is the immunological method by Shultheiss *et al.* [18, 19].

Investigation of the primary structure of AAC has revealed that it is a sulfhydryl (SH)-containing protein, and that the sulfhydryl group is necessary for its function. However, SH reagents such as disulfide or maleimide are thought not to inhibit AAC [20]. When incubated with ATP or ADP, AAC is able to react with N-ethylmaleimide. However, N-ethylmaleimide has a low molecular weight and high membrane permeability whereas EMA has a high molecular weight, releases fluorescence, and binds specifically to AAC. Therefore, the latter agent is more suitable for binding studies [6].

In the present study, the AAC content ratio of 10-week-old J-2-N hamsters was decreased compared to the control group regardless of the extent of myocardial damage. It is important to determine whether this finding is a cause or effect of cardiomyopathy. Our results suggested that degradation of membrane proteins in cardiomyopathic hamsters was induced by proteases in association with mitochondrial damage. However, the AAC content ratio was also significantly lower in young hamsters with no myocardial damage when compared to the control group. Moreover, the AAC content ratio was lower in the animals with more severe damage, suggesting that some disturbance of AAC synthesis had occurred.

The AAC content ratio in J-2-N hamsters surviving 1 year, was low compared to that of control Golden hamsters, but the decrease in the ratio was more marked in the young hamsters.

Our results support the report of Scultheiss *et al.* [2], who demonstrated a decrease of cytosolic ATP and an increase of mitochondrial ATP in animals to raise antibodies against AAC.

The rapid movement of ATP to the cytosol is controlled by AAC, so abnormalities of AAC could cause myocardial systolic or diastolic dysfunction, despite normal ATP production by the mitochondria.

Regarding the affinity of AAC for ADP or ATP, the Km value is related to the energy level of the mitochondria [21]. AAC is a very important protein in eucaryotic cells and its metabolism is considered to be relatively slow. It can thus be assumed that any abnormality of AAC would have a considerable influence in cardiomyopathic hamsters.

BI014.6 cardiomyopathic hamsters show a decrease of the myocardial ATP and creatine phosphate content, but show no evidence of a decrease in oxidative phosphorylation [22] that would suggest the possible involvement of AAC.

The activity of H^+-ATPase and oligomysin-sensitive H^+-ATPase was higher in J-2-N hamsters compared to the control group. This was probably because of impaired ATP transport and the retention of ATP in the mitochondria due to abnormalities of AAC. A decrease of the cytosolic ATP/ADP ratio and an increase of the mitochondrial ATP concentration could enhance the phosphorylation potential. Consequently, H^+-ATPase activity may increase in J-2-N hamsters as a compensatory mechanism for some deficiency of AAC.

In conclusion, the present study demonstrated that 1) The AAC content ratio was significantly lower in both 10-week old and 1-year old J-2-N hamsters when compared to control Golden hamsters. 2) Among 10-week-old J-2-N hamsters, the decrease in the AAC content ratio was more marked for the animals with more severe myocardial damage. 3) J-2-N hamsters which survived for 1 year showed a relatively high AAC content ratio. The above results suggested that some abnormality of AAC could be a possible cause of cardiomyopathy. 4) Since the H^+-ATPase level was higher in J-2-N hamsters than in Golden hamsters, the presence of a compensatory mechanism for abnormalities of AAC was also suggested.

This study demonstrated that the AAC content ratio was low in J-2-N cardiomyopathic hamsters, but that the decrease of AAC was not necessarily linked directly to cardiomyopathy because AAC have faculty in reserve for the movement of ATP or ADP under normal conditions.

However, it is still possible that the decrease of AAC plays an important role in the development and eventual progression of cardiomyopathy to heart failure in this animal model.

References

1. Klingenberg M, Rottenberg H: Relation of the gradient of the

ATP/ADP ratio and the membrane potential across the mitochondrial membrane. Eur J Biochem 73: 125–130, 1977
2. Schulze K, Becker BF, Schultheiss HP: Antibodies to the ADP/ATP carrier an autoantigen in myocarditis and dilate cardiomyopathy, penetrate into myocardial cells and disturb energy metabolism *in vivo*. Circ Res 64: 179–192, 1989
3. Kato M, Nagano M: Experimental animal models of cardiomyopathy. In: Opie LH and Sugimoto T (eds) Cardiomyopathy update-4. Metabolic and Molecular aspects of cardiomyopathy. University of Tokyo press, Tokyo, 1991, pp 69–81
4. Kato M, Takeda N, Yang J, Nagano M: The effects of angiotensin converting enzyme inhibitors and the role of Renin-Angiotensin-Aldosterone system in J-2-N cardiomyopathic hamsters. Jpn Circ J 56: 46–51, 1992
5. Tzagoloff A, Byington KH, Maclennan DH: Studies on the mitochondria adenosine triphosphatase system. II. The isolation and characterization of an oligomysin-sensitive adenosine triphosphatase from bovine heart mitochondria. J Biol Chem 243: 2405–2412, 1968
6. Müller M, Krebs JJR, Cherry RJ, Kawato S: Rotational diffusion of the ADP/ATP translocator in the inner membrane of mitochondria and in proteoliposomes. J Biol Chem 259: 3037–3043, 1984
7. Pande SU, Blanchaer MC: Reversible inhibition of mitochondria adenosine diphosphate phosphorylation by long chain acyl coenzyme A esters. J Biol Chem 246: 402–411, 1971
8. Lowry OH, Rosebrough NJ, Farr AL, Randall RJ: Protein measurement with the folin phenol reagent. J Biol Chem 193: 265–275, 1951
9. Moreadith RW, Jacobus WE: Creatine kinase of heart mitochondria functional coupling of ADP transfer to the adenine nucleotide translocase. J Biol Chem 257: 899–905, 1982
10. Schulze K, Becker BF, Schultheiss HP: Antibody to the ADP/ATP carrier, an autoantigen in myocarditis and dilated cardiomyopathy-Impair cardiac function. Circulation 81: 959–969, 1990
11. Nohl H, Klingenberg M: Kinetics of ADP, ATP transport in mitochondria as studied by the quench-flow method. Biochim Biophys Acta 503: 155–169, 1978
12. Munding A, Drees M, Beyer K, Klingenberg M: Probing of sulfhydryl groups in the adenosine 5′-diphosphate/adenosine 5′-triphosphate carrier by maleimide spin-labels. Biochemistry 26: 8637–8644, 1989
13. Winkler HH, Lenninger AL: Characterization of the atractyloside sensitive adenine nucleotide transport system in rat liver mitochondria. J Biol Chem 243: 3000–3008, 1968
14. Doussiere J, Ligeti E, Brandolin G, Vignais P: Control of oxidative phosphorylation in rat heart mitochondria. Biochim Biophys Acta 766: 492–500, 1984
15. Buchanan B, Eiermann W, Riccio P, Aquila H, Klingenberg M: Antibody evidence for different conformational states of ADP, ATP translocator protein isolated from mitochondria. Proc Natl Acad Sci 73: 2280–2284, 1976
16. Block MR, Lauguin JM, Vignais PV: Atractyloside and bongkrekic acid sites in the mitochondrial ADP/ATP carrier protein. FEBS Lett 131: 213–218, 1981
17. Klingenberg M, Mayer I, Appel M: Interaction of fluorescent 3′-[1,5-(Dimethylamino)naphthoyl]adenine nucleotides with the solubilized ADP/ATP carrier. Biochemistry 24: 3650–3659, 1985
18. Schultheiss HP, Klingenberg M: Immunoelectrophoretic characterization of the ADP/ATP carrier from heart, kidney, and liver. Arch Biochem Biophys 239: 273–279, 1985
19. Schultheiss HP, Bolte HD: Immunological analysis of auto-antibodies against the adenine nucleotide translocator in dilated cardiomyopathy. Mol Cell Cardiol 17: 603–617, 1985
20. Aquila H, Eiermann W, Klingenberg M: Incorporation of N-ethyl-maleimide into the membrane bound ADP/ATP translocator. Eur J Biochem 122: 133–139, 1982
21. Pfaff E, Klingenberg M: Adenine nucleotide translocation of mitochondria. Eur J Biochem 6: 66–79, 1968
22. Wrogemann K, Blanchaer MC, Jacobson BE: Oxidative phosphorylation at various stages of the genetically determine cardiomyopathy in the Syrian hamster. Recent Adv studies Cardiac Struct Metabol 3: 467–478, 1973

Molecular and Cellular Biochemistry **119**: 95–103, 1993.

Age-associated damage in mitochondrial DNA in human hearts

Mika Hayakawa, Satoru Sugiyama, Kazuki Hattori,[1]
Masaaki Takasawa[1] and Takayuki Ozawa
Departments of Biomedical Chemistry, and [1] Internal Medicine, Faculty of Medicine, University of Nagoya, Nagoya, Japan

Abstract

Damage to mitochondrial DNA seems to be involved in the etiology of age-associated degenerative diseases. The aim of this study is to elucidate effects of aging on human mitochondrial DNA. 8-Hydroxy-deoxyguanosine, a product of free radical damage to deoxyguanosine, is reported to cause random point mutations. In human mitochondrial DNA, 8-hydroxy-deoxyguanosine increased exponentially with age, and the population of mitochondrial DNA with deletion increased also exponentially with age. Furthermore, a clear correlation existed between the accumulation of 8-hydroxy-deoxyguanosine and that of mitochondrial DNA with deletion. We also determined the effects of aging on rat mitochondrial function together with 8-hydroxy-deoxyguanosine content in mitochondrial DNA. The activities of complexes I and IV of the mitochondrial electron transport chain decreased significantly in rats aged 100 weeks compared with those in rats aged 7 weeks. A concomitant increase in 8-hydroxy-deoxyguanosine was observed in mitochondrial DNA of rats aged 100 weeks. From our results, it is concluded that the age-associated accumulation of somatically acquired oxygen damage together with deletions in mitochondrial DNA might be important contributors to the deterioration of cardiac function associated with age. (Mol Cell Biochem **119:** 95–103, 1993)

Key words: aging, mitochondrial DNA, 8-hydroxy-deoxyguanosine, deletion, mitochondrial electron transport activity

Abbreviations: mtDNA – mitochondrial DNA; 8-OH-dG – 8-Hydroxy-Deoxyguanosine; dG – Deoxyguanosine; μHPLC/MS – Micro-High Performance Liquid Chromatography/Mass Spectrometry

Introduction

Since the elderly population is growing, the mechanism responsible for age-associated decline in organ function has attracted much interest. DNA damage, in particular that in mitochondrial DNA (mtDNA), accumulated with age seems to be a major contributor to this mechanism [1, 2]. Mitochondria, responsible organelle for cellular energy production, have their own DNA [3]. One of the features of mtDNA is that the rate of mutation in mtDNA is much higher than that of nuclear DNA [1], and mtDNA mutations accumulate during the whole life of an individuals [4]. Recent advances in gene technology emphasized the relationship between mtDNA mutations, resulting in deterioration of cellular energy metabolism, and the development of age-associ-

Address for offprints: T. Ozawa, Department of Biomedical Chemistry, Faculty of Medicine, University of Nagoya, Tsuruma, Showa-ku, Nagoya, 466, Japan

ated degenerative diseases [5–9]. We demonstrated that the incidence of mtDNA deletion increased with age [7]. If mutation of mtDNA is related to age-related organ dysfunction, the following points are of primary importance: 1. How large a population of mutant mtDNA is required to develop clinical signs and symptoms? Since another feature of mtDNA is that each mitochondrion has two or three copies of mtDNA and several thousands of mtDNA exist in a cell [10]. 2. What mechanism is operated in the genesis of mtDNA mutation? To resolve these problems, we developed a kinetic PCR method to quantify mtDNA mutations [11], and a micro-high performance liquid chromatography/mass spectrometry (μHPLC/MS) system which could detect 8-hydroxydeoxyguanosine (8-OH-dG) as low as one picomole [12], because the oxidative damage to DNA, in particular 8-OH-dG converted from deoxyguanosine (dG), is demonstrated to cause random point mutations in mtDNA [13]. In this paper, using these methods, we determined the effect of aging on human mtDNA mutation, and we also estimated the effect of aging on the activity of rat mitochondrial energy production system.

Methods

Coenzyme Q1 and coenzyme Q2 were kindly supplied by Eisai Co., Tokyo, Japan. Nicotinamide adenine dinucleotide, reduced from (NADH), and other chemicals were purchased from Sigma Chemical Co., St. Louis, MO, except where noted.

Clinical studies

Measurement of 8-OH-dG in human heart mtDNA

Subjects. Cardiac muscles from 20 human cadavers (8 male and 12 female) were obtained at random, regardless of disease at autopsy. The ages of subjects were 24 to 97 yr, and clinical features are shown in Table 1.

Preparation of mitochondrial fraction. The tissue was put into ice-cold 50 mM Tris-HCl, pH 7.4 containing 0.25 M sucrose, minced with a pair of scissors and homogenized using an Ultra Turrax Homogenizer (Janke & Kunkel, F.R.G.) at half maximal speed. Nagarse (2 mg/g tissue; Katayama Chemical Co., Osaka, Japan) was added and the mixture was homogenized using a Teflon homogenizer. The homogenates were centrifuged at 700 *g* for 20 min and the supernatants obtained were centrifuged at 9,000 *g* for 15 min. The pellets were suspended with 0.1 mM ethylenediaminetetraacetic acid (EDTA), 5 mM Tris-HCl, pH 7.4 containing 0.25 M sucrose (A buffer), and centrifuged at 16,000 *g* for 20 min. The washing medium was discarded, then the mitochondrial pellets were finally suspended in 0.1 M EDTA, 10 mM Tris-HCl, pH 7.4 to contain about 10 mg of mitochondrial protein per milliliter of suspension. The protein content of the suspension was determined by the biuret method [14]. This mitochondrial fraction contained no detectable nuclear DNA by ethidium bromide fluorescence after electrophoresis on 1% agarose gel.

Extraction of mtDNA. Freshly isolated mitochondria suspended in 0.1 M EDTA, 10 mM Tris-HCl, pH 7.4, were incubated in a shaking water bath for 4 hr at 37° C in the presence of 0.5% sodium dodecyl sulfate (SDS) and 100 μg/ml of proteinase K. After DNA was extracted twice with equal volumes of phenol/chloroform/isoamyl alcohol (25 : 25 : 1) and then once with chloroform/isoamyl alcohol (25 : 1), the aqueous phase was mixed and precipitated with 1/10 volume of 3 M sodium acetate, pH 7.4, and 2 volumes of absolute ethanol at − 20° C overnight, and then rinsed with 70% ethanol. The precipitated DNA was recovered in 500 μl of 10 mM Tris-HCl, pH 8.0, 1 mM EDTA. Then further

Table 1. Clincial features of subjects at autopsy

Subject No.	Sex	Age (yr)	Clinical diagnosis
1	F	24	Accident
2	F	28	Pulmonary embolism
3	M	36	Pulmonary embolism
4	F	46	Uterus cancer
5	F	48	Thymoma
6	F	49	Burkitt lymphoma
7	M	55	Gastric cancer
8	M	58	Gastric cancer
9	F	60	Lung cancer
10	M	60	Gastric cancer
11	M	72	Malignant lymphoma
12	F	73	Malignant lymphoma
13	F	75	Lung cancer
14	F	76	Subarachnoid hemorrhage
15	F	80	Pancreatic cancer
16	M	83	Respiratory failure
17	M	83	Hepatic failure
18	M	85	Rectal cancer
19	F	88	Lung cancer
20	F	97	Gastric cancer

purification was made by digestion with 10 μg/ml of pancreas ribonuclease for 30 min at 37° C followed by phenol/chloroform/isoamyl alcohol extraction.

Enzymatic hydrolysis of mtDNA to nucleosides. DNA samples (0.2–1 μg per 100 μl) in 40 mM Tris-HCl, pH 8.5, containing 10 mM $MgCl_2$ were incubated with DNase I (200 units per mg of DNA) spleen exonuclease (0.01 unit per mg of DNA), snake venom exonuclease (0.5 unit per mg of DNA), and alkaline phosphatase (10 units per mg of DNA) for 2 hr at 37° C. Complete DNA hydrolysis was monitored by agarose gel electrophoresis and staining with ethidium bromide.

μHPLC/MS analysis. Nucleosides obtained by mtDNA hydrolysis were analyzed by a μHPLC/MS system [12]. The μHPLC was performed using a Develosil-ODS column packed with 5 μm particles (0.3 × 150 mm, total volume 7 μl, Nomura Chem. Co., Seto, Japan). The solvent system used was a mixture of 7% methanol, 50 mM ammonium formate, and 0.5% glycerol, pH 4.0. The flow rate was 4 μl/min controlled by a Milton Roy Micro Metric Metering Pump. DNA hydrolysate, 50–100 μl, was subjected to pre-column concentration before injection into the column with the aid of Valco M4-N6W-T valves. Total eluent from the column was directly injected into a mass spectrometer (JMS-DX 303, JOEL) by a connecting capillary. Both selected ion monitoring and ionization mass spectrum of the sample were recorded.

Quantitative analysis of mtDNA with deletion in human heart

We have reported that the incidence of mtDNA with a 7.4 kb-deletion was increased age dependently [7]. Accordingly, in this paper, the ratio of deleted mtDNA relative to the total mtDNA was quantified with the rest of each mtDNA specimen, using a strategy illustrated in Fig. 1: one pair of PCR primers, L853-H38, could amplify a 1.0 kb fragment from the mtDNA with the 7.4 kb deletion, but not from the wild mtDNA. The other, L625-H726 for the total mtDNA, amplified a 1.0 kb fragment in the CO1 gene common to the wild and deleted mtDNA. The reaction was carried out on 1 μl of the DNA solution (approximately 10 ng of total DNA) in a final volume of 50 μl, which included 200 μM of each deoxyribonucleoside triphosphate (dNTP), 1 μM of each primer, 2.5 units of *Taq* DNA polymerase (Promega Co., Madison, WI), PCR buffer (50 mM Tris-HCl, pH 8.4, that contained 50 mM KCl, 1.5 mM $MgCl_2$, and 0.01% gelatin), and 7.4 KBq α[^{32}P]deoxycytosine triphosphate (30 TBq/mmol; Amersham International Ltd., Little Chalfont, Bucks, U.K.), using a Perkin-Elmer Cetus Thermal Cycler (Perkin-Elmer/Cetus Corp., Norwalk, CT). The cycle times were as follows: denaturation, 15 sec at 94° C; annealing, 15 sec at 50° C; and primer extension, 80 sec at 72° C. The reaction products were electrophoresed on 1% agarose gels at 100V for 30 min, transferred onto a nylon membrane (Hybond-N$^+$, Amersham), and autoradiographed. The radioactivities of the labeled 1.0 kb DNA fragments at increasing numbers of PCR amplification cycles were converted to digital computer memories by a laser image analyzer (Fujix BAS2000 system, Fuji Film, Tokyo, Japan). By plotting logarithms of radioactivity against the cycle number of PCR amplification, straight lines were obtained with different slopes due to differences in amplification efficiency of the DNA polymerase to different DNA fragments even of the same size. Extrapolation of the lines to zero amplification cycle gave the ratio of mtDNA with the 7.4 kb deletions to the total mtDNA in the original tissue specimen before PCR amplification.

Experimental studies

Measurement of 8-OH-dG in rat heart mtDNA

Experiments were carried out on male specific pathogen free Wistar rats. Rats aged 7 and 100 weeks old were cervically dislocated and hearts were removed. Segments from ten hearts were combined for preparation of the mitochondrial fraction, due to the large amount of protein required for measurement of 8-OH-dG. Mitochondrial DNA was extracted using the same methods as described above, and 8-OH-dG in mtDNA was quantitatively determined using the μHPLC/MS system.

Rat mitochondrial electron transport activity

Preparation of rat tissue mitochondria. The age of the rats when they were cervically dislocated was 7 (n = 7) and 100 (n = 7) weeks old. Rats were fasted for 24 hr before the experiments and allowed free access only to water. After cervically dislocated, their hearts were removed and put into ice-cold 50 mM Tris-HCl, pH 7.4 containing 0.25 M sucrose, minced with a pair of scissors and homogenized using a Physcotron Handy Micro Homogenizer (Niti-on, Tokyo, Japan) for 30 sec. The

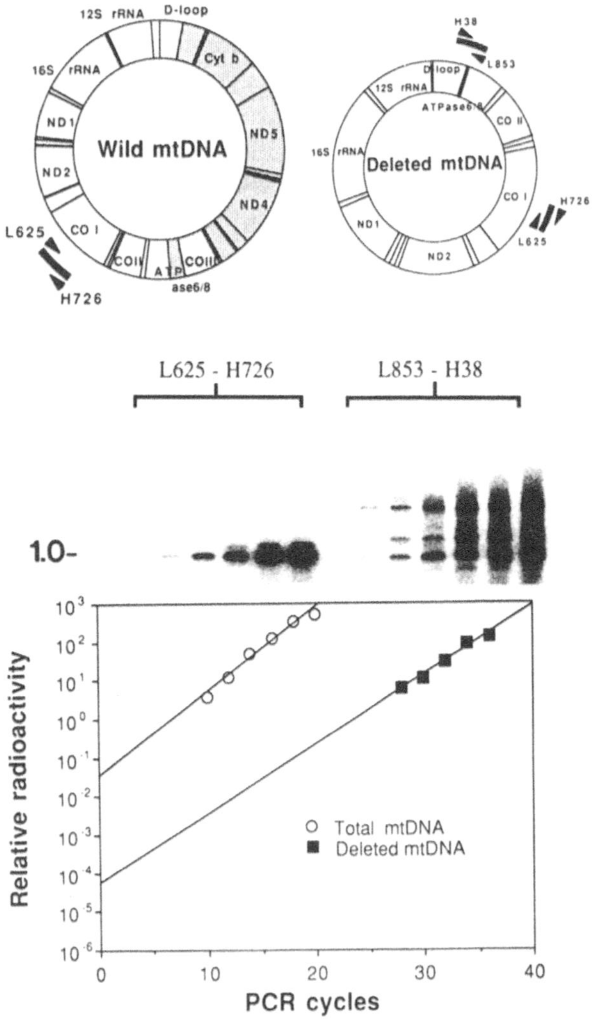

Fig. 1. The kinetic PCR method for the quantitative determination of deleted mtDNA.

Upper panel: As schematically illustrated, two different pairs of PCR primers were synthesized: One, ACGAAAATCTGTTCGCTTCA nucleotide position (nt) 8531 to 8550 [L853] and AAATTTGAAATC TGGTTAGG nt400 to 381 [H38], could amplify a 1.0 kb fragment from mtDNA with 7.4 kb deletion [hatched area between the ATPase6 gene and the displacement loop (D-loop), but not from the wild mtDNA. The other, TATAGTGGAGGCCGGAGCAG nt6251 to 6270 [L625] and GAATGAGCCTACAGATGATA nt7280 to 7261 [H726], was for the total mtDNA amplifying a 1.0 kb fragment in CO 1 gene common to the wild and deleted mtDNA.

Middle panel: PCR amplification from the total mtDNA was stopped after 10, 12, 14, 16, 18, and 20 cycles, and that from the deleted mtDNA after 28, 30, 32, 34, 36, 38, and 40 cycles. DNA fragments were electrophoresed on 1% agarose gel. With the L625-H726 primer pair, fragments with 1.0 kb size (illustrated as 1.0) were amplified. With L853-H38, fragments with 1.0 kb and others with multiple sizes were amplified.

Lower panel: By plotting logarithms of radioactivity of the labeled 1.0 kb DNA fragment against the cycle number of PCR amplification, straight lines with different slopes were obtained. Extrapolation of the lines to zero amplification cycle gave the ratio of mtDNA with deletion to the total DNA in an original DNA specimen (Subject No. 6 in Table 1) before PCR amplification.

homogenates were differentially centrifuged using the method described above. Then the pellets were washed with 10 mM Tris-HCl, pH 7.8 containing 0.25 M sucrose and finally resuspended in the same buffer. Protein in the mitochondrial preparations were measured by the bicinchoninic acid method according to Smith *et al.* [15].

Measurement of electron transport chain activity. The specific activity of rotenone-sensitive NADH-ubiquinone oxidoreductase (complex I) was determined by a modification of the method of Hatefi and Rieske [16]. The reaction mixture consisted of 20 μl of potassium phosphate buffer (1.0 M, pH 8.0), 20 μl NaN_3 (0.1 M), 50 μl coenzyme Q_1 (1.0 mM), 10 μl asolectin suspension (15 mg/ml in 1 mM EDTA) and 880 μl distilled water. The reaction mixture was incubated at 38° C before use. The reaction was initiated by adding 12 μl NADH (10 mM) and 40 μl mitochondrial suspension, sequentially. After 15 sec, 10 μl rotenone was added. The reaction rate was followed for 1 min by recording the decrease in absorbance at 340 nm, using a Shimadzu UV-visible recording spectrophotometer, model UV-250 (Shimadzu, Kyoto, Japan). The specific activity of NADH-ubiquinone oxidoreductase was deduced from the rate of decrease in absorbance.

The specific activity of succinate-ubiquinone oxidoreductase (complex II) was determined by a modification of the method of Hatefi and Stiggall [17]. The reaction mixture consisted of 20 μl sodium succinate (1.0 M, pH 7.4), 0.5 μl EDTA (0.2 M, pH 7.0), 20 μl NaN_3 (0.1 M) and 920 μl potassium phosphate buffer (50 mM, pH 7.4). The reaction mixture was incubated at 38° C before use. The reaction was initiated by adding 16 μl 2,6-dichloroindophenol (4.65 mM, DCIP), 20 μl coenzyme Q_1 (2.5 mM) and 10 μl mitochondrial suspension. The reaction rate was followed for 3 min by recording the decrease in absorbance at 600 nm. The specific activity of succinate-ubiquinone oxidoreductase was deduced from the rate of decrease in absorbance.

The specific activity of ubiquinol-cytochrmoe *c* oxidoreductase (complex III) was determined by the method of Shimomura *et al.* [18]. Reduced coenzyme Q_2 was prepared by the method of Rieske [19]. The reaction mixture consisted of 200 μl cytochrome *c* (Type III, 30 mM), 100 μl soybean phospholipid mixture (containing 21% L-α-phosphatidylcholine), 700 μl reaction buffer containing 25 mM potassium phosphate, pH 7.5, 25 μM EDTA and 0.02% Tween# 20. Before starting the enzymic reaction, reduced coen-

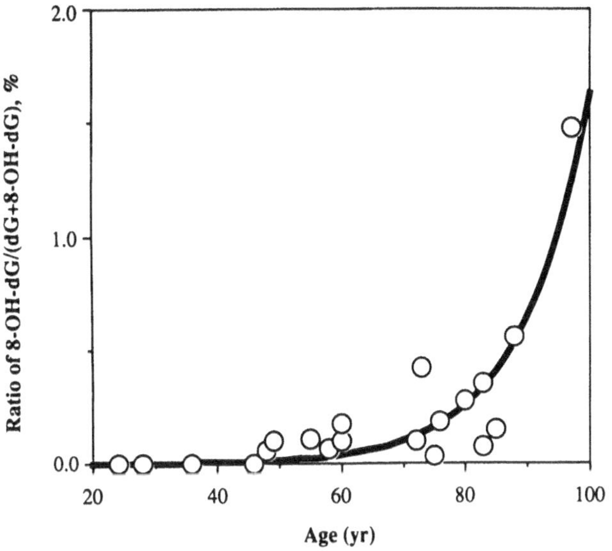

Fig. 2. The ratio of 8-OH-dG/(dG + 8-OH-dG) in human heart mtDNA from various ages. The ratio is plotted against the ages of subjects.

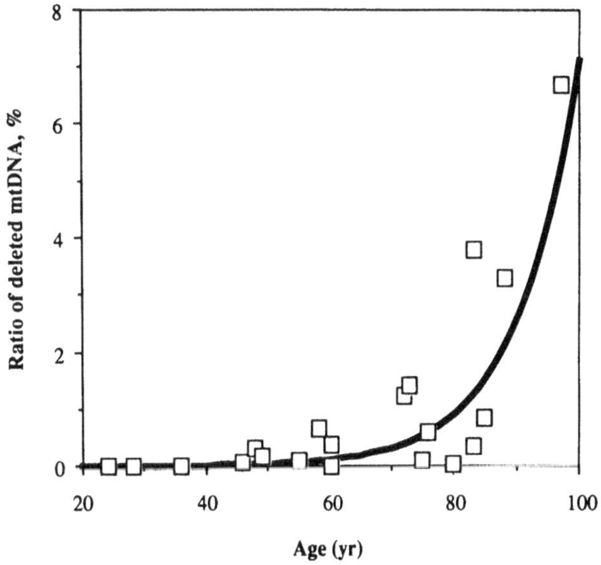

Fig. 3. The ratio of deleted mtDNA to the total mtDNA. The ratio is plotted against the ages of subjects.

zyme Q_2 was added to a concentration of 63 mM from a concentrated ethanolic solution, and nonenzymic reduction of cytochrome *c* was measured at 550 nm for 30 sec. Then the reaction was started by adding the enzyme and the reaction rate was followed by recording the increase in absorbance at 500 nm. The specific activity of ubiquinol-cytochrome *c* oxidoreductase was estimated from the rate of increase in absorbance.

The specific activity of cytochrome *c* oxidase (complex IV) was determined by a modification of the method of Wharton and Tzagoloff [20]. To prepare ferrocytochrome *c*, 1% ferricytochrome *c* was reduced completely by dithionite, and excess dithionite was removed by passing the solution through a column of Sephadex G-25 (fine). Potassium phosphate buffer (2.67 ml; 50 mM, pH 7.0) and 30 μl 10% Triton X-100 were added to 200 μl ferrocytochrome *c* solution. Immediately after the addition of 100 μl mitochondrial suspension, the reaction was followed for 15 sec by recording the rate of decrease in absorbance at 550 nm. The specific activity of cytochrome *c* oxidase was estimated from the rate of decrease in absorbance.

All analyses were performed in duplicate, and mean values were recorded. Since each group consisted of 7 samples, 14 measurements were made for each complex.

This study was performed in accordance with the guidelines for animal experimentation set by University of Nagoya.

Statistical analysis

All results are shown as mean ± SD, and ANOVA with Dunnett's test was used for statistical analysis of the data. Differences were considered significant when probability (P) values were less than 0.05.

Results

Clinical studies

The ratio of 8-OH-dG/(dG + 8-OH-dG) in human heart mtDNA

The ratio of 8-OH-dG/(dG + 8-OH-dG) in mtDNA increased exponentially with age [$\log f$ (%) = $-3.79 + 0.0400 \times$ age, $r = 0.85$, $P < 0.01$], reaching 0.26% at age 80 yr and 0.65% at age 90 yr (Fig. 2).

Quantitative analysis of mtDNA with deletion in human heart

The proportion of mtDNA with a 7.4 kb-deletion to the total mtDNA increased exponentially with age [$\log f$ (%) of deleted mtDNA) = $-3.59 + 0.0444 \times$age, $r = 0.83$, $P < 0.01$], and the population was estimated to be 0.92% and 2.55% in subjects of age 80 yr and 90 yr, respectively (Fig. 3). Clear correlation existed between the ratio of 8-OH-dG/(dG + 8-OH-dG) in mtDNA and the population of mtDNA with deletion [(% of deleted mtDNA) = $0.0178 + 4.68 \times$ 8-OH-dG/(dG + 8-OH-dG), $r = 0.92$, $P < 0.01$].

Experimental studies

The content of 8-OH-dG in rat mtDNA

Table 2 shows the ratio of 8-OH-dG/(dG + 8-OH-dG) in rat heart mtDNA. The ratio of 8-OH-dG/(dG + 8-OH-dG) in mtDNA in the 7 week group was 0.49%. Whereas, in the 100 week group, the ratio was four times higher than those in the other two groups, reaching 2.19%.

Rat mitochondrial electron transport activity

The specific activities of complexes, I, II, III and IV of rat heart mitochondria are shown in Fig. 4A–4D, respectively. The activities of complex I of rats aged 100 weeks (881 ± 56 nmol/min/mg protein) were significantly lower than those of rats aged 7 weeks (1273 ± 208). The specific activities of complex IV of rats aged 100 weeks (1889 ± 223 nmol/min/mg protein) were significantly lower than those of rats aged 7 weeks (2420 ± 184). In contrast, there were no significant differences in the specific activities of complex II (7 and 100 weeks; 513 ± 70 and 488 ± 46, respectively), and complex III (1438 ± 128 and 1356 ± 123, respectively) among the three age groups.

Discussion

The resting cardiac output falls steadily with age, as does the cardiac output with exercise. The fall in cardiac output, particularly with exercise, appears to be due to a decrease in stroke volume as well as a decrease in maximal heart rate. The decline in stroke volume is, at least in part, attributed to diminished contractility of cardiac muscle. Energy metabolism is closely related to cardiac muscle contraction. Effects of aging on the mitochondria have been widely studied [21–24], and morphologically, concerning the subcellular particles of the cells, age dependent alterations have been found only in mitochondria [25]. Trounce *et al.* [26] reported that activity of cytochrome *c* oxidase in mitochondria decreases with age, but mitochondrial monoamine oxidase activity does not. Some subunits of the former enzyme are encoded by mtDNA, however, the latter enzyme is biosynthesized entirely according to the information of nuclear DNA. In this study, the specific activities of complexes I and IV of heart mitochondria decreased significantly in rats aged 100 weeks (69% of rats aged 7 weeks). No significant decreases were observed in complexes II and III. Several subunits of complexes I and IV are encoded by mtDNA, whereas, no or few subunits of complexes II and III are encoded by mtDNA. These results suggest the high susceptibility of mtDNA to aging. In addition, Pikó *et al.* [27] reported that the frequency of the mutation rate of mtDNA in the rat and mouse was increased during the aging process.

Oxygen free radicals are proposed to be involved in aging [28]. MtDNA is constantly exposed to oxygen free radicals leaked from the mitochondrial energy producing system, and the occasion of oxidative damage to mtDNA might be expected to be much higher than that to nuclear DNA. Recently, it was revealed that the formation of 8-OH-dG from dG by active oxygen might play an important role in age-related damage to DNA [29]. Our study demonstrated that, in human heart mtDNA, the content of 8-OH-dG accumulated exponentially with age, reaching 0.26% at age 80 yr and 0.65% at age 90 yr. The level of 8-OH-dG in rat heart mtDNA in the age 7 weeks group was 0.49%. Wherease, in the age 100 week group, it was found to increase about four-fold, reaching 2.19%.

In this study, using the kinetic PCR method, the population of deleted mtDNA in human hearts increased exponentially with age, and in subjects over age 80 yr, the deleted mtDNA population markedly increased. A recent study [13] on the mutagenesis of 8-OH-dG in a mammalian cell clearly demonstrated that a synthetic c-Ha-*ras* protooncogene containing 8-hydroxy-guanine induces random point mutations at the modified site and adjacent positions. Both the conformation and the electron negativity of a 8-hydroxy-guanine were found to be quite different from those of normal guanine, inducing local DNA structure alternation such as B→ Z transition. Accumulation of 8-hydroxy-guanine in mtDNA reported here, up to 1.5% of dG (one per 150 bp of mtDNA ≈ 15 turns of double helix), could induce double-strand break leading to the generation of long stretches of single-stranded DNA

Table 2. The ratio of 8-OH-dG/(dG + 8-OH-dG) in rat heart mtDNA

Groups	mtDNA ng/mg protein	dG pmol/mg protein	8-OH-dG pmol/mg protein	8-OH-dG %*
7 weeks	8.9	5.0	0.023	0.49
100 weeks	3.8	2.4	0.053	2.19

(mean)

* Values express % of 8-OH-dG/(dG + 8-OH-dG).

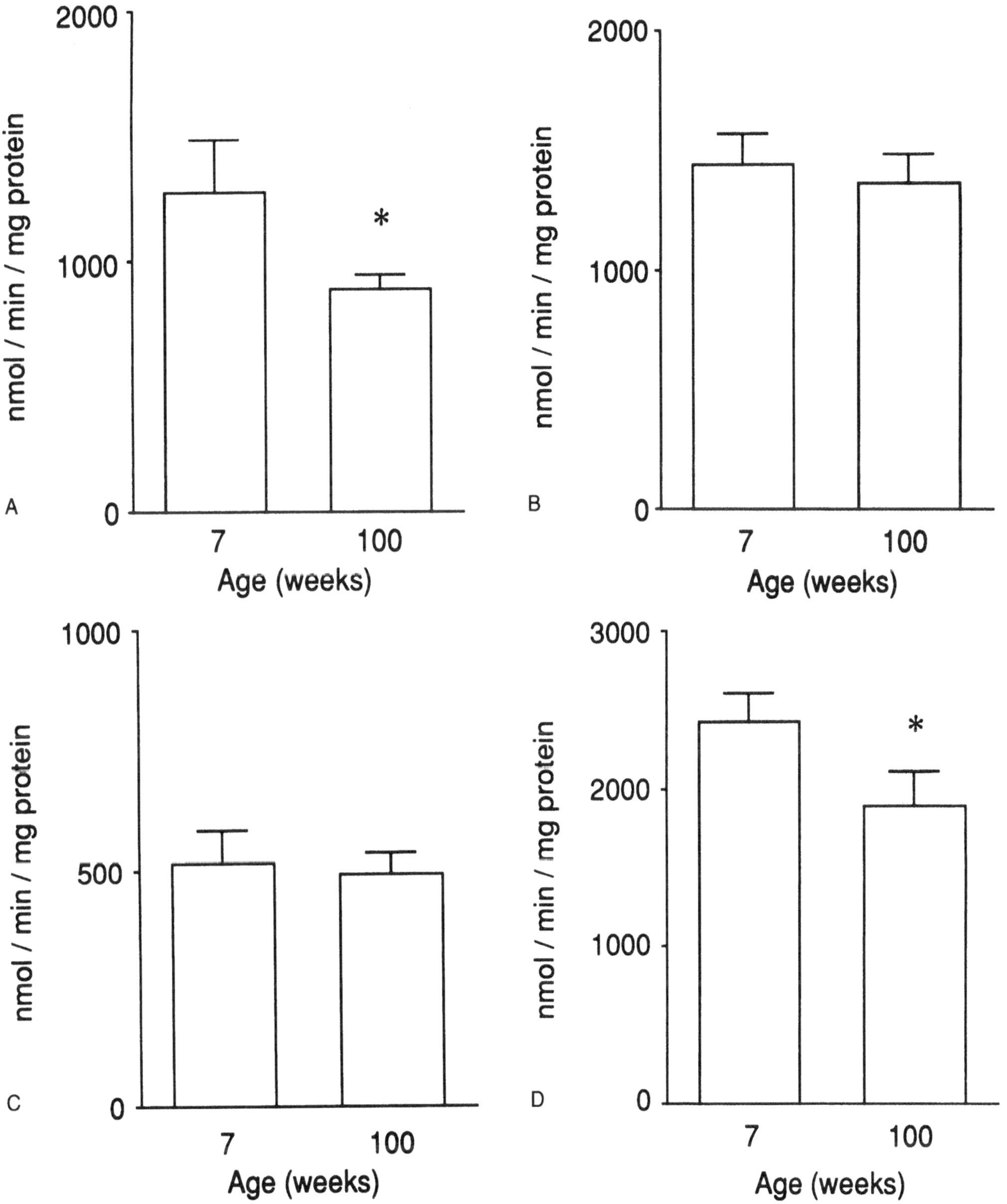

Fig. 4(A). The specific activity of complex I of the electron transport chain of rat heart mitochondria. The activity of rats aged 100 weeks was significantly low compared with those of rats aged 7 weeks. * $P < 0.01$.
(B) The specific activity of complex II of the electron transport chain of rat heart mitochondria. There were no significant differences between two groups.
(C) The specific activity of complex III of the electron transport chain of rat heart mitochondria. There were no significant differences between two groups.
(D) The specific activity of complex IV of the electrol transport chain of rat heart mitochondria. The activity of rats aged 100 weeks was significantly decreased compared with those of rats aged 7 weeks. * $P < 0.01$.

which is prerequisite of a large deletion. Conversely, defected mitochondrial respiratory chain encoded by the deleted mtDNA would enhance oxygen free-radical formation resulting more accumulation of 8-OH-dG. Such a vicious cycle of the oxygen damage and deletion in mtDNA seems to result in those changes to be synergistic and exponential with a significant correlation, $r = 0.92$, as reported here. MtDNA with a 5.0 kb-deletion was observed in striatal mtDNA in patients with Parkinsons's disease, a well known age-associated disease, and the proportion of mtDNA with deletion to the total mtDNA was 5% in patients with Parkinson's disease, while 0.3% in the controls [11]. From our results, heart mtDNA with deletion is estimated to be over 1.0% at age 80's, up to 6.7% in a 97 yr-old subject. Accordingly, mtDNA mutation, at least in a part, is involved in the genesis of heart failure in patients over 80 yr-old, and usefulness of drugs, such as digitalis, catecholamines, and phosphodiesterase inhibitor to improve cardiac performance in highly aged persons cannot be expected. Indeed, pharmacological treatment is hampered clinically by a decreased responsiveness to inotropic agents in aged individuals. Protection of mtDNA against oxidative damage is of primary importance in preventing age-associated decline in cardiac performance.

Actual measurements with human heart mtDNA *in viov* demonstrated here the synergistic accumulation of 8-hydroxy-guanine and the large deletion of mtDNA associated with age. Consequence of mtDNA mutations will result in defect of the respiratory chain enzyme, leading to energy deficit in a cell and tissue. Human muscle mitochondrial enzymes encoded by mtDNA has been reported to lose their activities with age [26], which was actually demonstrated here using rat hearts. These facts could explain the development of non-atherosclerotic heart dysfunction that seems to be an inevitable progressive change in human life.

References

1. Linnane AW, Marzuki S, Ozawa T, Tanaka M: Mitochondrial DNA mutations as an important contributor to ageing and degenerative diseases. Lancet i: 642–645, 1989
2. Ames BN: Endogenous oxidative DNA damage, aging, and cancer. Free Radic Rex Commun 7: 121–128, 1989
3. Anderson S, Bankier AT, Barrell BT, de Bruijn MHL, Coulson AR, Drouin J, Eperon IC, Nierlich DP, Roe BA, Sanger F, Schreier PH, Smith AJH, Staden R, Young IG: Sequence and organization of the human mitochondrial genome. Nature 290: 457–465, 1981
4. Ozawa T, Yoneda M, Tanaka M, Ohno K, Sato W, Suzuki H, Nishikimi M, Yamamoto M, Nonaka I, Horai S: Maternal inheritance of deleted mitochondrial DNA in a family with mitochondrial myopathy. Biochem Biophys Res Commun 154: 1240–1247, 1988
5. Cortopassi GA, Arnheim N: Detection of a specific mitochondrial DNA deletion in tissues of older humans. Nucleic Acids Res 18: 6927–6933, 1990
6. Linnane AW, Baumer A, Maxwell RJ, Preston H, Zhange C, Marzuki S: Mitochondrial gene mutation: The ageing process and degenerative diseases. Biochem Int 22: 1067–1076, 1990
7. Hattori K, Tanaka M, Sugiyama S, Obayashi T, Ito T, Satake T, Hanaki Y, Asai J, Nagano M, Ozawa T: Age-dependent increase in deleted mitochondrial DNA in the human heart: Possible contributory factor to presbycardia. Am Heart J 121: 1735–1742, 1991
8. Yen TC, Su JH, King KL, Wei YH: Ageing-associated 5 kb deletion in human liver mitochondrial DNA. Biochem Biophys Res Commun 178: 124–131, 1991
9. Torii K, Sugiyama S, Tanaka M, Takagi K, Hanaki Y, Iida K, Matsuyama M, Hirabayashi N, Uno Y, Ozawa T: Aging-associated deletions of human diaphragmatic mitochondrial DNA. Am J Respir Cell Mol Biol 6: 543–549, 1992
10. Robin ED, Wong R: Mitochondrial DNA molecules and virtual number of mitochondria per cell in mammalian cells. J Cell Physiol 136: 507–513, 1988
11. Ozawa T, Tanaka M, Ikebe S, Ohno K, Kondo T, Mizuno Y: Quantitative determination of deleted mitochondrial DNA relative to normal DNA in parkinsonian striatum by a kinetic PCR analysis. Biochem Biophys Res Commun 172: 483–489, 1990
12. Hayakawa M, Ogawa T, Sugiyama S, Tanaka M, Ozawa T: Massive conversion of guanosine to 8-hydroxy-guanosine in mouse liver mitochondrial DNA by administration of azidothymidine. Biochem Biophys Res Commun 176: 87–93, 1991
13. Kamiya H, Miura K, Ishikawa H, Inoue H, Nishimura S, Ohtsuka E: *c*-Ha-*ras* containing 8-hydoxyguanine at codon 12 induces point mutations at the modified and adjacent positions. Cancer Res 52: 3483–3485, 1992
14. Gornall AG, Bardawill CJ, David MM: Determination of serum proteins by means of the biuret reaction. J Biol Chem 177: 751–766, 1949
15. Smith PK, Krohn RI, Hermanson GT, Mallia AK, Gartner FH, Provenzano MD, Fujimoto EK, Goeke NM, Olson BJ, Klenk DC: Measurement of protein using bicinchoninic acid. Anal Biochem 150: 76–85, 1985
16. Hatefi Y, Rieske JS: Preparation and properties of DPNH-coenzyme Q reductase (complex I of the respiratory chain). In: RW Estabrook, ME Pullman (eds.) Methods in Enzymology. Vol 10. Academic Press, New York, 1967, pp 235–239
17. Hatefi Y, Stiggall DL: Preparation and properties of succinate: Ubiquinone oxidoreductase (complex II). In: S Fleischer, L Packer (eds.) Methods in Enzymology. Vol 53. Academic Press, New York, 1978, pp 21–27
18. Shimomura Y, Nishikimi M, Ozawa T: Isolation and reconstitution of the iron-sulfur protein in ubiquinol-cytochrome *c* oxidoreductase complex. J Biol Che 259: 14059–14063, 1984
19. Rieske JS: Preparation and properties of reduced coenzyme Q-cytochrome *c* reductase (complex III of the respiratory chain).

In: RW Estabrook, ME Pullman (eds.) Methods in Enzymology. Vol 10, Academic Press, New York, 1967, pp 239–245
20. Wharton DC, Tzagoloff A: Cytochrome oxidase from beef heart mitochondria. In: RW Estabrook, ME Pullman (eds.) Methods in Enzymology. Vol 10. Academic Press, New York, 1967, pp 245–250
21. Asano K, Amagase S, Matsuura ET, Yamagishi H: Changes in the rat liver mitochondrial DNA upon aging. Mech Ageing Dev 60: 275–284, 1991
22. Bandy B, Davison AJ: Mitochondrial mutations may increase oxidative stress: Implications for carcinogenesis and aging? Free Radic Biol Med 8: 523–539, 1990
23. Gadaleta MN, Petruzzella V, Renis M, Fracasso F, Cantatore P: Reduced transcription of mitochondrial DNA in the senescent rat. Tissue dependence and effect of L-carnitine. Eur J Biochem 187: 501–506, 1990
24. Tauchi H, Sato T: Some micromeasuring studies of hepatic cells in senility. J Gerontol 17: 254–259, 1962
25. Tauchi H, Sato T: Age changes in size and number of mitochondria of human hepatic cells. J Gerontol 23: 454–461, 1968
26. Trounce I, Byrne E, Marzuki S: Decline in skeletal muscle mitochondrial respiratory chain function: Possible factor in ageing. Lancet ii: 637–639, 1989
27. Pikó L, Hougham AJ, Bulpitt KJ: Studies of sequence heterogeneity of mitochondrial DNA from rat and mouse tissues: Evidence for an increased frequency of deletions/additions with aging. Mech Ageing Dev 43: 279–293, 1988
28. Harman D: The aging process. Proc Natl Acad Sci USA 78: 7124–7128, 1981
29. Fraga CG, Shigenaga MK, Park JW, Degan P, Ames BN: Oxidative damage to DNA during aging: 8-hydroxy-2'-deoxyguanosine in rat organ DNA and urine. Proc Natl Acad Sci USA 87: 4533–4537, 1990

Molecular and Cellular Biochemistry **119**: 105–108, 1993.

Mitochondrial DNA deletion in human myocardium

Nobuakira Takeda, Akira Tanamura, Takaaki Iwai, Izuru Nakamura, Mitsutoshi Kato, Tadanari Ohkubo and Kenji Noma
Department of Internal Medicine, Aoto Hospital, Jikei University School of Medicine, Katsushika-ku, Tokyo, Japan

Abstract

Mutation of myocardial mitochondrial DNA was investigated in human left ventricles obtained at autopsy using the polymerase chain reaction (PCR). Seventeen autopsy cases were examined, including patients with diabetes mellitus, myocardial infarction, cardiomyopathy, cancer, and other diseases. Two cases of diabetes mellitus, 2 of myocardial infarction, and 1 of pulmonary fibrosis showed a 7.4 kb deletion of myocardial mitochondrial DNA. Primer shift PCR confirmed that an amplified DNA fragment had not been obtained by misannealing of the primers. It is unclear how much these findings are related to the severity or prognosis of the various diseases, but they indicate that mutation of myocardial mitochondrial DNA can occur in other diseases besides cardiomyopathy, although the influence of aging could not be excluded. (Mol Cell Biochem **119:** 105–108, 1993)

Key words: mitochondrial DNA, PCR, diabetes mellitus, myocardial infarction, cardiomyopathy

Introduction

Mitochondrial cytopathy in cardiac lesions has recently been reported as a result of gene analysis, e.g., point mutations of mitochondrial DNA in patients with mitochondrial myopathy, encephalopathy, lactic acidosis, and stroke-like episodes (MELAS) [1–4]. In various cardiovascular diseases, new interesting evidence has also been obtained using molecular biological methods, e.g., point mutations or deletions of myocardial mitochondrial DNA, have been found in hypertrophic or dilated cardiomyopathy [5, 6]. However, it is unclear how these mitochondrial DNA mutations are related to the pathogenesis of cardiomyopathy. It is also unclear whether or not these alterations exist in patients with other acquired heart diseases. Therefore, we investigated whether myocardial mitochondrial DNA deletion could be detected in patients with diseases other than cardiomyopathy using the polymerase chain reaction (PCR) [7, 8].

Materials and methods

Human left ventricles were obtained at autopsies performed within 3 h of death.

Left ventricular myocardium (ca. 20 mg) was homogenized for 30 seconds using a Physcotron Handy Micro-Homogenizer (Niti-on, Tokyo), and was digested for 12 h at 37° C in 10 mM Tris-HCl and 0.1 M ethylenedia-

Address for offprints: N. Takeda, Department of Internal Medicine, Aoto Hospital, Jikei University School of Medicine, Aoto 6-41-2, Katsushika-ku, Tokyo 125, Japan

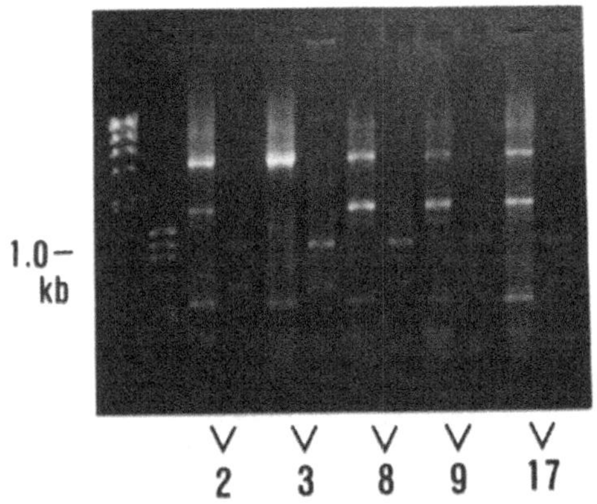

Fig. 1. Result of the standard PCR. The left lane of each sample shows PCR amplification using primers L116-H617 and the right lane shows amplification with primers L853-H38. Right lanes indicate the existence of a 7.4 kb deletion. The leftmost two lanes show markers. The number of each lane corresponds to the patient numbers shown in Table 1.

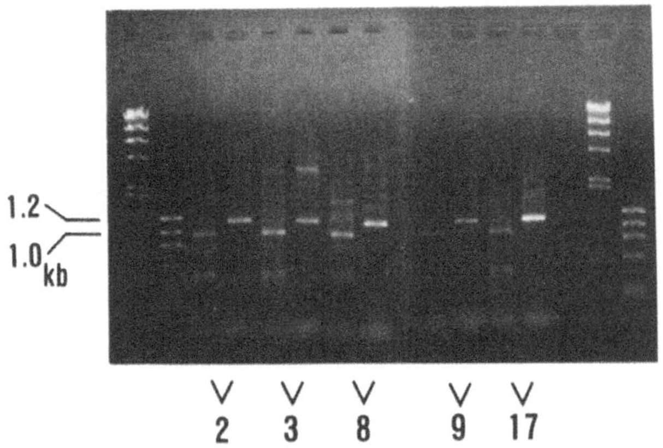

Fig. 2. Primer shift PCR. The left lane of each sample shows PCR amplification using primers L853-H38 and the right lane shows amplification with primers L853-H60.

minetetraacetic acid (EDTA) (pH 7.4) containing 0.1 mg/ml proteinase K and 0.5% sodium dodecyl sulfate (SDS). Then DNA was extracted with phenol, chloroform, and isoamylalcohol, after which it was precipitated with 3 M sodium acetate (pH 7.4) and ethanol at − 80° C for 30 min. The pellet was obtained by centrifugation (13,000 rpm, 5 min), was rinsed with 70% ethanol, and was dissolved in 30 μl of 10 mM Tris-HCl and 0.1 M EDTA (pH 8.0).

Oligonucleotide primers for the PCR were synthesized and purified on cartridges obtained from Sci-Media Ltd. (Tokyo).

PCR amplification was carried out with a reaction mixture of template DNA (10 ng/μl), two primers, dNTPs, PCR buffer (500 mM KCl, 100 mM Tris-HCl, pH 8.3, 15 mM $MgCl_2$, 0.01% gelatin), and Taq DNA polymerase (Takara, Kyoto). The apparatus used for PCR was a Program Temp Control System PC-700 (Astec, Tokyo). Amplification was carried out for 30 cycles under the following conditions: denaturation at 94° C for 15 sec, annealing at 50° C for 15 sec, and primer extension at 72° C for 80 sec. Amplified DNA fragments were separated by agarose gel electrophoresis, stained with ethidium bromide, and detected by photofluorography.

To ensure that amplified DNA fragments were not obtained by misannealing of the primers to an unexpected site, primer shift PCR was also carried out [9].

Results

Seventeen autopsy cases (6 males and 11 females) were examined. They had the following diagnoses: cardiomyopathy, 1; diabetes mellitus, 2; myocardial infarction, 3; pulmonary fibrosis, 2; apoplexy, 1; solid cancer, 6; leukemia, 1; and toxicosis, 1 (Table 1). The mean age of the patients was 65.3 years.

Table 1. Profiles of patients

No.	Age (y)	Gender	Diagnosis	No.	Age (y)	Gender	Diagnosis
1	57	male	Dilated cardiomyopathy	10	92	male	Gastric cancer
2	76	male	Diabetes mellitus	11	84	female	Acute myocardial infarction
3	78	female	Acute myocardial infarction	12	60	female	Pulmonary fibrosis
4	54	male	Gastric cancer	13	89	male	Hepatoma
5	47	female	Apoplexy	14	28	female	Gastric cancer
6	29	female	Gestational toxicosis, DIC	15	54	female	Pancreas cancer
7	78	female	Lung cancer	16	66	female	Acute myelogenous leukemia
8	69	male	CHF, OMI	17	72	female	Diabetes mellitus
9	77	female	Pulmonary fibrosis, OMI				

DIC: disseminated intravascular coagulation, CHF: congestive heart failure, OMI: old myocardial infarction.

Table 2. Primers used for PCR

Primers	Sequence (5′→3′)
L116	AACTCAAAGGACCTGGCGGT
L853	ACGAAAATCTGTTCGCTTCA
H38	AAATTTGAAATCTGGTTAGG
H60	AAACATTTTCAGTGTATTGC
H617	CGGGGAAACGCCATATCGGG

A 4 kb band was detected by the combination of primers L116–H617 in all samples, indicating that the PCR worked well (Fig. 1). With the combination of primers L853-H38, a single band of ca. 1.0 kb was detected in five samples, indicating that there was a mitochondrial DNA deletion of 7.4 kb (Fig. 1). The deletion was found in 2 cases of diabetes mellitus, 2 cases of myocardial infarction, and 1 case of pulmonary fibrosis. Figure 2 shows that this band obtained with primers L853-H38 was not the result of misannealing. Figure 3 shows a representative gel which reveals no band between these two primers, indicating that there was no mitochondrial DNA deletion.

Discussion

The molecular biological approach to investigation of the pathogenesis of cardiomyopathy has brought to light some interesting evidence, e.g., mutations of myocardial mitochondrial DNA and hybrid or mutant cardiac myosin heavy chain in familial hypertrophic cardiomyopathy [10, 11]. It is still unclear whether these myocardial mitochondrial DNA mutations influence cardiac performance or affect the severity or prognosis of the underlying diseases. It is also unclear whether these mutations exist in diseases besides cardiomyopathy. In the present study, deletion of myocardial mitochondrial DNA was found in patients with diabetes mellitus and myocardial infarction. In diabetes, not only vascular involvement but also the metabolic disorder itself could induce cardiac lesions [12–16]. We have previously reported on alterations of myocardial contractility and the shifting of ventricular myosin isoenzymes towards V_3, which has the lowest ATPase activity, in experimental diabetic rats [17–19]. This change of the ventricular myosin isoenzyme distribution can be considered as an adaptation to maintain myocardial force development with low energy utilization in the face of deteriorating myocardial energy metabolism due to diabetes. Thus, the metabolic disorder caused by diabetes might be related to the myocardial mitochondrial DNA deletion detected in this study. Myocardial ischemia produces free radicals which cause DNA damage. Chronic myocardial ischemia might have induced mitochondrial DNA deletion in the patients with myocardial infarction in our study. Since oxygen radicals increase with aging, this might also influence the results of the present study [20]. Only one case of cardiomyopathy was examined and no deletions were detected. This patient might have had no deletion of myocardial mitochondrial DNA or it might have existed in a part of the DNA which was not examined.

Table 3. The combination of primers

Combination of primers	Distance between primers (kb)	Amplified fragment (kb)
L116 + H617	5.0	5.0
L853 + H38	8.4	1.0
L853 + H60	8.6	1.2

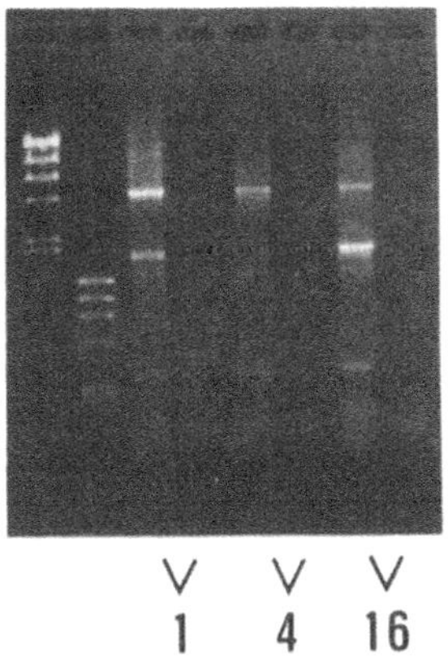

Fig. 3. Result of the standard PCR. The conditions were the same as for the PCR shown in Fig. 1. The right lanes show no band, indicating that there was no deletion between primers L853 and H38.

In conclusion, mutations of myocardial mitochondrial DNA can also occur in other diseases besides cardiomyopathy, but further examination is necessary to clarify the influence of aging and whether such mutations are related to the severity or prognosis of the underlying diseases.

References

1. Pavlakis S, Phillips PC, DiMauro S, DeVivo DC, Rowland LP: Mitochondrial myopathy, encephalopathy, lactic acidosis, and stroke-like episodes. Ann Neurol 16: 481–488, 1984

2. Ito T, Hattori K, Tanaka M, Sugiyama S, Ozawa T: Mitochondrial cytopathy. Jpn Circ J 54: 1214–1220, 1989
3. Tanaka M, Ino H, Ohno K, Obayashi T, Ikebe S, Sano T, Ichiki T, Kobayashi M, Wada Y, Ozawa T: Mitochondrial DNA mutations in mitochondrial myopathy, encephalopathy, lactic acidosis, and stroke-like episodes (MELAS). Biochem Biophys Res Commun 174: 861–868, 1991
4. Goto Y, Horai S, Matsuoka T, Koga Y, Nihei K, Kobayashi M, Nonaka I: Mitochondrial myopathy, encephalopathy, lactic acidosis, and stroke-like episodes (MELAS): a correlative study of the clinical features and mitochondrial DNA mutation. Neurology 42: 545–550, 1992
5. Ozawa T, Tanaka M, Sugiyama S, Hattori K, Ito T, Ohno K, Takahashi A, Sato W, Takada G, Mayumi B, Yamamoto K, Adachi K, Koga Y, Toshima H: Multiple mitochondrial DNA deletions exist in cardiomyocytes of patients with hypertrophic or dilated cardiomyopathy. Biochem Biophys Res Commun 170: 830–836, 1990
6. Obayashi T, Takasawa M, Hattori K, Ito T, Tanaka M, Sugiyama S, Ozawa T, Tanaka T, Itoyama S, Deguchi H, Kawamura K, Koga Y, Toshima H, Takeda N, Nagano M: Point mutations in mitochondrial DNA in patients with hypertrophic cardiomyopathy. Cardiac Structure and Metabolism (in Japanese) 14: 407–412, 1991
7. Saiki RK, Scharf S, Faloona F, Mullis KB, Horn GT, Erlich HA, Arnheim N: Enzymatic amplification of β-globin genomic sequences and restriction site analysis for diagnosis of sickle cell anemia. Science 230: 1350–1354, 1985
8. Mullis K, Faloona F, Scharf S, Saiki R, Horn G, Erlich H: Specific enzymatic amplification of DNA *in vitro*: the polymerase chain reaction. Cold Spring Harbor Symp. Quant Biol 51: 263–273, 1986
9. Sato W, Tanaka M, Ohno K, Yamamoto T, Tanaka G, Ozawa T: Multiple populations of deleted mitochondrial DNA deleted by a novel gene amplification method. Biochem Biophys Res Commun 162: 664–672, 1989
10. Tanigawa G, Jarcho JA, Kass S, Solomon SD, Vosberg H-P, Seidman JG, Seidman CE: A molecular basis for familial hypertrophic cardiomyopathy: An α/β cardiac myosin heavy chain hybrid gene. Cell 62: 991–998, 1990
11. Geisterfer-Lowrance AAT, Kass S, Tanigawa G, Vosberg H-P, McKenna W, Seidman CE, Seidman JG: A molecular basis for familial hypertrophic cardiomyopathy: A β cardiac myosin heavy chain gene missense mutation. Cell 62: 999–1006, 1990
12. Regan TJ, Lyons MM, Ahmed SS: Evidence for cardiomyopathy in familial diabetes mellitus. J Clin Invest 60: 885–889, 1977
13. Penpargkul S, Fein F, Sonnenblick EH, Scheuer J: Depressed cardiac sarcoplasmic reticular function from diabetic rats. J Mol Cell Cardiol 13: 303–309, 1981
14. Fein FS, Stroheck TE, Malhotra A, Scheuer J, Sonnenblick EH: Reversibility of diabetic cardiomyopathy with insulin in rats. Circ Res 49: 1251–1261, 1981
15. Pierce GN, Dhalla NS: Sarcolemmal Na^+-K^+-ATPase activity in diabetic rat heart. Am J Physiol 245: C241–C247, 1983
16. Kawamura M, Suzuki H, Nagata E, Kawanishi M, Watari S, Ohkubo T: Altered Ca^{2+} transport of myocardial sarcolemma in diabetes mellitus. Jikeikai Med J 37 (Suppl): 61–70, 1990
17. Takeda N, Nakamura I, Hatanaka T, Ohkubo T, Nagano M: Myocardial mechanical and myosin isoenzyme alterations in streptozotocin-diabetic rats. Jpn Heart J 29: 455–463, 1988
18. Takeda N, Nakamura I, Ohkubo T, Hatanaka T, Nagano M: Effects of physical training on the myocardium of streptozotocin-induced diabetic rats. Basic Res Cardiol 83: 525–530, 1988
19. Takeda N, Ohkubo T, Tanamura A, Iwai T, Nakamura I, Hatanaka T, Nagano M: Myocardial mechanics and myosin isoenzymes in streptozotocin-induced diabetic rats. In: M. Nagano and N.S. Dhalla (eds.) The Diabetic Heart. Raven Press, New York, 1991, pp 291–299
20. Hattori K, Tanaka M, Sugiyama S, Obayashi T, Ito T, Satake T, Hanaki Y, Asai J, Nagano M, Ozawa T: Age-dependent increase in deleted mitochondrial DNA in the human heart: Possible contributory factor to presbycardia. Am Heart J 121: 1735–1742, 1991

Molecular and Cellular Biochemistry **119**: 109–120, 1993.

Ischemic changes in myocardial ionic contents of the isolated perfused rat hearts as studied by NMR

Takaharu Ishibashi, Mikio Nakazawa and Shoichi Imai
Department of Pharmacology, Niigata University School of Medicine, Niigata, 951, Japan

Abstract

Using ^{31}P-, ^{23}Na- and ^{39}K-NMR, we assessed ischemic changes in high energy phosphates and ion contents of isolated perfused rat hearts continuously and systematically. To discriminate intra- and extracellular Na^+, a shift reagent ($Dy(TTHA)^{3-}$) was used in ^{23}Na-NMR study. In ^{39}K-NMR study, the extracellular K^+ signal was suppressed by inversion recovery pulse sequence in order to obtain intracellular K^+ signal without using shift reagents. During the early period of ischemia, increases in intracellular Na^+ and inorganic phosphate (Pi) were observed in addition to the well-documented decreases in creatine phosphate and ATP and a fall of intracellular pH, suggesting an augmented operation of Na^+-H^+ exchange triggered by a fall of the intracellular pH resulted from breakdown of ATP. At around 15 min of ischemia, a second larger increase in intracellular Na^+ and a decrease in intracellular K^+ were observed in association with a second increase in Pi. This was accompnanied by an abrupt rise of the ventricular end-diastolic pressure. As there was a depletion of ATP at this time, the increase in intracellular Na^+ and associated decrease in intracellular K^+ may be explained by inhibition of the Na^+-K^+ ATPase due to the depletion of ATP. A longer observation with ^{31}P-NMR revealed a second phosphate peak (at lower magnetic field to ordinary Pi peak) which increased its intensity as ischemic time lengthened. The pH of this 2nd peak changed in parallel with the changes in pH of the bathing solution, indicating the appearance of a compartment whose hydrogen concentration is in equilibrium with that of the external compartment. Thus, the peak could be used as an index of irreversible membrane damage of the myocardium. (Mol Cell Biochem **119:** 109–120, 1993)

Key words: ^{31}P-NMR, ^{23}Na-NMR, ^{39}K-NMR, ischemia, rat heart

Introduction

Changes in myocardial energy metabolism and ionic permeability of the membrane during ischemia have been the subject of intensive studies [1–6]. During myocardial ischemia, there occurs a shift of energy metabolism from aerobic to anaerobic glycolysis with resultant decreases in intracellular pH and ATP. These phenomena are well-documented together with the changes in mechanical function [6, 7]. Early efflux of K^+ during ischemia has been shown with ion-selective microelectrode [3, 8] or with myocardial extract [1]. However, there is still controversy as to the changes in intracellular Na^+ [3, 6, 8] during ischemia. Activation and/or inactivation of various ion pumps, channels and exchange systems are assumed to occur during ischemia in response to changes in myocardial energy state. However, due to the technical difficulties inherent in non-

Address for offprints: T. Ishibashi, Department of Pharmacology, Niigata University School of Medicine, Niigata, 951, Japan

invasive and continuous recording of the events occurring in the isolated functioning hearts there have been only a few systematic studies that endeavoured to find out correlation between the changes in myocardial ionic contents and the energy state of the myocardium. In view of the circumstances a non-invasive, continuous recording of changes in energy state, pH, Na^+ and K^+ of the perfused rat heart during ischemia was attempted with a nuclear magnetic resonance spectroscopy (NMR) with the aim of exploring the contributions of Na^+-K^+ pump and Na^+-H^+ exchange system to ischemic changes in intracellular Na^+, K^+ and pH as well as to those in high energy phosphates and of finding some indices of irreversible damage of myocardium detectable with NMR. Preliminary reports of this work were presented elsewhere in parts]9, 10].

Materials and methods

Isolated perfused heart preparation

Hearts excised from male Wistar rats (250–350 g) under ether anesthesia were perfused according to the method of Langendorff (perfusion pressure of 75 cm H_2O). The perfusion medium was a modified Krebs-Ringer solution (37° C). Unless otherwise noted, the composition of the perfusion medium is as shown in Table 1 (K-R I). Coronary flow was measured with an electromagnetic flowmeter (Statham SP 2201) equipped with a cannulating type flow probe (2 mm in diameter) that is placed on the in-flow circuit leading to the aortic cannula. Left ventricular pressure was measured with a saline or distilled water (in the case of ^{23}Na-NMR) filled latex balloon coupled to a pressure transducer (Statham P-50) with a polyethylene tube. The balloon was inflated to give an end diastolic pressure of 0–5 mmHg. Heart rate was measured continuously with a tachometer triggered by left ventricular pressure signals. After completion of the necessary procedures for starting perfusion, the hearts were placed in a glass NMR tube (diameter of 15 mm for both ^{31}P-NMR and ^{39}K-NMR, and of 20 mm for ^{23}Na-NMR) in a superconducting magnet. Coronary effluent was evacuated from the NMR tube with a vacuum pump.

All experimental procedures were conducted after a preliminary perfusion period of about 40 min. Throughout the experiments, all animals were dealt with in a humane way in accordance with the guidelines for animal experimentation set by the Japanese Association for Laboratory Animal Science.

Spectra measurement

Spectra of free induction decays were obtained using a FX270 NMR spectrometer (JEOL Ltd, Tokyo, Japan), operating at desired wave lengths as described below, and were accumulated and stored automatically at preprogrammed intervals. Before Fourier transformation, free induction decays were multiplied by the desired factor with an exponential time constant.

^{31}P-NMR spectra measurement and data analysis

^{31}P-NMR spectra were obtained with the NMR spectrometer operating at 109.16 MHz. The spectral parameters were 45° pulse, 0.5 sec intervals, and 600 transients (5 min).

Adenosine triphosphate (β-ATP), creatine phosphate (CrP), and intracellular inorganic phosphate (Pi) were quantitated by integrating the areas under each peak with a personal computer (HP-85, Hewlett Packard) connected to a digitizer (Graphic Tablet 911A, Hewlett-Packard). A capillary filled with methylene diphosphonate (MDP) and placed within the NMR tube served as the external standard, and the relative intensities of each peak were used for quantitative analysis. Intracellular pH was estimated from the pH-dependent chemical shift of the Pi peak relative to the CrP peak that is pH independent. For determination of intracellular pH, the titration curve was established in this laboratory using Pi and CrP solution, the ionic compositions of which resembled those found *in vivo* [11].

^{23}Na-NMR spectra measurement and data analysis

^{23}Na-NMR spectra were obtained at 71.30 MHz. The spectral parameters were 90° pulse, 0.5 sec intervals, and 360 transients (3 min). Capillaries filled with a mixture of 1 M NaCl and 100 mM $(TrisH)_3Dy(TTHA)$ ·3TrisHCl were placed within the NMR tube and were served as the external standard. In order to discriminate the intracellular Na^+ from the extracellular one, we introduced a shift reagent $(Dy(TTHA)^{3-})$ into perfusion- and bathing-solutions (see below). The relative intensities of each peak were used for quantitative analysis.

^{39}K-NMR spectra measurement and data analysis

^{39}K-NMR spectra were obtained at 12.54 MHz. The spectral parameters were 90° pulse 250 ms intervals (2000–4000 transients during 8.3–16.7 min) for traditional observation (single pulse). Inversion recovery pulse sequence (180°-PD-90°) with intervals of 360 ms for 20 min (3333 transients) was used for observation of intracellular K^+ (described below). The relative changes in intensity of the peak to that obtained during control period (normoxic perfusion) was used for quantitative analysis.

Agents

Dysprosium chloride ($DyCl_3$) was purchased from Nakarai Chemicals, Ltd. (Kyoto, Japan) and treithylenetetraminehexaacetic acid (TTHA) from Dojin Laboratories (Kumamoto, Japan).

$(TrisH)_3Dy(TTHA)\cdot 3TrisHCl$ (100 mM) for external standard and for bathing solution (Table 1, B III) was prepared by dissolving TTHA with 1 M Tris followed by addition of $DyCl_3$ under stirring. Tris of 2 M was used to adjust pH at 7.4 at 37° C and the volume was adjusted with distilled water. The stock solution of $Na_3Dy(TTHA)\cdot 3Na$ (100 mM) for perfusion medium [12] (Table 1, K-R II) was prepared by dissolving TTHA and $DyCl_3$ with 1 N NaOH, and pH was adjusted by further addition of 1 N NaOH. The amount of Na^+ required to prepare the stock solution was subtracted from the amount of NaCl that should otherwise be added to perfusion medium (Table 1).

Statistical analysis

All values are expressed as mean ± SEM. Statistical assessment of the data along the time course was made by Student's t test. Differences were considered significant at a probability value of less than 0.05.

Results

Ischemic changes observed with ^{31}P-NMR

Ischemia was induced by completely stopping the aortic inflow and drop infusion (80 ml/h) of Krebs-Ringer solution (37° C) equilibrated with 95% N_2 and 5% CO_2 (B I solution in Table 1) was begun to avoid the uptake of oxygen from the bathing solution (Fig. 1A). Upon induction of ischemia a rapid disappearance of creatine phosphate (CrP) was observed with a more gradual decay of ATP (β). Intracellular inorganic phosphate (Pi) showed a rapid increase and the peak shifted to higher magnetic field as the duration of ischemia lengthened (Fig. 2), indicating the acidification (pH about 6.0) of the myocardium. The shift became stable (about

Table 1. Composition of solutions (all components are expressed in mM)

	K-R I	K-R II	K-R III	B I	B II	B III
NaCl	125.2	125.2[a]	119.3	125.2	149.0	
KCl	4.7	4.7		4.7	5.9	
$CaCl_2$	2.5	2.5	2.5	2.5	2.5	
KH_2PO_4	1.2	1.2		1.2		
$NaHCO_3$	24.9	24.9	24.9	24.9	1.0	
NaH_2PO_4			1.2			
glucose	5.5	5.5	5.5	5.5	5.5	
pyruvate sodium	2.0	2.0	2.0	2.0	2.0	
$Na_3Dy(TTHA)3NaCl$		10.0				
$trisH_3Dy(TTHA)3tris$						10.0
mannitol						290.0
gas mixture	95% O_2 + 5% CO_2	95% O_2 + 5% CO_2	95% O_2 + 5% CO_2	95% N_2 + 5% CO_2	95% N_2 + 5% CO_2	
pH	7.4	7.4	7.4	7.4	6.0	7.4

K-R I: Modified Krebs-Ringer bicarbonate solution. K-R II: Modified Krebs-Ringer solution with shift reagent. K-R III: Modified Krebs-Ringer solution without potassium. B I: Bathing solution (pH 7.4). B II: Bathing solution (pH 6.0). B III: Bathing solution with shift reagent. a: Depending on the amount of NaOH that was used to adjust the pH of shift reagent to 7.4, necessary amount of NaCl added to the solution was changed so as to keep the final concentration of Na^+ at 125.2 mM.

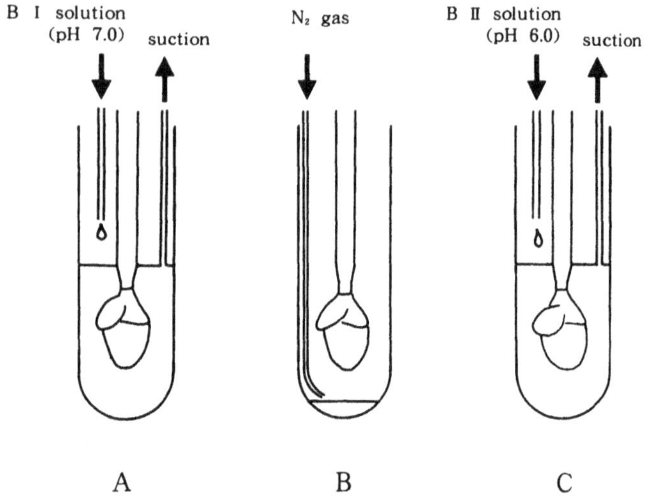

Fig. 1. Schematic drawings showing the conditions of perfusion during ischemia of the isolated heart preparation subjected to ^{31}P-NMR studies. A: Standard situation. Hearts were soaked in normal bathing solution. B: Hearts were placed in atmosphere of nitrogen. C: Hearts were soaked in the bathing solution of pH 6.0. The composition of the bathing solution is shown in Table 1. See text for details.

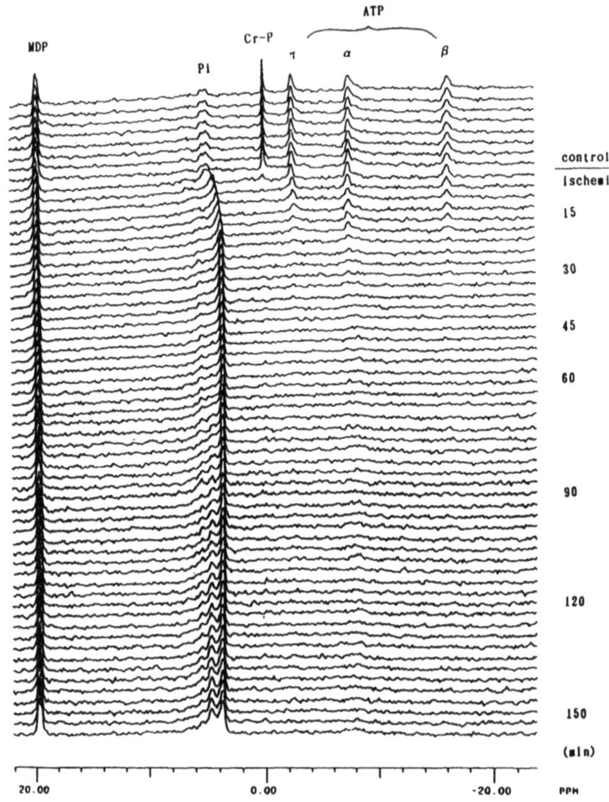

Fig. 2. A representative record of a serial ^{31}P-NMR spectra taken during ischemia under the condition as depicted in Fig. 1A. MDP: methylene diphosphonate as an external standard. Pi: inorganic phosphate. Cr-P: Creatine phosphate. ATP (α, β, γ): adenosine triphosphate. The initial ischemic changes completed about 30 min after induction of ischemia. A new peak (second Pi peak) developed and grew up after about 90 min of ischemia.

1.8 ppm to the higher field) after 20–30 min of ischemia and the peak stayed at that position thereafter. The changes in these parameters of 6 preparations observed during 1 h of ischemia were summarized in Fig. 3.

After about 90 min of ischemia, another change was observed: a new peak emerged at the left side of the ordinary Pi peak (lower field or more alkaline side (pH about 7.0) than the peak of Pi (pH about 6.0)). The intensity of the new peak increased as ischemia lengthened, while the ordinary Pi peak decreased in proportion to the growth of the new peak (Figs 2 and 4) leaving the sum total intensity of the ordinary Pi and the new peaks constant. Whereas reperfusion after 3 h of ischemia resulted in a gradual decrease in the ordinary Pi peak, no change was observed in the new peak after reperfusion (Fig. 4). To identify the substance responsible for the new peak, 6% perchloric acid extract of the heart from which the new peak was recorded was subjected to ^{31}P-NMR examination. Only one Pi peak was noted in the extract (Fig. 5), indicating that Pi was responsible not only for the ordinary Pi peak, but also for the new peak (thus, it is the second Pi peak) which became manifest after 90 min of ischemia.

As is shown in Fig. 1A, the heart was surrounded by Krebs-Ringer solution of pH 7.4 (B I) to improve s/n ratio. As the second Pi peak developed at lower field, i.e., more alkaline side, than the ordinary Pi peak, we examined the effect of pH of the bathing solution on the pH of the second peak.

In the first series of the experiment, the bathing solution (K-R I) was removed upon induction of ischemia and moistened nitrogen gas was infused continuously (100 ml/min, Fig. 1B) into the NMR tube. The space between the tube and the heart was refilled with B I (pH 7.4, equilibrated by 95% O_2 and 5% CO_2) only during the period of observation of NMR spectra (for 3 min after 30 and 180 min). As shown in Fig. 6, this procedure prevented the development of the second Pi peak and the intensity of the first peak no longer decreased (n = 6). In the second series of the experiment, pH of the bathing solution was changed from 7.4 to 6.0 (from B I to B II in Table 1) after 30 min of ischemia at a time at which the decrease in intracellular pH was maximal (pH about 6.0). This procedure again inhibited the development of the second Pi peak and kept the intensity of the first Pi peak constant (Fig. 7).

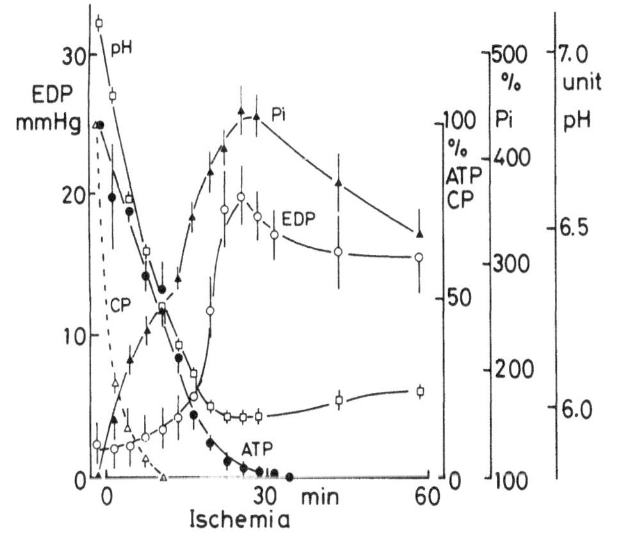

Fig. 3. Changes in intracellular pH, end diastolic pressure and high energy phosphates during 1 h of global ischemia. Results of 6 experiments are shown. EDP: end diastolic pressure of the left ventricle. CP: creatine phosphate. Pi: inorganic phosphate. ATP: adenosine triphosphate (as assessed using the signal of β-ATP).

Ischemic changes observed with ^{23}Na-NMR

Model experiments

In a model experiment conducted with a glass ball filled with various concentrations of NaCl and placed in NMR tube (the space between the ball and the wall of the NMR tube was filled with solution B I), addition of 10 mM of $Dy(TTHA)^{3-}$ to the bathing solution resulted in a good separation of the ^{23}Na peak of the bathing solution from that of the glass ball (about 4 ppm) and a good correlation was found between the intensity of the ^{23}Na spectra of the glass ball and the Na^+ concentration of the solution within it (r = 0.997).

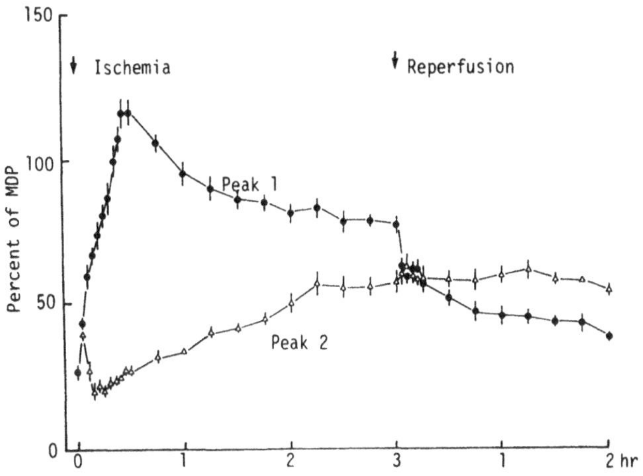

Fig. 4. Changes in intensities of the ordinary Pi peak and the second Pi peak during 3 h of ischemia and after reperfusion. Results of 6 experiments are shown. Peak 1: the ordinary Pi peak. Peak 2: a new (the second) peak found in the present study, which is clearly visible after about 90 min of ischemia (also see Fig. 2).

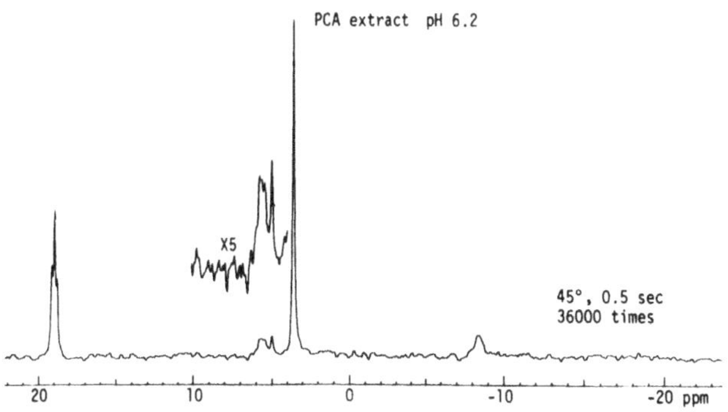

Fig. 5. ^{31}P-NMR spectrum taken from the PCA extract of the heart subjected to 3-h ischemia. Inset is an enlargement of the spectra between 4 and 10 ppm. Only one peak of Pi is obvious. A peak at about 19 ppm represents MDP, an external standard.

Experiments in isolated perfused hearts

Hearts were perfused with a Krebs-Ringer solution with 10 mM of the shift reagent (K-R II) and the space between the heart and the wall of the NMR tube was filled with a solution containing 10 mM of the shift reagent (B III in Table 1) by continuous infusion (6 ml/min) through a tubing the tip of which was advanced to the bottom of the NMR tube. The perfusate was sucked from the NMR tube with a tubing the orifice of which was just above the heart to exclude the interference by Na^+ signals originating from the perfusate effectively.

A typical serial record is shown in Fig. 8. Under control condition, the intracellular Na^+ spectrum was not visible due to a broadening of extracellular Na^+ spectrum. To get a distinct separation between the two peaks, the concentration of the shift reagent was increased which resulted in a decrease in developed tension of the left ventricle that could not be overcome with increase in Ca^{2+} in the perfusion medium. Induction of ischemia resulted in an increase in the intracellular Na^+ signal. The height of the signal increased as the time of ischemia lengthened (Fig. 8). Figure 9 summarized the relative changes in intracellular Na^+ concentration that occurred during the course of ischemia (n = 6). The increase in the intracellular Na^+ was biphasic, the beginning of the second phase corresponding to a rapid rise in the end diastolic pressure of the left ventricle (Fig. 9). To obtain the absolute value of Na^+, several hearts were subjected to determination of Na^+ content with atomic absorption spectrometer [1] after ischemia of 1 h. Extracellular Na^+ was washed out with ice-cold

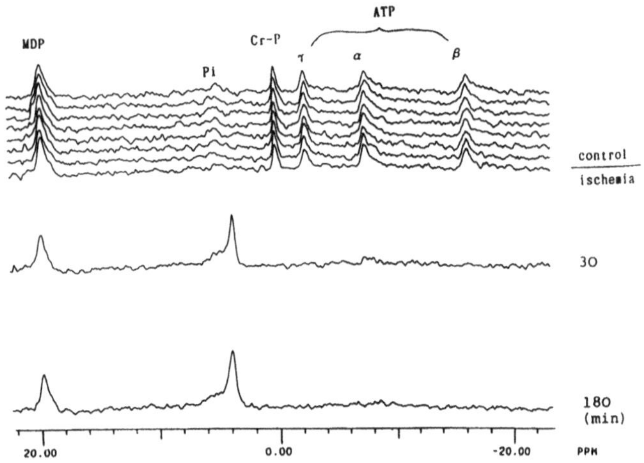

Fig. 6. A representative record of ^{31}P-NMR spectra during ischemia taken under the condition as described in Fig. 1B. MDP: methylene diphosphonate as an external standard. Pi: inorganic phosphate. Cr-P: creatine phosphate. ATP (α, β, γ): adenosine triphosphate. At 30 and 180 min of ischemia, the space between the wall of the NMR tube and the heart was filled with B I solution for 3 min to take ^{31}P-NMR spectra.

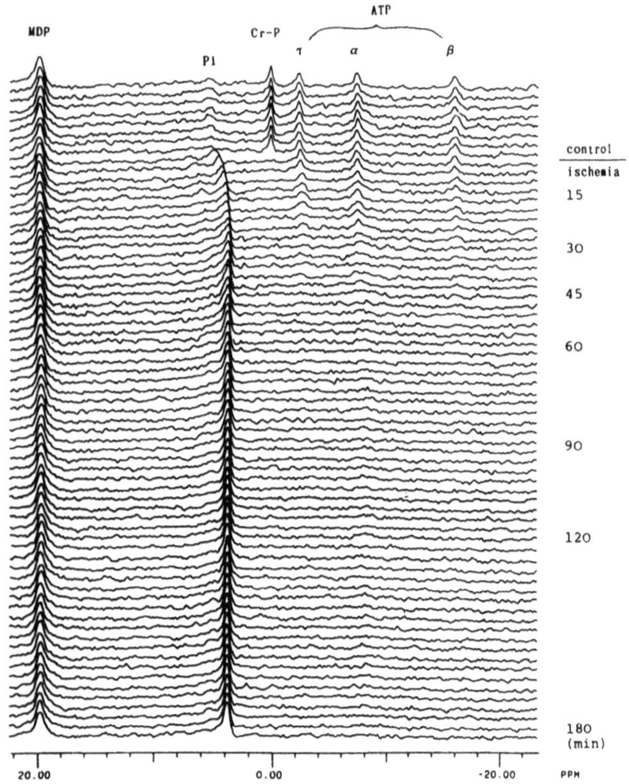

Fig. 7. A representative record of serial ^{31}P-NMR spectra taken during ischemia under the condition as described in Fig. 1C. MDP: methylene diphosphonate as an external standard. Pi: inorganic phosphate. Cr-P: creatine phosphate. ATP (α, β, γ): adenosine triphosphate. After 30 min of ischemia, bathing solution was changed from B I solution (pH 7.4) to B II solution (pH 6.0).

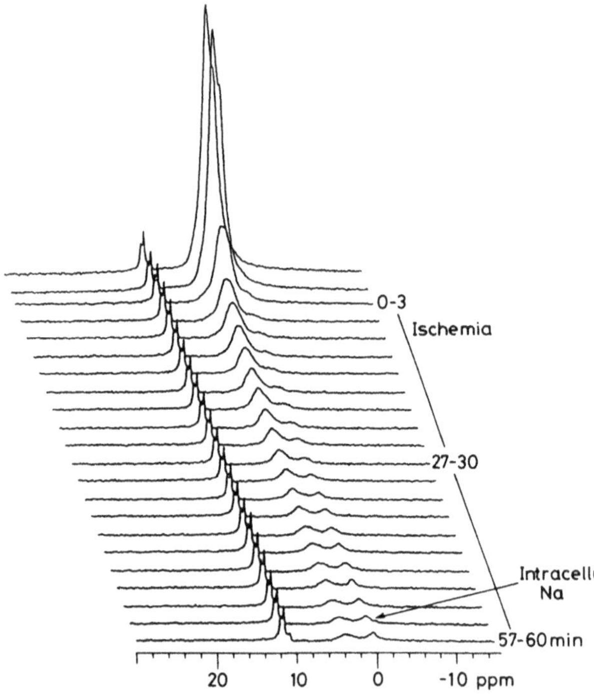

Fig. 8. A representative record of serial ^{23}Na-NMR spectra taken during ischemia. A spectrum indicated by an arrow represent intracellular Na^+. The spectrum at about 12 ppm represents an external standard.

sucrose-histidine buffer [13]. The value of Na^+ concentration in myocardium estimated using the values thus obtained was 11.17 μmole/ml after 1 h of ischemia.

Ischemic changes observed with ^{39}K-NMR

In this series of experiment, an inversion recovery pulse sequence was used to nullify the signal of extracellular K^+. The spectrum obtained from perfused heart with single pulse sequence (90°) was composed of a broad (dome shaped) spectrum and a sharp one. The sharp spectrum was susceptible to shift reagent in the perfusion medium, while the broad one was not (Fig. 10), indicating that the sharp one represented extracellular K^+. Having confirmed this, an inversion recovery pulse sequence (180°-PD-90°) was applied to estimate T_1s of extracellular and intracellular K^+. Real recording with pulse delay (PD) of 10 ms in Fig. 11 (upper figure) clearly showed a sharp cleavage in the middle of a dome-shaped spectrum. This implies that T_1 of extracellular K^+ is longer than that of intracellular one. The difference was confirmed by curve fit analysis (Simplex method) which revealed the T_1 of extracellular K^+ to be 45 msec and that of intracellular K^+ to be 1 msec (Fig.

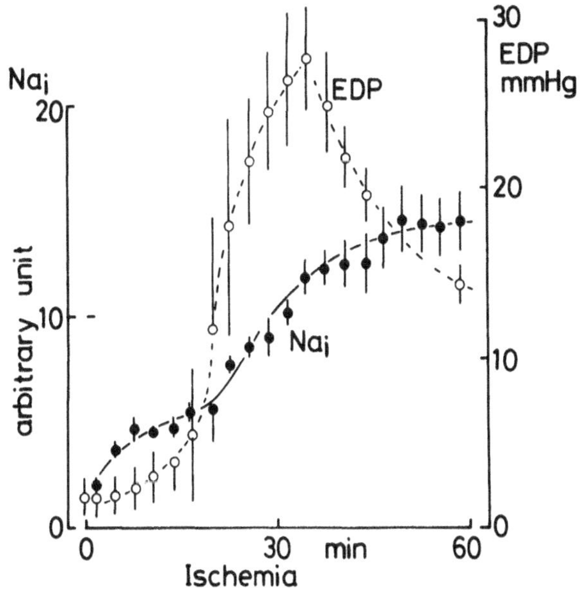

Fig. 9. Changes in intracellular Na$^+$ and end diastolic pressure during 1 h of ischemia. Results of 6 experiments are shown. EDP: end diastolic pressure of the left ventricle. Na$_i$: intracellular Na$^+$.

11, lower). Thus, pulse sequence necessary to nullify the extracellular K$^+$ was decided to be 180°-30 msec-90° (30 msec was derived from 45 msec × 0.69).

As shown in Fig. 12 (upper record), this sequence completely abolished the extracellular component of ^{39}K spectra. Spectra of similar intensities were obtained by omission of K$^+$ from the perfusion medium (K-R III in Table 1) by the use of both single pulse sequence and the inversion recovery pulse sequence (Fig. 12, lower).

^{39}K-NMR spectra before induction of ischemia and during the early ischemic period are shown in Fig. 12. It is obvious from the figure that the width of the spectra

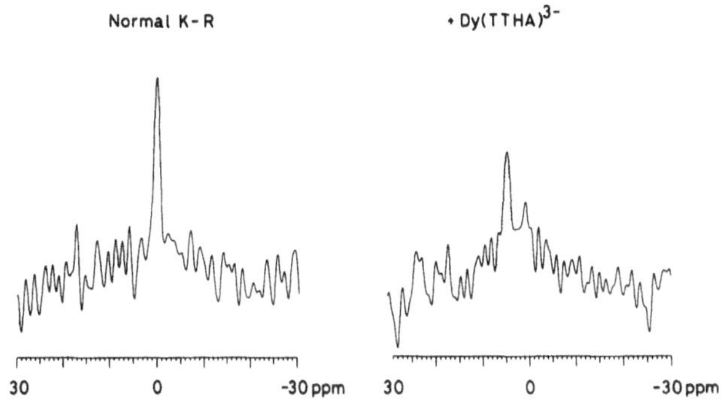

Fig. 10. ^{39}K-NMR spectra from normoxic perfused rat heart. Spectra were obtained with a single pulse sequence (90°, 2000 scans) under the condition of normoxic perfusion (Normal K-R: left figure) and in the presence of the shift reagent (+ Dy(TTHA)$^{3-}$: right figure). The shift reagent produced a shift of the sharp peak (dotted area).

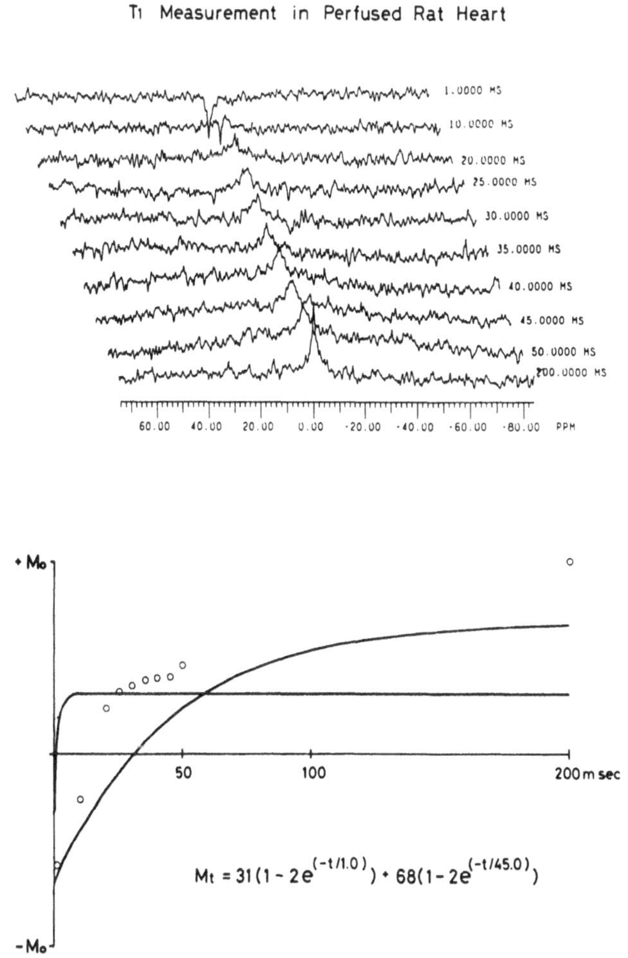

Fig. 11. Measurement of T$_1$ of K$^+$ in perfused rat heart. The upper panel represents a record of serial application of inversion pulse sequence (180°-PD-90°). PD (pulse delay) is indicated in the right of each tracing. Curve fit analysis of the spectra is shown in the lower figure. The height of each spectrum (open circle) is plotted against PD (in the horizontal axis), and the relation is analyzed according to the theoretical equation on the assumption that measured points are composed of two elements. Estimated relation is indicated as an equation and the two components are drawn in the figure. + Mo: positive height of the spectra. The maximum positive element (PD = 200 msec) was taken as 100%. − Mo: negative height of the spectra. The absolute value of the maximum positive element was taken as 100%.

decreased after ischemia. Summarized data (Fig. 13) revealed a transient, significant decrease in intracellular K$^+$ of about 25% of the control.

Discussion

Use of NMR enables us to record metabolic changes in the biological system in a continuous and non-invasive manner [14]. However, some technical problems re-

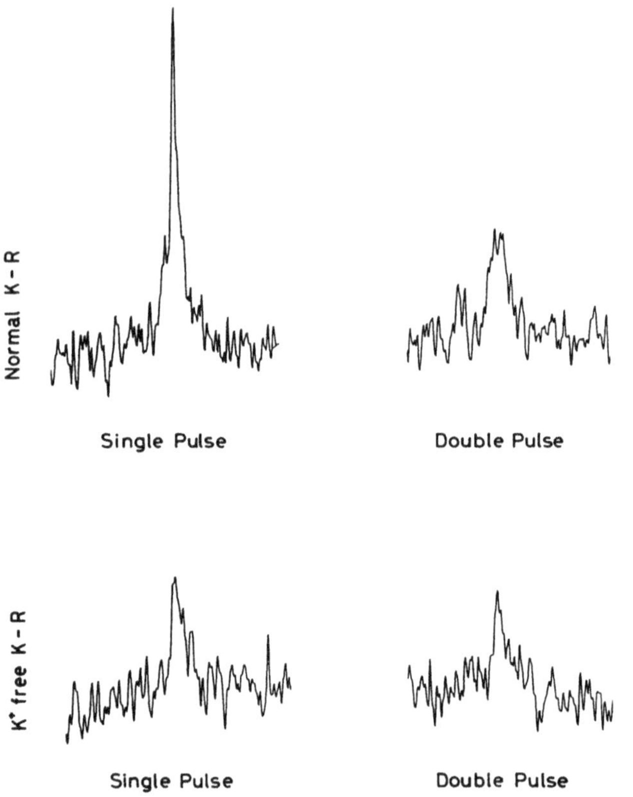

Fig. 12. Validation of inversion recovery pulse sequence used to nullify the extracellular K^+ signal. Upper figures: Spectra from the heart perfused with K-R I solution (Normal K-R) obtained with single pulse sequence (90°: Single pulse) and with inversion recovery pulse sequence (180°-30 msec-90°: Double pulse). Inversion recovery pulse sequence abolished extracellular K^+ signal (sharp component). Lower figures: The heart was perfused with K-R III solution (K^+ free K-R) and spectra were obtained with single (Single pulse) and inversion recovery pulse sequence (Double pulse). Similar signals were obtained with these two pulse sequences.

main unsolved [15]. For example, repetition time interval presently used with ^{31}P-NMR was not long enough to avoid saturation of signals. Yet, we had to chose this repetition time interval to increase the numbers of accumulation feasible within a limited time interval. This is necessary to overcome the problem of poor s/n ratio while catching up the rapid changes in metabolic parameters such as intracellular pH. Under the circumstances we gave up the idea of estimating the absolute values of ATP, CrP and Pi, although there were reports on the absolute values from other laboratories [16–18]. Assessment of the relative changes of these compounds seems to be sufficient to evaluate the ischemic derangement in question in the present study [5, 19–21]. Furthermore, in view of the possibility that T_1 of high energy phosphates may change during ischemia [22], we have to develop a different protocol to acquire NMR signals in the ischemic hearts, if the absolute values are to be obtained. Anyway, it is known that the values of high energy phosphates measured by NMR are not the same as those obtained by traditional method of measurement [16].

The concentration of $Dy(TTHA)^{3-}$ used as a shift reagent in the present study has been shown not to be toxic to the heart [12]. For example, it was shown not to produce any significant change in the ratio of intensities of the CrP to Pi (a sensitive indicator of the vitality of the heart [23]). However, broadening of the peaks brought about by the shift reagent presented a new problem in that it makes the separation of the peaks of intra- and extracellular Na^+ less satisfactory. To avoid the problem, we adopted a new method for observation of intracellular K^+ that does not require shift reagent [24–26]. Although it is in principle not impossible to use this procedure for observation of intracellular Na^+ [27], the error in estimation would be larger because of the fact that the intracellular Na^+ content is much lower than that of K^+. Furthermore, the procedure is not free from error if there is alteration of T_1 of intracellular ions by ischemic derangement.

In spite of the above problems, NMR is an important technique that cannot be dispensed with for observation of ischemic derangement because it can be used in *in vivo* (or near *in vivo*) conditions and is the only method that enables the recording of the time course of changes in the same preparation and thus to reduce the number of animals sacrificed, etc. Therefore, we used ^{31}P-, ^{23}Na- and ^{39}K-NMR to study the effect of ischemia on the myocardial ionic contents and energy metabolism in the perfused rat heart.

Observations with ^{31}P-NMR revealed a marked increase in Pi in association with rapid decreases in ATP and CrP during early phase of ischemia, indicating a severe oxygen deficiency even in the early phase of ischemia. Simultaneous decrease in intracellular pH as was indicated by a chemical shift of Pi may be ascribable mainly to the breakdown of ATP and the accumulation of lactate resulted from a shift to anaerobic metabolism. Though Gevers [28] has indicated no proton generation from lactate under circumstances of pH 7.2 where all ATP and ADP form complexes with Mg^{2+}, correction by Seelye [29] suggested that the production of lactate coupled with ATP regeneration system would produce enough proton to explain the observed fall in the intracellular pH. An abrupt increase in end diastolic pressure (EDP) which occurred after 15–20 min of ischemia

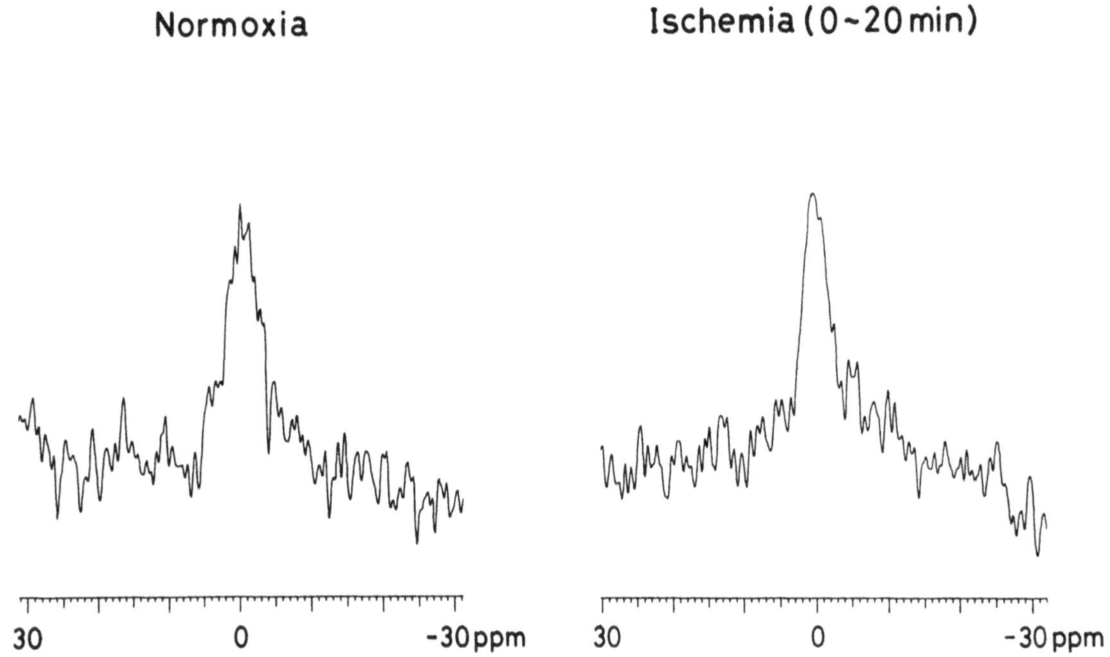

Fig. 13. ^{39}K-NMR spectra obtained from isolated perfused rat heart during control perfusion and during the early phase of ischemia. After ischemia, the peak of intracellular K^+ became slender.

corresponded to the depletion of ATP and a further rise of Pi. Thus, the phenomenon may be due to the formation of rigor complex as was indicated by Murray and Weber [30].

In the present study a new peak of Pi was observed after 60–90 min of ischemia when the heart was bathed in the solution of pH 7.4. When the heart was bathed in the solution of pH 6.0 (a pH similar to the intracellular pH at an early phase of ischemia) or when the bathing solution was withdrawn during ischemia, the new peak of Pi did not appear. The new peak differs from the second peak observed in an intact perfused rat heart and considered to represent mitochondrial phosphate peak [31] in the following two points: First, the position of the peak is different, i.e., the peak was from a compartment with different pH. Secondly, there is a difference in susceptibilities to pH changes of the external (bathing) solutions. The susceptibility of our second peak to pH of the external solution indicates that our second phosphate peak was the product of a special experimental condition used, i.e., the heart was bathed in a bathing solution to obtain a good resolution of spectral data (see methods). Reperfusion after 3 h of ischemia did not change the pH and the intensity of this second peak. These results suggest that the second Pi peak found in the present study is in a compartment that is severely damaged to allow free exchange of protons between the intra- and extracellular spaces, but was disconnected from the circulation. As such, it would serve as an experimental index of the advanced damage of myocardium.

In accordance with other reports [32, 33], the increase in intracellular Na^+ was observed during ischemia. However, the value before ischemia estimated in the present study (about 1/20 of that after 1 h of ischemia) was considerably lower than the values obtained with intracellular ion-selective microelectrodes [8] or with NMR by other researchers [34]. This discrepancy may be explained, at least in part, by wash out of the extracellular fluid inclusive of intracavity fluid, a possible major source of the extracellular ^{23}Na signal, with the method conducted in the present study as a preparatory step for determination of intracellular Na^+ content with atomic absorption spectrometer.

^{23}Na-NMR conducted in the present study indicated the biphasic nature of the increase in intracellular Na^+ during ischemia. The early increase (the first phase) corresponded to a rapid decrease in intracellular pH, indicating an activation of Na^+-H^+ exchange system [4, 35]. This exchange system is not operative under low-flow ischemic conditions, as we have previously demonstrated [5]. However, the system seems to be active under global ischemic conditions. The operation of this system in global ischemia [36] as well as on reperfusion [37] has already been reported. The rise of intracellular Na^+ paused when decrease in pH became stable (pH

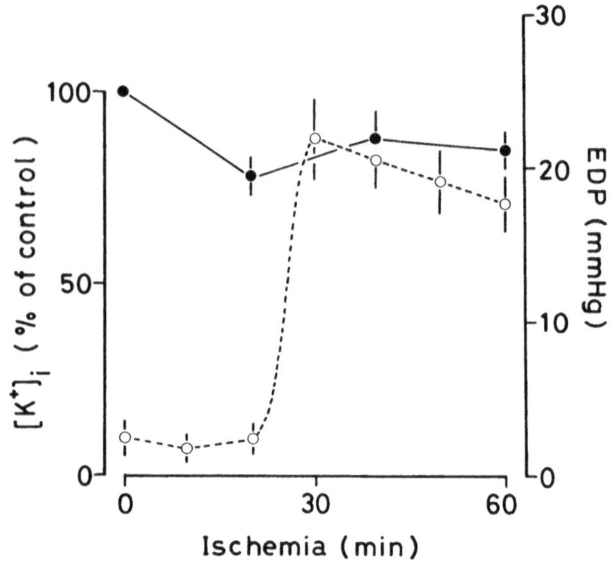

Fig. 14. Changes in intracellular K^+ and end diastolic pressure during 1 h of ischemia. Results of 6 experiments are shown. EDP: end diastolic pressure of the left ventricle. $[K^+]_i$: intracellular K^+.

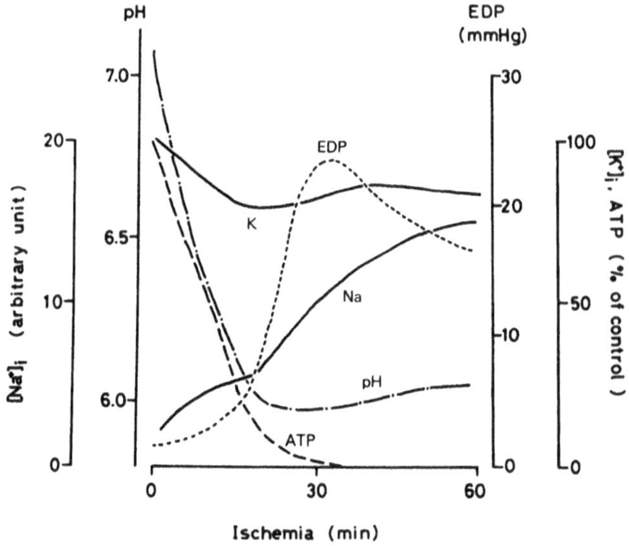

Fig. 15. Schematic diagram showing the ischemic changes in the ionic contents, pH and ATP of the isolated perfused rat hearts observed with ^{31}P-, ^{23}Na- and ^{39}K-NMR as related to the changes in ventricular end diastolic pressure.

about 6.0). This phenomenon would be explained by an inactivation of Na^+-H^+ exchange system by gradual acidification of extracellular fluid [4]. On the depletion of ATP, a second marked rise of Na^+ started, indicating an inhibition of Na^+-K^+ pump as the K_m of ATP for the pump is known to be very low (0.1 mM) [6].

Before depletion of ATP, intracellular K^+ decreased and stayed at that low level thereafter. This is in agreement with the results obtained by Kléber [2] as regards the extracellular K^+ accumulation. Incomplete inhibition of Na^+-K^+ pump [2] may account for the initial decrease because myocardial ATP-sensitive K^+ channel has been indicated not to be open at least during this period of ischemia [38]. According to Knopf *et al.*, [3] the plateau phase of K^+ accumulation at the later phase of ischemia is due to stimulation of Na^+-K^+ pump. Three possible mechanisms for the activation of Na^+-K^+ pump were proposed: 1) release of adrenaline during ischemia, 2) increase in intracellular Na and 3) depolarization of the membrane due to the increase in extracellular K^+. However, in view of a secondary large increase in intracellular Na^+ observed in the present study corresponding to this period, it would be reasonable to assume the activation of ATP-sensitive K^+ channel rather than to assume the activation of Na^+-K^+ pump. Ischemic changes in high energy phosphates, ions and end diastolic pressure found in the present study is summarized with a schematic drawing in Fig. 15.

In conclusion, the present experiments conducted in the isolated perfused rat heart with ^{31}P-, ^{23}Na- and ^{39}K-NMR demonstrated the enhancement of Na^+-H^+ exchange and the inhibition of Na^+-K^+ pump during the early phase of global ischemia for which a marked decrease in ATP is responsible. A rapid increase in end diastolic pressure observed at a mid phase of ischemia may probably be explained by rigor complex formed through depletion of ATP. Further increase in intracellular Na^+ as well as increase in end diastolic pressure may be taken to denote the irreversible damage of myocardium. After ischemia of a very long duration, the second Pi peak appeared, indicating a collapse of membrane barrier for free exchange of proton between the intracellular and extracellular spaces.

Acknowledgements

This work was supported in parts by a Grant-in-Aid for Co-operative Research (A) to S.I. (No. 60304061), a Grant-in-Aid for Scientific Research (B) to S.I. (No. 60480126) and a Grant-in-Aid for Encouragement of Young Scientists to T.I. (No. 63770123) from the Ministry of Education, Science and Culture in Japan.

References

1. Kawada T, Yoshida Y, Sakurai H, Imai S: Myocardial Na^+ during ischemia and accumulation of Ca^{2+} after reperfusion: A study with momensin and dichlorobenzamil. Jpn J Pharmacol 59: 191–200, 1992
2. Kléber AG: Extracellular potassium accumulation in acute myocardial ischemia. J Mol Cell Cardiol 16: 389–394, 1984
3. Knopf H, Theising R, Moon CH, Hirche H: Continuous determination of extracellular space and changes of K^+, Na^+, Ca^+, and H^+ during global ischaemia in isolated rat hearts. J Mol Cell Cardiol 22: 1259–1272, 1990
4. Lazdunski M, Frelin C, Vigne P: The sodium/hydrogen exchange system in cardiac cells: Its biochemical and pharmacological properties and its role in regulating internal concentrations of sodium and internal pH. J Mol Cell Cardiol 17: 1029–1042, 1985
5. Imai S, An-Yun S, Ishibashi T, Nakazawa M: Na^+/H^+ Exchange is not operative under low-flow ischemic conditions. J Mol Cell Cardiol 23: 505–517, 1991
6. Allen DG, Orchard CH: Myocardial contractile function during ischemia and hypoxia. Circ Res 60: 153–168, 1987
7. Steenbergen C, Deleeuw G, Rich T, Williamson JR: Effects of acidosis and ischemia on contractility and intracellular pH of rat heart. Circ Res 41: 849–858, 1977
8. Kléber AG: Resting membrane potential, extracellular potassium activity, and intracellular sodium activity during acute global ischemia in isolated perfused guinea pig hearts. Circ Res 52: 442–450, 1983
9. Ishibashi T, Nakazawa M, Imai S: Changes in myocardial energy metabolism and Na^+ permeability produced by a long ischemia as studied by ^{31}P- and ^{23}Na-NMR. J Mol Cell Cardiol 20 (suppl. I): S45, 1988 (Abstract)
10. Ishibashi T, Matsubara T, Nakazawa M, Imai S: A continuous noninvasive observation of the myocardial K^+ content during ischemia and reperfusion with ^{39}K-NMR without shift reagents. Jpn J Pharmacol 49 (Suppl): 124P, 1989 (Abstract)
11. Garlick PB, Radda G, Seeley J: Studies of acidosis in the ischaemic heart by phosphorus nuclear magnetic resonance. Biochem J 184: 547–554, 1979
12. Pike MM, Frazer JC, Dedrick DF, Ingwall JS, Allen PD, Springer Jr. CS, Smith TW: ^{23}Na and ^{39}K nuclear magnetic resonance studies of perfused rat hearts: Discrimination of intra- and extracellular ions using a shift reagent. Biophys J 48: 159–173, 1985
13. Shiga T, Nakazawa M, Imai S: Ventricular pressure-heart rate products before induction of ischemia as a determinant of the reperfusion-induced accumulation of calcium within myocardium. Jpn J Pharmacol 45: 379–387, 1987
14. Lundberg P, Harmsen E, Ho C, Vogel HJ: Nuclear magnetic resonance studies of cellular metabolism. Anal Biochem 191: 193–222, 1990
15. Ishibashi T, Imai S: Fundamental technique for analysis of NMR spectra (in Japanese). Med Philosoph 11: 87–93, 1992
16. Jeffrey FM, Storey CJ, Nunnally RL, Malloy CR: Effect of ischemia on NMR detection of phosphorylated metabolites in the intact rat heart. Biochemistry 28: 5323–5326, 1989
17. Murphy E, Gabel SA, Funk A, London RE: NMR observability of ATP: Preferential depletion of cytosolic ATP during ischemia in perfused rat liver. Biochemistry 27: 526–528, 1988
18. Clarke K, Willis RJ: Energy metabolism and contractile function in rat heart during graded, isovolumic perfusion using ^{31}P nuclear magnetic resonance spectroscopy. J Mol Cell Cardiol 19: 1153–1160, 1987
19. Matsubara T, Ishibashi T, Nakazawa M, Yamazoe M, Izumi T, Shibata A, Imai S: Effects of lidocaine on ischemic myocardial metabolism assessed by ^{31}P-NMR in the isolated perfused rat heart. Jpn Heart J 32: 493–504, 1991
20. Ishibashi T, Matsubara T, Nakazawa M, Katsumata N, Imai S: On the mode of cardioprotection produced by a new bradycardic agent, FR76830, during ischaemia and after reperfusion in the isolated perfused rat heart: A ^{31}P-NMR study. Cardiovasc Res 24: 1008–1012, 1990
21. Tanaka M, Ishibashi T, Imai S: Effects of two angiotensin converting enzyme inhibitors on the mechanical function and energy metabolism of isolated rat hearts: A nuclear magnetic resonance study with an active form of benazeprilat and captopril. Arzneim -Forsch 40: 1082–1086, 1990
22. Sako EY, Kingsley-Hickman PB, From AHL, Foker JE, Ugurbil K: ATP synthesis kinetics and mitochondrial function in the postischemic myocardium as studied by ^{31}P NMR. J Biochem Tokyo 263: 10600–10607, 1988
23. Ingwall JS: Phosphorus nuclear magnetic resonance spectroscopy of cardiac and skeletal muscles. Am J Physiol 242: H729–H744, 1982
24. Seo Y, Murakami M, Suzuki E, Watari H: A new method to discriminate intracellular and extracellular K by ^{39}K NMR without chemical-shift reagents. J Magn Res 75: 529–533, 1987
25. Seo Y, Murakami M, Suzuki E, Kuki S, Nagayama K, Watari H: NMR characteristics of intracellular K in the rat salivary gland: A ^{39}K NMR study using double-quantum filtering. Biochemistry 29: 599–603, 1990
26. Kuki S, Suzuki E, Watari H, Takami H, Matsuda H, Kawashima Y: Potassium-39 nuclear magnetic resonance observation of intracellular potassium without chemical shift reagents during metabolic inhibition in the isolated perfused heart. Circ Res 67: 401–405, 1990
27. Hotta Y, Ando H, Naruse M, Takeya K: Shift reagent free estimation of intracellular sodium content in the working heart by ^{23}Na NMR. Jpn J Pharmacol 52 (Suppl. I): 280P, 1990
28. Gevers W: Generation of protons by metabolic processes in heart cells. J Mol Cell Cardiol 9: 867–874, 1977
29. Seelye RN: Proton generation and control during anaerobic glycolysis in heart cells. J Mol Cell Cardiol 12: 1483–1486, 1980
30. Murray JM, Weber A: The cooperative action of muscle proteins. Scientific American 230: 58–71, 1974
31. Garlick PB, Brown TR, Sullivan RH, Ugurbil K: Observation of a second phosphate pool in the perfused heart by ^{31}P NMR: Is this the mitochondrial phosphate? J Mol Cell Cardiol 15: 855–858, 1983
32. Regan TJ, Broisman L, Haider B, Eaddy C, Oldewurtel HA: Dissociation of myocardial sodium and potassium alterations in mild versus severe ischemia. Am J Physiol 238: H575–H580, 1980
33. Pridjian AK, Levitsky S, Krukenkamp I, Silverman NA, Feinberg H: Intracellular sodium and calcium in the postischemic myocardium. Ann Thrac Surg 43: 416–419, 1987
34. Wittenberg BA, Gupta RK: NMR studies of intracellular sodium ions in mammalian cardiac myocytes. J Biol Chem 260: 2031–2034, 1985
35. Balschi JA, Frazer JC, Fetters JK, Clarke K, Springer CS, Smith

TW, Ingwall JS: Shift reagent and Na^{23} NMR discriminates between extra and intra cellular sodium pools in ischemic heart. Circulation 72: 355, 1985
36. Scholz W, Albus U, Linz W, Martorana P, Lang HJ, Schölkens BA: Effects of Na^+/H^+ exchange inhibitors in cardiac ischemia. J Mol Cell Cardiol 24: 731–740, 1992
37. Karmazyn M: The 1990 Merck Frosst Award. Ischemic and reperfusion injury in the heart. Cellular mechanisms and pharmacological interventions. Can J Physiol Pharmacol 69: 719–730, 1991
38. Sargent CA, Smith MA, Dzwonczyk S, Sleph PG, Grover GJ: Effect of potassium channel blockade on the anti-ischemic actions of mechanistically diverse agents. J Pharmacol Exp Ther 259: 97–103, 1991

Molecular and Cellular Biochemistry **119**: 121–127, 1993.

^{31}P-NMR magnetization transfer study of reperfused rat heart

Akira Kobayashi, Yoshimi Okayama, Noboru Yamazaki
The Third Department of Internal Medicine, Hamamatsu University School of Medicine, 3600, Handa-cho, Hamamatsu, 431-31, Japan

Abstract

The relationships between pressure rate product (PRP) and flux(PCr→ ATP) or flux(Pi→ ATP) were studied in isolated perfused rat hearts by the method of saturation transfer using ^{31}P-NMR during the preischemic and reperfusion periods. The hearts were made ischemic for 15 min, followed by 60 min of reperfusion. PRP was almost completely depressed, and recovered to 60% of the control level (preischemic period) after reperfusion. The ATP level during reperfusion was significantly decreased, whereas there was no significant change in PCr level. Pi level of reperfused hearts was significantly higher than that in the control. Both flux(PCr→ ATP) and flux(Pi→ ATP) were significantly decreased during the reperfusion period (both $p< 0.05$). However, the flux(PCr→ ATP)/PRP ratio during reperfusion did not differ from that of the control. This result indicates that the decrease in flux(PCr→ ATP) was matched by a similar decrease in cardiac performance. In contrast, the flux(Pi→ ATP)/PRP ratio during reperfusion was significantly decreased compared to that of control. These results suggest that the stunned heart needs less ATP turnover in proportion to its depressed contractile activity, and flux(Pi→ ATP) may limit the recovery of postischemic performance. (Mol Cell Biochem **119:** 121–127, 1993)

Key words: saturation transfer, nuclear magnetic resonance, reperfused heart, high-energy phosphate, ATP, creatine phosphate, inorganic phosphate, cardiac performance

Introduction

High-energy phosphate metabolism plays an important role in myocardial function, and its production by oxidative phosphorylation is directed toward contractile activity. Recently, Ambrosio *et al.* [1] showed a discrepancy between ATP levels and cardiac performance using nuclear magnetic resonance techniques (^{31}P-NMR). Neely and Grotyhann [2] also could find no relation between tissue ATP levels and postischemic recovery. Thus, it is now clear that measurements of tissue ATP levels cannot substitute for direct measurements of irreversible ischemic injury, such as ultrastructural changes or ischemic contracture.

The steady-state content of high energy phosphate, ATP plus creatine phosphate (PCr), presumably reflects the overall balance between metabolic supply and contractile utilization [3]. Mitochondrial oxidative phosphorylation synthesizes most of the ATP in the heart, and the creatine kinase reaction resynthesizes the ATP that is broken down during cardiac contraction [4]. Creatine kinase resynthesizes ATP so rapidly that direct measurement of the reaction rates in the heart has not

Address for offprints: A. Kobayashi, The Third Department of Internal Medicine, Hamamatsu University School of Medicine, 3600, Handa-cho, Hamamatsu, 431-31, Japan

been possible using conventional techniques. This limitation, however, has been partially overcome with the relatively recent expansion of nuclear magnetic resonance (NMR) spectroscopy to studies of intact cells and tissues [5].

The ^{31}P-NMB technique of magnetization transfer now provides the unique to study the kinetics of enzymes in situ, and it has been used to define the rate of ATP synthesis in isolated hearts by oxidative phosphorylation and from PCr via the creatine kinase reaction [6–8]. Bittl *et al.* [9] and Neubauer *et al.* [10] reported using this technique that flux (PCr→ ATP), reflecting the creatine kinase reaction, correlation with cardiac performance. Perry *et al.* [11] suggested that energy transfer plays an important role in the recovery of postischemic function.

In the normal heart, 90% of ATP is formed by oxidative phosphorylation while 10% is produced via glycolysis [12]. The glycolytic enzymes, glyceraldehyde-3-phosphate dehydrogenase and 3-phosphoglycerate kinase, together can catalyze Pi→ ATP exchange through the phosphorylated intermediate 3-phosphoglyceroyl phosphate and contribute to the NMR-measured Pi→ ATP rate [flux(Pi→ ATP)][13, 14]. However, no investigation of flux (Pi→ ATP) in the reperfused heart model has been reported. The present study investigated the exchange rates of flux(PCr→ ATP) and flux (Pi→ ATP) in ischemia/reperfused isolated rat hearts.

Materials and methods

1. Isolated perfused rat hearts

Male Wistar rats were anesthetized with pentobarbital (30 mg/kg, i.p.) and thoracotomy was performed. The hearts were rapidly excised and immediately immersed in an ice-cold buffer solution. The aorta was dissected free, and mounted onto a cannula, which was attached to a perfusion apparatus.

Retrograde perfusion of the heart was begun in the Langendorff mode under a constant perfusion pressure of 100 cmH_2O with modified Krebs-Henseleit buffer containing (mM/l): NaCl 118.0, KCl 4.7, $CaCl_2$ 2.5, $MgSO_4$ 1.2, KH_2PO_4 0.3, $NaHCO_3$ 25.0, Na_2EDTA 0.5, and glucose 10. The perfusate buffer was equilibrated with 95% O_2 and 5% CO_2. All experiments were performed at 37° C.

2. Experimental schedule

After an equilibrium period (45 min), total normothermic (37° C) ischemia was induced for 15 min followed by reperfusion for 60 min. Six rats were perfused with modified Krebs-Henseleit buffer solution. Twelve rats were infused with propranolol (0.4–11.2 μM) via the perfusion line for 30 min during the preischemic period, in order to change the pressure rate product. Indexes of cardiac performance of development pressure (DP), beating rate (BR), and pressure rate product (PRP) were measured 7 times; at 45 min, 15 min, and just before the start of ischemia, just after ischemia, just before the start of reperfusion, and 45 min, and 60 min after reperfusion. Since a complete saturation transfer measurement could be obtained in 30 min, saturation transfer measurements were carried out twice: once during the pre-ischemic period and again during the reperfusion period (from 30 min to 60 min).

3. Measurement of cardiac performance

A water-filled latex balloon was inserted into the left ventricle through the left atrium via the mitral valve. The balloon was connected to a pressure transducer (Nihon Koden, Japan) via a small-bore polyethylene tube for continuous measurement of left ventricular development pressure (LVEDP) and beating rate (BR), in order to observe changes in the pressure rate product (PRP).

4. ^{31}P-NMR spectra

^{31}P-NMR spectra of perfused hearts were obtained using a JEOL GSX 270 WBNMR spectrometer (Nihon Denshi, Japan) operating at 109.16 MHz. The probe consisted of a vertically-mounted Helmholtz coil (17 mm in diameter) tuned to the phosphorus frequency. Spectra were recorded using 45° pulses and a 1 sec repetition time, and were accumulated 200 times to obtain the steady-site concentrations of ATP, creatine phosphate (PCr), and inorganic phosphate (Pi). Quantification was performed by comparison of the peak height of interest with that of the external standard (hexamethylphosphoric triamide: 100 mM) within the radiofrequency coil. Peak height value of the external standard was defined as 100 U.

Figure 1 shows typical ^{31}P-NMR spectra obtained in

this study. Immediately following, 200 additional scans were again accumulated with selective irradiation applied at the γ-phosphate of ATP. The intensity of the PCr [PCr(a)] and Pi [Pi(a)] signals were then measured. Separate studies showed that the narrow-band pulse directly attenuated the magnetization of PCr and Pi by less than 5% when the carrier frequency was placed in a symmetrical position to the PCr or Pi signal.

5. *Saturation transfer measurement*

Saturation recovery experiments utilized of a 90° pulse in the presence of selective irradiation of the γ-P peak of ATP, followed by a variable time delay (0.3, 0.6, 1.2, 2.4 and 4.8 sec) and a 90° observation pulse (Fig. 1). This cycle was repeated 50 times. T_1 was calculated using T_1 analyzing software (JEOL). A complete saturation transfer measurement was acquired in 30 min in this study. Magnetization transfer was used to study the pseudo-first order rate constants ($\varkappa_1$ and $\varkappa_2$) and the forward fluxes of the Pi to ATP exchange reaction (Pi to γ-P of ATP) and the creatine kinase reaction in the direction of PCr to γ-P of ATP, respectively. The theoretical principles underlying this method have been described previously [6, 14–16]. Briefly, two equations were used to determine $\varkappa$:

$$M^+/M^0 = (1 + \varkappa T_1)^{-1} \quad (1)$$

$$1/\tau = 1/T_1 + \varkappa \quad (2)$$

where M^+ is the magnetization intensity of the Pi or PCr resonance in the presence of selective irradiation of the γ-P of ATP, and M^0 is the magnetization intensity of these resonances in the presence of a control irradiation. $1/T_1$ represents the intrinsic longitudinal relaxation rate (in the absence of chemical exchange), and $1/\tau$ is the measured or apparent longitudinal relaxation rate. Rate constants ($\varkappa$) for the forward and reverse reactions may potentially be derived from NMR data. The pseudo-first-order constants of $\varkappa_1$ ($\varkappa_{Pi \rightarrow \gamma-ATP}$) and $\varkappa_2$ ($\varkappa_{PCr \rightarrow \gamma-ATP}$) were calculated using the following equations:

$$\varkappa_1 (\varkappa_{Pi \rightarrow ATP} = \Delta M_{Pi}/M^0_{Pi} \times 1/\tau_{Pi} \quad (4)$$
$$\varkappa_2 (\varkappa_{PCr \rightarrow ATP}) = \Delta M_{PCr}/M^0_{PCr} \times 1/\tau_{PCr} \quad (5)$$

These equations are derived from equations (1) and (2). The exchange fluxes for the PCr→ ATP and Pi→ ATP reactions were then calculated by the equations:

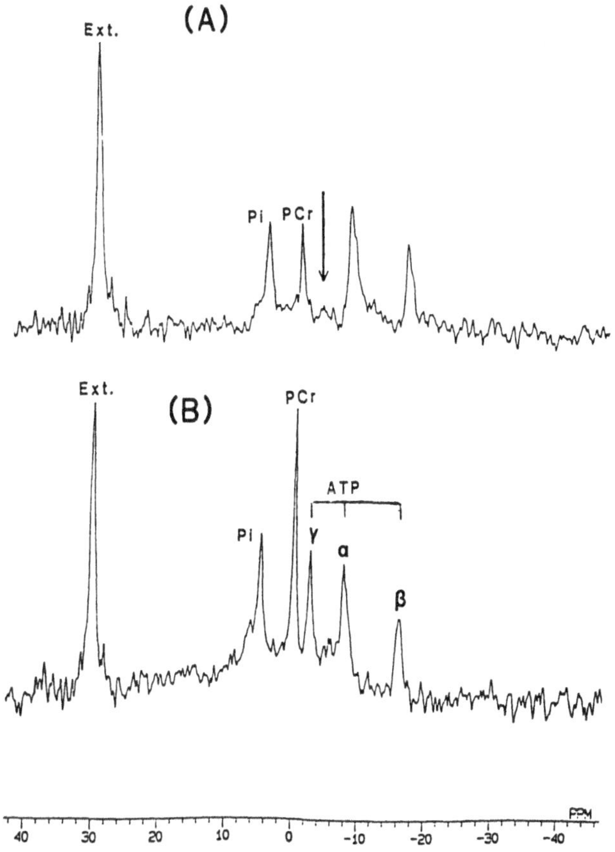

Fig. 1. ^{31}P-NMR spectra.
(A) represents the selective saturating irradiation at the γ-phosphate resonance of ATP. The arrow (↓) indicates the position of the saturating radiofrequency and the γ-P of ATP was made invisible.
(B) represents ^{31}P-NMR spectra of isolated perfused rat heart collected under isometric pressure at 100 cmH_2O perfusion pressure. Spectra were accumulated using 200 transients with a 45° pulse and 1.0 sec interpulse delay. Ext.: external standard peak.

$$\text{Flux} = \varkappa_1[\text{Pi}] \text{ or } \varkappa_2[\text{PCr}] \quad (6)$$

using the corresponding concentrations with the appropriate $\Delta M/M^0$ measurement.

6. *Statistical analysis*

All data are presented as mean ± SE. To test for significant differences between the control (preischemia) and reperfusion data, the paired t-test was used. The relation between flux(PCr→ ATP) or flux(Pi→ ATP) and PRP rate was tested using linear regression analysis. Values of $p < 0.05$ were considered to be statistically significant.

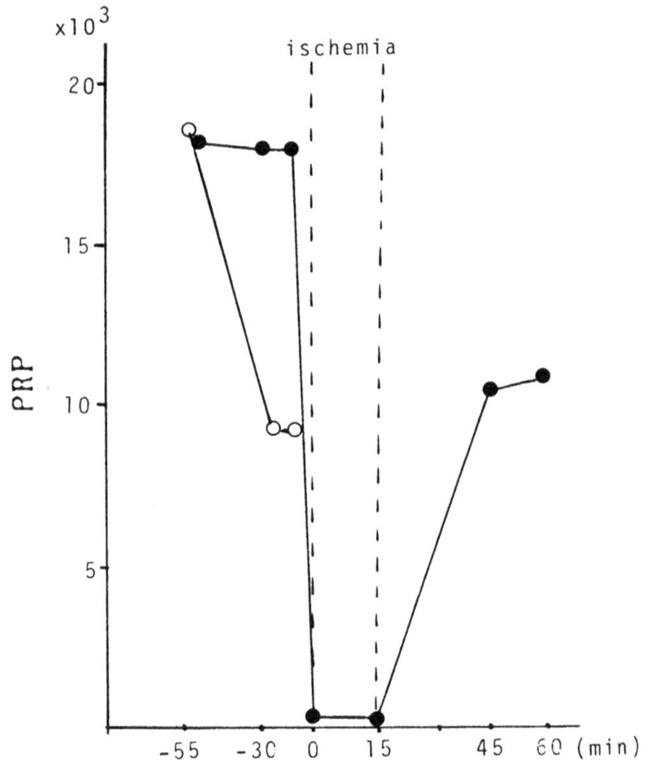

Fig. 2. Time course of changes in pressure rate product.
Closed circles represent the non-treated group, and open circles represent the propranolol-treated group. Propranolol was administered to change the pressure rate product during the preischemic period. Ischemia was induced for 15 min, followed by reperfusion for 60 min.

Results

1. Cardiac performance

Development pressure was significantly decreased during reperfusion. Propranolol-treated rats showed significant change in beating rate during the preischemic period. However, non-treated rats showed no significant change in beating rate. The pressure rate product (PRP) was significantly decreased during reperfusion. Thus, cardiac work (PRP) was almost completely depressed, and recovered to approximately 60% of the control level after reperfusion (Fig. 2).

2. Changes in phosphometabolites

The levels of phosphorous metabolites in preischemic and reperfused hearts are shown in Table 1. The ATP level during the reperfusion period was significantly lower than that in the preischemic period ($p < 0.05$). On the other hand, there was no significant change in PCr level. Pi level during the reperfusion period was significantly higher than that in the preischemic period ($p < 0.05$).

3. Flux(PCr→ ATP) and flux(Pi→ ATP)

Table 2 shows the data of flux(PCr→ ATP) and flux (Pi→ ATP) during the preischemic and reperfusion periods. Figures 3 and 4 show the relationship between flux(PCr→ ATP) or flux(Pi→ ATP) and PRP. Flux (PCr→ ATP) and flux(Pi→ ATP) during the reperfusion period were significantly decreased compared with preischemic values (both $p < 0.05$). Since flux(PCr→ ATP) and flux(Pi→ ATP) change to various degrees with cardiac work, it is necessary to investigate the relationship between flux(PCr→ ATP) or flux(Pi→ ATP) and cardiac performance (PRP).

The relationship between flux(PCr→ ATP) and PRP showed a close positive linear correlation during the preischemic and reperfusion periods ($Y = 1.1 \times 10^{-3}X + 8.5$, $r = 0.76$, $p < 0.05$), indicating that the decrease in flux(PCr→ ATP) was matched by a similar decrease in cardiac work during the preischemic and reperfusion periods. On the other hand, similar changes in flux(Pi→ ATP) were not observed between the preischemic and reperfusion periods. First, flux(Pi→ ATP) showed a strong correlation with PRP during the preischemic period ($Y = 0.44 \times 10^{-3}X + 3.6$, $r = 0.70$, $p < 0.05$). However, the value of flux(Pi→ ATP) during the reperfusion period was lower in proportion to PRP, with no linear correlation with PRP during the reperfusion period. Thus, flux(Pi→ ATP) during the reperfusion period did not match to cardiac work.

Table 1. Changes in phosphometabolites

	Preischemia	Reperfusion
ATP	41.7 ± 2.7 U	29.7 ± 1.1 U*
PCr	79.1 ± 3.6 U	78.1 ± 5.7 U
Pi	29.1 ± 2.4 U	37.3 ± 3.0 U*

Data represent mean ± SE.
* denotes statistically significant difference ($p < 0.05$).

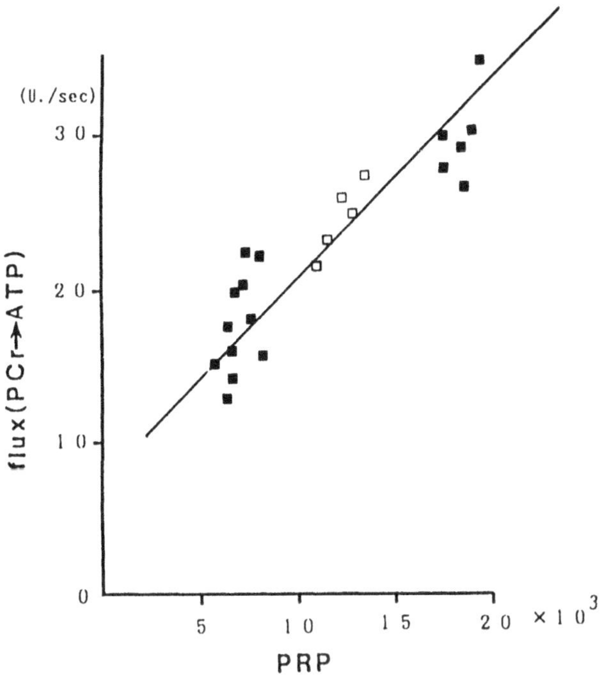

Fig. 3. Relationship between flux(PCr→ ATP) and cardiac work (PRP).

Closed squares represent flux(PCr→ ATP) during the preischemic period. Open squares represent flux(PCr→ ATP) during the reperfusion period. The relationship between flux(PCr→ ATP) and PRP shows a close positive linear correlation ($Y = 1.1 \times 10^{-3}X + 8.5$, $r = 0.76$, $p < 0.05$).

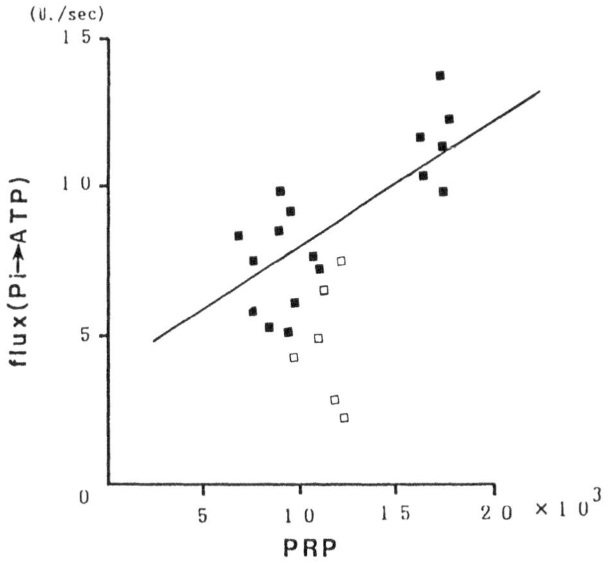

Fig. 4. Relationship between flux(Pi→ ATP) and cardiac work (PRP).

Closed squares represent flux(Pi→ ATP) during the preischemic period. Open squares represent flux(Pi→ ATP) during the reperfusion period. Flux(Pi→ ATP) during the preischemic period shows a close positive linear correlation to PRP ($Y = 0.44 \times 10^{-3}X + 3.6$, $r = 0.70$, $p < 0.05$), whereas values during the reperfusion period are lower in proportion to PRP.

Discussion

Recovery of postischemic function may be limited by energy production (oxidative phosphorylation), energy utilization (actomyosin ATPase), or energy transfer (via creatine kinase) [10]. Mitochondrial oxidative phosphorylation produces most of the ATP in the heart, and the creatine kinase reaction resynthesizes the ATP that is broken down during cardiac contraction. The steady-state content of high energy phosphate, ATP plus PCr, presumably reflects the overall balance between metabolic supply and contractile utilization [3]. As ATP is utilized, creatine kinase catalyzes the transfer of a high-energy phosphate group from creatine to ADP to ensure a constant supply of ATP. In the normal heart, 90% of ATP is formed by oxidative phosphorylation in the mitochondria, while 10% is produced via glycolysis [12]. The saturation transfer technique has been used to study the forward flux of the Pi to ATP exchange reaction [flux(Pi→ ATP)] and the creatine kinase reaction in the direction from PCr to ATP [flux(PCr→ ATP)]. ATP synthesized in the mitochondria is used preferentially for creatine phosphorylation, and this PCr is subsequently used to produce the ATP that is cleaved by the myofibrillar ATPase. Therefore, flux(PCr→ ATP) reflects the rate of ATP production through oxidative phosphorylation in the mitochondria. Flux(Pi→ ATP), however, reflects the rate of ATP production through the glycolytic pathway generated by the activation of both glycogenolysis and glycolysis [16]. Glyceraldehyde-3-phosphate dehydrogenase and phosphoglycerate kinase can catalyze significant Pi to ATP exchange. These enzymes are key regulatory enzymes in glycolysis [17].

Table 2. Flux(PCr→ ATP) and Flux(Pi→ ATP)

		Preischemia	Reperfusion
T_1	PCr	2.1 ± 0.1 sec	2.1 ± 0.2 sec
	Pi	0.8 ± 0.1 sec	0.7 ± 0.2 sec
flux(PCr→ ATP)		27.9 ± 1.5 U/sec	23.0 ± 2.4 U/sec*
flux(Pi→ ATP)		10.5 ± 1.1 U/sec	4.5 ± 1.6 U/sec*

Results are expressed as mean ± SE.
* denotes statistically significant difference from the preischemic group ($p < 0.05$).

We observed a close correlation between cardiac performance (PRP) and flux(PCr→ ATP) during the preischemic period. This was interpreted as evidence for coupling of energy transfer, production, and utilization under normoxic conditions. This result is in agreement with the results of Perry *et al.* [11] and Neubauer *et al.* [10]. A 15 min period of ischemia followed by 60 min of reperfusion produced mild, reversible injury, or stunning. PCr was close to preischemic control levels, while ATP was significantly decreased and inorganic phosphate (Pi) was significantly increased. Cardiac performance was almost completely depressed, and recovered to approximately 60% of the control level after reperfusion. Thus, we examined myocardium that was stunned (15 min of ischemia). We observed a close correlation between cardiac performance and flux(PCr→ ATP) during the reperfusion period as well as in the preischemic period. This indicates that the decrease in cardiac performance was matched by a similar decrease of flux(PCr→ ATP). That is, stunned myocardium requires less ATP turnover to maintain its depressed contractile activity. Neubauer *et al.* [10] also reported that flux(PCr→ ATP) correlates with cardiac performance. Bittl *et al.* [9] demonstrated that in the isolated rabbit heart reperfused after 10–60 min of ischemia, mitochondria creatine kinase activity is lost in proportion to the duration of ischemia, and that recovery of postischemic function correlated well with mitochondrial creatine kinase activity. In addition, Perry *et al.* [11] showed that mitochondrial creatine kinase activity correlated with flux(PCr→ ATP) velocity. These results are consistent with the hypothesis that energy production and transfer are compromised and limit mechanical function in the postischemic heart, and then indicate that flux(PCr→ ATP) plays an important role in the recovery of postischemic function. On the other hand, Neubauer *et al.* [10] observed no decrease of mitochondrial creatine kinase activity after 20, 40, or 60 min of ischemia, and have concluded that flux(PCr→ ATP) cannot be limiting in the recovery of function in the mildly or severely injured postischemic myocardium.

In addition, we measured the unidirectional exchange flux(PCr→ ATP) during the preischemic and reperfused periods, and observed a close correlation between cardiac performance (PRP) and flux(Pi→ ATP) during the preischemic period. On the contrary, there was no correlation between cardiac performance and flux(Pi→ ATP) during the reperfusion period. Flux (Pi→ ATP)/PRP ratio significantly decreased during the reperfusion period. Mitochondrial ATP production and glycolysis are important energy sources for cardiac contraction. Flux(Pi→ ATP) reflects the rate of ATP production through a variety of components of this process [17]. Seymour *et al.* [16] demonstrated that flux(Pi→ ATP) was markedly elevated in the hyperthyroid state, and suggested that this increased exchange reflects an increased flux through the glycolytic pathway generated by the activation of both glycogenolysis and glycolysis. Phosphoglycerate kinase is the rate-limiting enzyme in the Pi→ ATP exchange reaction [14]. Since GAPDH and PGK, the key regulatory enzymes in glycolysis, are the rate-limiting enzymes in the Pi→ ATP exchange reaction, the reduction of flux(Pi→ ATP) is interpreted as a reflection of decreased glycolytic flux [16]. Therefore, this result indicates that glycolytic flux (glycolytic ATP production) was significantly depressed in the stunned myocardium, and corresponds well to previous reports that mitochondrial ATP production and glycolysis are disturbed by ischemia/reperfusion, and that treatment with propranolol or diltiazem can protect against reperfusion injury of the myocardium [18, 19].

The mechanisms responsible for the depression of postischemic function (stunning) have not been identified. Proposed mechanisms include impaired energy production, utilization, or transfer, abnormalities of electromechanical coupling related to Ca^{2+} metabolism, activation of phospholipases, and free radical formation [21]. Another mechanism, changes in coronary flow or perfusion pressure due to occlusion of small vessels by postischemic edema formation (no-reflow phenomenon) [20], with increased Pi[22] and decreased pH, can contribute to depression of cardiac function [23]. Glycolytically-produced ATP is thought to be preferentially used for pumping sodium, potassium, and calcium ions across the sarcolemmna [24]. Therefore, the decrease in flux(Pi→ ATP) may reduce the activity of the Na^+/K^+-ATPase and ATP-dependent Ca^{2+} pumps, which cause Ca^{2+} extrusion, resulting in Ca^{2+} overload in the reperfused myocardium. In addition, decreased subsarcolemmal ATP may help to open ATP sensitive potassium channels and thereby also act to inhibit contraction [25]. There is considerable evidence that increased concentrations of inorganic phosphate (Pi) inhibit the contractile force by decreasing the number of cross-bridges in the force-generating state [26]. Pi levels significantly increased during the reperfusion period in the present study.

In conclusion, we investigated the relationship between flux(PCr→ ATP) or flux(Pi→ ATP) and cardiac performance (PRP) during the preischemic and reperfusion periods. Flux(PCr→ ATP) showed a close positive correlation with cardiac performance, whereas the decrease in velocity of flux(Pi→ ATP) was lower in proportion to the decrease of cardiac performance during the reperfusion period. These results suggest that the stunned heart needs less ATP turnover to maintain its depressed contractile activity, and flux(Pi→ ATP) may limit recovery of postischemic performance.

References

1. Ambrosio G, Jacobus WE, Bergman CA, Weismann HF, Becker LC: Preserved high energy phosphate metabolic reserve in globally 'stunned' hearts despite reduction of basal ATP content and contractility. J Mol Cell Cardiol 19: 953–964, 1987
2. Neely JR, Grotyohann LW: Role of glycolytic products in damage to ischemic myocardium. Dissociation of adenosine triphosphate levels and recovery of function of reperfused ischemic hearts. Circ Res 55: 816–824, 1984
3. Zweier J, Jacobus WE: Substrate-induced alterations of high energy phosphate metabolism and contractile function in the perfused heart. J Biol Chem 262: 8015–8021, 1987
4. Bessman SP, Geiger PJ: Transport of energy in muscle: The phosphorylcreatine shuttle. Science 211: 448–452, 1981
5. Foresen S, Hoffman RA: Study of moderately rapid chemical exchange reactions by means of nuclear magnetic double resonance. J Chem Phys 39: 2892–2901, 1963
6. Nunnally RL, Hollis DP: Adenosine triphosphate compartmentation in living hearts: A phosphorus nuclear magnetic resonance saturation transfer study. Biochem 18: 3642–3646, 1979
7. Degani H, Laughlin M, Campbell S, Shulman RG: Kinetics of creatine kinase in heart: A ^{31}P NMR saturation- and inversion-transfer study. Biochem 24: 5510–5516, 1985
8. Bittl JA, Ingwall JS: Reaction rates of creatine kinase and ATP synthesis in the isolated rate heart. A ^{31}P NMR magnetization transfer study. J Biol Chem 260: 3512–3517, 1985
9. Bittl JA, Baischi JA, Ingwall JS: Effects of norepinephrine infusion on myocardial high-energy phosphate content and turnover in the living rat. J Clin Invest 79: 1852–1859, 1987
10. Neubauer S, Hamman BL, Perry SB, Bittl JA, Ingwall JS: Velocity of the creatine kinase reaction decreases in postischemic myocardium: A ^{31}P-NMR magnetization transfer study of the isolated ferret heart. Circ Res 63: 1–15, 1988
11. Perry SB, McAuliffe J, Balschi JA, Hickey PR, Ingwall JS: Velocity of the creatine kinase reaction in the neonatal rabbit heart: Role of mitochondrial creatine kinase. Biochem 27: 2165–2172, 1988
12. Kobayashi K, Neely JR: Control of maximum rates of glycolysis in rat cardiac muscle. Circ Res 44: 166–175, 1979
13. Matthews PM, Bland JL, Gadian DG, Radda GK: A ^{31}P-NMR saturation transfer study of the regulation of creatine kinase in the rat heart. Biochim Biophys Acta 721: 312–320, 1982
14. Brindle KM, Radda GK: ^{31}P-NMR saturation transfer measurements of exchange between Pi and ATP in the reactions catalysed by glyceraldehyde-3-phosphate dehydro genase and phosphoglycerate kinase *in vitro*. Biochim Biophys Acta 928: 45–55, 1987
15. Ugurbil K, Petein M, Maidan R, Michurski S, From AHL: Measurement of an individual rate constant in the presence of multiple exchanges: Application to myocardial creatine kinase reaction. Biochem 25: 100–107, 1986
16. Seymour AML, Eldar H, Radda GK: Hyperthyroidism results in increased glycolytic capacity in the rat heart. A ^{31}P-NMR study. Biochim Biophys Acta 1055: 107–116, 1990
17. Brindle KM: ^{31}P NMR magnetization-transfer measurement of flux between inorganic phosphate and adenosine 5'-triphosphate in yeast cell genetically modified to overproduce phosphoglycerate kinase. Biochem 27: 6187–6196, 1988
18. Diederichs F, Wittenberg H, Sommerfield U: Myocardial cell damage and breakdown of cation homeostasis during conditions of ischaemia and reperfusion, the oxygen paradox, and reduced extracellular calcium. J Clin Chem Clin Biochem 28: 139–148, 1990
19. Piper HM, Sezer O, Schleyer M, Schwartz P, Hutter JF, Spieckermann PG: Development of ischemia-induced damage in defined mitochondrial subpopulations. J Mol Cell Cardiol 17. 885–896, 1985
20. Apstein CS, Deckelbaum L, Hagopian L, Hood WB: Acute ischemia and reperfusion: Contractility, relaxation and glycolysis. Am J Physiol 235: H637–H648, 1978
21. Bolli R: Mechanism of myocardial 'stunning'. Circulation 82: 723–738, 1990
22. Kusuoka H, Koretune Y, Chacko VP, Weisfeldt ML, Marban E: Excitation-contraction coupling in postischemic myocardium. Circ Res 66: 1268–1276, 1990
23. Fabiato A, Fabiato F: Effects of pH on the myofilaments and the sarcoplasmic reticulum of skinned cells from cardiac skeletal muscles. J Physiol 276: 233–255, 1978
24. MacDonald TF, Hunter EG, Macleod DP: ATP partition in cardiac muscle with respect to transmembrane electrical activity. Pflugers Arch 322: 95–108, 1971
25. Niki I, Ashcroft FM, Ashcroft SJH: The dependence on intracellular ATP concentration of ATP-sensitive K-channels and of Na^+,K^+-ATPase in intact HIT-TI5 beta-cells. FBBS Lett 257: 361–364, 1989
26. Herzig JW, Peterson JW, Ruegg JC, Solaro RJ: Vanadate and phosphate ions reduce tension and increase cross-bridge kinetics in chemically skinned heart muscle. Biochim Biophys Acta 672: 191–196, 1981.

Molecular and Cellular Biochemistry **119**: 129–132, 1993.

Release of 6-KETO-PGF$_{1\alpha}$ and thromboxane B$_2$ in late appearing cardioprotection induced by the stable PGI analogue: 7-OXO-PGI

László Szekeres[1] and Árpád Tósaki[2,*]

*Institute of Pharmacology, Albert Szent-Györgyi Medical University, Szeged, Hungary[1] and Institut Henry Beaufour, Les Ulis, France[2] * Present address: University of Connecticut Health Center, Farmington, USA*

Abstract

We have shown earlier that prostacyclin (PGI_2) and its stable analogue: 7-oxo-prostacyclin(7-OXO) may induce a prolonged, late appearing (24–48 h after drug administration), dose dependent protection of the heart from harmful consequences of a subsequent severe ischaemic stress, such as myocardial ischaemia, life-threatening ventricular arrhythmias and early ischaemic morphological changes. In an other study we observed that a similar but shortlived (less than 1 h) cardioprotection, induced by 'preconditioning' brief coronary artery occlusions, is greatly reduced by blockade of the cyclooxygenase pathway, suggesting that prostanoids might play a role in this shortlasting protection.

Objective of our present study was to elucidate the importance of some arachidonic acid (AA) metabolites, such as PGI_2 and thromboxane A_2 (TXA_2) in the mechanism of the late appearing, prolonged cardioprotection. Estimation of the metabolites: 6-keto-PGF$_{1\alpha}$ (6-KETO) and thromboxane B_2 (TXB_2) was made from the perfusate of isolated Langendorff hearts of guinea-pigs pretreated with 50 μg/kg 7-OXO, 24 and 48 h before preparation. Pretreatment alone produced a slight, but significant elevation of 6-KETO (from 206 ± 11 to 284 ± 19 pg/ml/min after 24 h, and to 261 ± 18 pg/ml/min after 48 h). No change was seen in TXB_2 production. Global ischaemia for 25 min (followed by 25 min reperfusion) markedly increased the release of both AA metabolites; maximal values were observed in the third min of reperfusion (6-KETO from 206 ± 11 to 1275 ± 55 pg/ml/min and TXB_2 from 29 ± 4 to 172 ± 12 pg/ml/min). All values returned to the preischaemic level by the 25th min of reperfusion. Ischaemic increase in 6-KETO level was significantly higher in the perfusate of hearts from pretreated animals (1507 ± 73 pg/ml/min after 24 h, and 1398 ± 54 pg/ml/min after 48 h) that in those of untreated controls. There was no difference in TXB_2 values. Thus both basal and ischaemic release of PGI_2 increased 24 and 48 h after pretreatment with 7-OXO but not TXA_2 production. Results suggest that endogenous prostanoids might play a role in late appearing cardioprotection. (Mol Cell Biochem **119:** 129–132, 1993)

Key words: late appearing cardioprotection, PGI_2, TXA_2, ischaemia, reperfusion

Introduction

To protect from consequences of myocardial ischaemia and/or reperfusion is an old ambition of cardiovascular research. In the last years even more attention was focussed on this problem, particularly when it was recognized that short periods of ischaemic stress may initiate in the heart mechanisms protecting against the

Address for offprints: L. Szekeres, Institute of Pharmacology, A. Szent-Györgyi Medical University, Dóm tér 12, H-6701, Szeged, Hungary

harmful consequences of a subsequent, more severe ischaemic stress. The initial ischaemic stimulus, called 'preconditioning' [1] induces protective (e.g. antiarrhythmic) mechanisms which last for about 30 min in rats [2] and less than 1 h in dogs [3]. It was shown [4] that the protective effect is greatly reduced after blockade of the cyclooxygenase pathway by meclophenamate. Thus this short-term cardioprotection can be partly attributed to the release of prostanoids, such as prostacyclin (PGI_2).

In 1983 we have described a later appearing and longer-lasting protection of the heart after administration of PGI_2 or of its stable analogue: 7-oxo-prostacyclin (7-OXO [5]. We found protection against the consequences of prolonged ischaemia, such as ischaemic ST-segment elevation after coronary artery occlusion, early postocclusion and reperfusion arrhythmias in anaesthetised dogs and late arrhythmias in conscious dogs. Moreover the postischaemic rise of the left ventricular end-diastolic pressure (LVEDP) in dogs and isolated working guinea-pig hearts was dose- and time dependent, with optimal protection 24 and 48 h after 7-OXO administration. Under similar conditions 7-OXO prolonged ventricular effective refractory period (VERP) and repolarisation-time (QT) in normoxic conscious rabbits and guinea-pigs, and in anaesthetised dogs [Reviewed in 6].

The aim of the present study was to elucidate the importance of some arachidonic acid (AA) metabolites, such as PGI_2 and thromboxane A_2 (TXA_2) in the mechanism of late appearing, prolonged cardioprotection. Therefore the stable metabolites of PGI_2 and TXA_2 namely 6-KETO and TXB_2 were estimated from the perfusate of isolated Langendorff guinea-pig hearts. The animals were pretreated with 7-OXO, 24 and 48 h before preparation. A part of the isolated heart preparations was subjected to global ischaemia, followed by reperfusion.

Methods

Animals

27 male guinea-pigs of 370–420 g body weight were used for all studies.

Perfusion technique and medium

Guinea-pigs were heparinized with 500 U disodium heparin i.p. After five minutes the animals were killed with a blow on the head and the heart excised and placed in cold (4° C) medium until contractions had ceased (usually within 15 s). Each heart was then cannulated via the aorta and perfused by the Langendorff method at a constant perfusion pressure equivalent to 100 cm water (10 kPa). Bicarbonate buffer containing 11.1 mmol glucose, 118 mmol NaCl, 5.8 mmol KCl, 2.5 mmol $CaCl_2$, 25 mmol $NaHCO_3$, 1.2 mmol NaH_2PO_4 and 1.2 mmol $MgCl_2$ was the perfusion fluid. This was gassed with 95% oxygen and 5% carbon dioxide (pH = 7.4 at 37° C). Before use, all solutions were filtered through a 5-μm porosity filter to exclude any particulate contaminants.

Induction of ischaemia and reperfusion

After 10-min aerobic perfusion of the heart, the aortic inflow line was clamped at a point close to the origin of the aortic cannula. Global ischaemia was maintained for 25 min. Then reperfusion was initiated by unclamping and removing the occluder. Duration of the reperfusion period was 25 min. To prevent the myocardium from drying out during normothemic global ischaemia, the thermostated glassware (in which hearts were suspended) was covered and vapor content was kept at a constant level (between 90 and 100%).

Measurement of stable metabolites of PGI_2 and TXA_2

Coronary effluents in 1-min aliquots sampled before starting global ischaemia, then at the end of the 1., 3., 5., 7., 10., and 25.min. of reperfusion were collected in an ice bath and then extracted for determination of 6-KETO, a stable metabolite of PGI_2, and TXB_2, a stable metabolite of TXA_2. The perfusate collected was immediately applied to Sep-Pak C_{18} columns (Waters), and 6-KETO and TXB_2 metabolites were eluted with 80% ethanol/20% distilled water. Samples were evaporated in a Speed Vac concentrator and 6-KETO and TXB_2 were determined by radioimmunassay (RIA, New England Nuclear, Stevenage, UK).

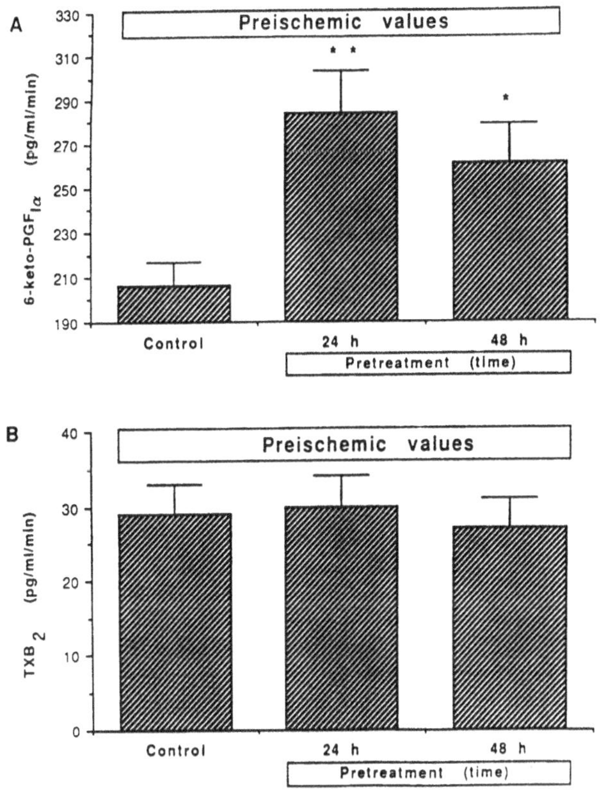

Fig. 1. Effect of 7-OXO pretreatment on AA metabolites in the nonischaemic heart. 6-KETO (panel A) and TXB_2 (panel B) in the coronary effluent of aerobically perfused, isolated guinea-pig hearts. Samples are from the perfusate of hearts from non-treated animals, and from those 24 and 48 h after pretreatment with 7-OXO. Ordinates are 6-KETO and TXB_2 levels in pg/ml/min.

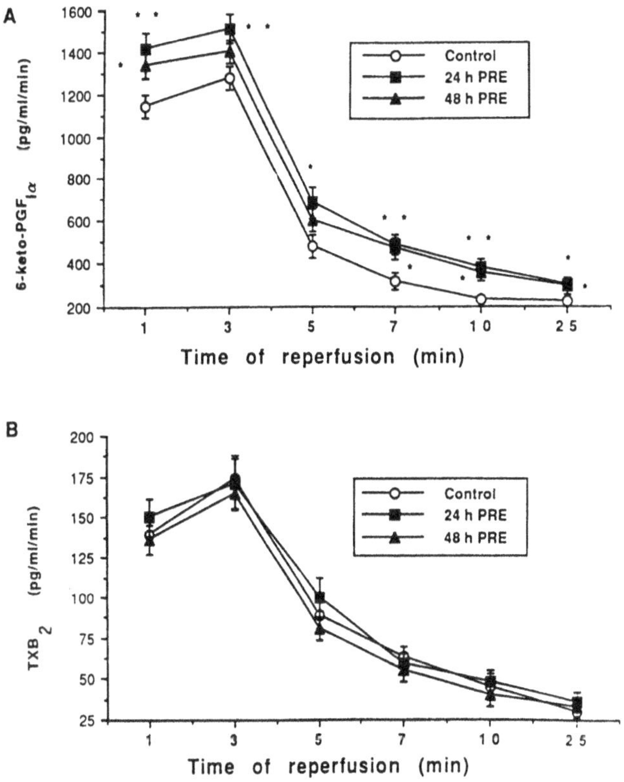

Fig. 2. Effect of 7-OXO pretreatment on AA metabolites during reperfusion. 6-KETO (panel A) and TXB_2 (panel B) levels in the coronary effluent of hearts of untreated, control animals and from those 24 and 48 h after pretreatment with 7-OXO. Samples were collected at different times during reperfusion, following a period of 25-min global ischaemia. Ordinates: 6-KETO and TXB_2 values in pg/ml/min. Abscissa: time in minutes.

Experimental protocol

Three groups of animals ($n = 9$ in each group) were formed. The first group, serving as control was not pretreated with 7-OXO. The hearts were isolated, subjected to 25-min global ischaemia, followed by 25-min reperfusion. The AA metabolites were determined as described above. The same procedure was made with the second and third group, in addition both groups were pretreated with 50 μg/kg 7-OXO i.p. but preparation of the isolated Langendorff heart was carried out 24 h (second group) and 48 h (third group) after drug administration.

Statistical analysis

The values of 6-KETO and TXB_2 were expressed as means ± SEM. Difference from corresponding values of the control group were calculated by using analysis of variance (ANOVA) followed by Fisher's test. Significant difference was assumed when p was less than 0.05.

Results

Effect of 7-OXO-pretreatment on AA-metabolites in the nonischaemic heart

Determinations made during aerobic perfusion have shown a slight but significant increase in 6-KETO release, induced by pretreatment with 7-OXO. The values of 206 ± 11 pg/ml/min in untreated aerobic controls increased to 284 ± 19 pg/ml/min 24 h after 7-OXO pretreatment, and to 261 ± 18 pg/ml/min 48 h after the same pretreatment. No change was seen in TXB_2 production (Fig. 1).

Effect of 7-OXO-pretreatment on AA metabolites during reperfusion

Global ischaemia followed by reperfusion markedly increased the release of both AA metabolites; maximal values were observed in the third minute after starting reperfusion, therefore the following postischaemic values are related to the third minute of reperfusion. Thus 6-KETO increased from 206 ± 11 pg/ml/min to 1275 ± 55 pg/ml/min, and TXB_2 from 29 ± 4 pg/ml/min to 172 ± 12 pg/ml/min in control hearts, not pretreated with 7-OXO. After the third minute of reperfusion there was a marked decline in the release of AA metabolites, and all values returned to near-preischaemic level by the 25th min of reperfusion (Fig. 2).

The trend of changes was similar in hearts of animals pretreated with 7-OXO, however, the postischaemic release of 6-KETO was significantly higher than in the untreated controls (1507 ± 73 pg/ml/min after 24 h, and 1398 ± 54 pg/ml/min after 48 h). There was no difference in TXB_2 values (Fig. 2).

Discussion

The results obtained demonstrate that both aerobic and ischaemic release of PGI_2 increased 24 and 48 h after pretreatment with 7-OXO. Nevertheless TXA_2 production was unaffected by 7-OXO pretreatment. At present we have no direct evidence, whether release of endogenous PGI_2 contributes to the late appearing, longer lasting cardioprotection induced by pretreatment with 7-OXO.

There are, however, studies suggesting that increased PGI_2 release days after pretreatment is not merely a casual coincidence. Thus Coker *et al.* [7] have shown that PGI_2 is released following coronary artery occlusion; this release is from the ischaemic area; and the amount released correlates inversely with the number of ectopic beats occurring early in ischaemia and at the time of sampling. Furthermore inhibitors of thromboxane synthetase, such as low dose aspirin and dazoxiben, that act, in part, by diverting endoperoxides towards increased PGI_2 synthesis were shown to protect against both ischaemia and reperfusion induced arrhythmias [8, 9].

Moreover nafazatrom, a substance promoting prostacyclin release in the ischaemic myocardium proved to be also protective in such situations [10].

Another line of evidence is that blockade of the cycloxygenase pathway prevents the protective effects of 'preconditioning' [4]. This suggests that the protective (anti-ischaemic, antiarrhythmic) effect of 'preconditioning' is due, at least in part, to the generation of an endogenous substance derived from the cyclooxygenase pathway. Since prostacyclin has a very short half-life, much shorter than the duration of the protection, further studies are needed to elucidate, which mechanism is responsible for the increased release of PGI_2 to the ischaemic stimulus 24 h and 48 h after pretreatment with 7-OXO. In addition more evidence is needed in order to prove that this enhanced release of PGI_2 is related to the late appearing, prolonged cardioprotection.

References

1. Murry CE, Jennings RB, Reimer KA: Preconditioning with ischaemia: A delay of lethal cell injury in ischaemic myocardium. Circulation 74: 1124–1136, 1986
2. Komori S, Parratt JR, Szekeres L, Végh Á: Preconditioning reduces the severity of ischaemia and reperfusion-induced arrhythmias in both anaesthetized rats and dogs. J Physiol Lond 423: 16p, 1990
3. Végh Á, Komori S, Szekeres L, Parratt JR: Antiarrhythmic effects of preconditioning in anaesthetised dogs and rats. Cardiovasc Res 26: 487–495, 1992
4. Végh Á, Szekeres L, Parratt JR: Protective effects of preconditioning of the ischaemic myocardium involve cyclo-oxygenase products. Cardiovasc Res 24: 1020–1023, 1990
5. Szekeres L, Krassói I, Udvary É: Delayed antiischaemic effect of PgI_2 and of a new stable PgI_2 analogue: 7-oxo-Prostacyclin-Na in experimental model angina in dogs. J Mol Cell Cardiol 15: Suppl I, 394, 1983
6. Szekeres L, Pataricza J, Szilvássy Z, Udvary É, Végh Á: Cardioprotection: Endogenous protective mechanism promoted by prostacyclin. In: RG Gülch, G Kissling (eds.) Current Topics in Heart Failure. Steinkopff Verl. Darmstadt. Basic Res Card 86: Suppl 3: 215–221, 1991
7. Coker SJ, Parratt JR, Ledingham IMcA, Zeitlin IJ: Thromboxane and prostacyclin release from ischaemic myocardium in relation to arrhythmias. Nature 291: 323–324, 1981
8. Coker SJ, Parratt JR, Ledingham IMcA, Zeitlin IJ: Aspirin inhibits the early myocardial release of thromboxane B_2 and ventricular ectopic activity following acute coronary artery occlusion in dogs. Br J Pharmacol 72: 593–595, 1981
9. Coker SJ, Parratt JR: Effects of dazoxiben on arrhythmias and ventricular fibrillation induced by coronary artery occlusion and reperfusion in anaesthetised greyhounds. Br J Clin Pharmacol 15: 87–96, 1983
10. Coker SJ, Parratt JR: The effects of nafazatrom on arrhythmias and prostanoid release during coronary artery occlusion and reperfusion in anaesthetised greyhounds. J Mol Cell Cardiol 16: 43–52, 1984

Molecular and Cellular Biochemistry **119**: 133–141, 1993.

Effects of iloprost, a PGI_2 derivative, on ischemic myocardial energy and carbohydrate metabolism in dogs

Kazuo Ichihara, Kiminobu Yamamoto and Yasushi Abiko
Department of Pharmacology, Asahikawa Medical College, 4-5 Nishikagura, Asahikawa 078, Japan

Abstract

Effects of iloprost, which is a stable prostacyclin analogue, on the ischemic myocardium were examined in the open-chest dog heart in terms of biochemical parameters. Ischemia was initiated by ligating the left anterior descending coronary artery. When the coronary artery was ligated for 3 min, the levels or glycogen, fructose-1,6-diphosphate (FDP), adenosine triphosphate and creatinephosphate decreased, and the levels of glucose-6-phosphate (G6P), fructose-6-phosphate (F6P), lactate, adenosine diphosphate and adenosine monophosphate increased. During ischemia, therefore, energy charge potential was significantly decreased from 0.89 ± 0.01 to 0.82 ± 0.01, and ([G6P] + [F6P])/[FDP] and [lactate]/[pyruvate] ratios were significantly increased from 1.75 ± 0.30 to 29.05 ± 5.70 and 13 ± 3 to 393 ± 112, respectively. Iloprost (0.1, 0.3, or 1 $\mu g \cdot kg^{-1}$) was injected intravenously 5 min before the onset of ischemia. Iloprost (0.1, 0.3, and 1 $\mu g \cdot kg^{-1}$) reduced the ischemia-induced decrease in energy charge potential to 94, 74, and 86%, respectively, the increase in ([G6P] + [F6P])/[FDP] to 38, 29, 32%, respectively, and the increase in [lactate]/[pyruvate] to 67, 45, 65%, respectively. These results suggest that iloprost lessens the myocardial metabolic derangements produced by ischemia, and the most potent effect was obtained at the dose of 0.3 $\mu g \cdot kg^{-1}$. (Mol Cell Biochem **119:** 133–141, 1993)

Key words: iloprost, ischemia, energy metabolism, carbohydrate metabolism

Introduction

It is known that the prostacyclin (PGI_2) – thromboxane A_2 (TXA_2) system influences the function of the myocardium and of the coronary arteries in pathophysiological conditions such as ischemia. For example, PGI_2 inhibits the increase in the levels of lactate and cyclic AMP in the ischemic myocardium [1], and decreases ultimate infarct size and the death rate [2]. Therefore, it is considered that PGI_2 is an effective drug for therapy of ischemic heart diseases [3]. However, PGI_2 is unstable and has a very short biological half life. Schrör *et al.* [4] have demonstrated that ZK36374 (iloprost or E1030), which is a stable PGI_2 analogue, dilates the coronary artery, inhibits platelet aggregation, and inhibits release of enzymes from the infarcted myocardium. Aksulu *et al.* [5] have reported that iloprost has an antiarrhythmic property. Therefore, iloprost, instead of PGI_2, can be one of the effective drugs on the ischemic heart disease. The myocardial cell contains a lot of mitochondria, which produce energy for contraction by means of an aerobic metabolism using oxygen. If the coronary artery is obstructed, the myocardium can not produce energy aerobically, because of decrease in cor-

Address for offprints: K. Ichihara, Department of Pharmacology, Asahikawa Medical College, 2–5 Nishikagura, Asahikawa 078, Japan

onary blood flow and/or oxygen supply to the myocardium. Under these conditions, an anaerobic metabolism takes place and is accelerated in order to produce energy for contraction anaerobically [6]. In fact, when the coronary artery is ligated in the dog heart, the anaerobic metabolism is accelerated [7, 8]. The anaerobic metabolism supplies energy to the heart. Nevertheless, the levels of adenosine triphosphate (ATP) and creatinephosphate (CrP) are low in the ischemic myocardium, because the energy produced by anaerobic metabolism is not enough to maintain the function of the myocardium.

We [9] have been trying evaluation of the anti-ischemic effect of various drugs by the use of the biochemical parameters, which reflect a balance between anaerobic and aerobic metabolism. For example, in the myocardium where ischemia causes conversion of the myocardial metabolism from aerobic to anaerobic, the drug that improves oxygen balance in the ischemic myocardium can switch the metabolism back from anaerobic to aerobic [10]. The present study, therefore, was undertaken to examine the effect of iloprost on changes in the myocardial energy and carbohydrate metabolism caused by ischemia.

Methods

Animal preparation

Healthy mongrel dogs (53 animals) of either sex weighing 6.5–16.7 kg were anesthetized with sodium pentobarbital (30 $mg \cdot kg^{-1}$, iv), and endotracheally intubated and ventilated with a respirator. A left thoracotomy was performed between the fourth and fifth ribs to expose the left ventricular wall. A main trunk of the left anterior descending coronary artery (LAD) was dissected free from the adjacent tissues at a distal portion to the first diagonal branch, and was loosely encircled with a 2–0 silk thread ligature. Ischemia was initiated by ligating the LAD. An ischemic region of the myocardium was assessed by visible cyanosis. Heart rate was counted from ECG taken in the standard limb lead II. Arterial blood pressures were measured via a cannula introduced into the left carotid artery. Coronary blood flow was measured by an electromagnetic flow probe positioned at just proximal to the ligature.

After a 60-min period of stabilization, either vehicle or iloprost (E3010) (0.1, 0.3, or 1 $\mu g \cdot kg^{-1}$) was injected iv over a period of 30 sec into the left femoral vein. An ampoule containing 10 $\mu g \cdot 2 ml^{-1}$ of iloprost and that containing vehicle were supplied by Eisai Pharmaceutical Co. Ltd. Japan. After 5 min. of injection, the ligature around the LAD was not tied in vehicle (n = 6) or iloprost 0.1 $\mu g \cdot kg^{-1}$ (n = 5), 0.3 $\mu g \cdot kg^{-1}$ (n = 6), or 1.0 $\mu g \cdot kg^{-1}$ (n = 7) treated groups, or tied for 3 min. in vehicle (n = 6), or iloprost 0.1 $\mu g \cdot kg^{-1}$ (n = 10), 0.3 $\mu g \cdot kg^{-1}$ (n = 7), or 1.0 $\mu g \cdot kg^{-1}$ (n = 6) treated groups. When the ischemia had been produced for 3 min, a full thickness sample of the myocardium was taken from the center of the ischemic area. The samples were immediately pressed and frozen with clamps previously chilled in liquid nitrogen. An equivalent non-ischemic control sample was taken from the heart in which the LAD was not tied. The subendocardial portion was collected and used for analysis [7].

Biochemical analysis

The frozen endocardial tissue sample was pulverized in a mortar with a pestle precooled with liquid nitrogen and extracted with 6% perchloric acid. The levels of glycogen, glucose, glucose-6-phosphate (G6P), fructose-6-phosphate (F6P), fructose-1,6-diphosphate (FDP), pyruvate, lactate, ATP, adenosine diphosphate (ADP), adenosine monophosphate (AMP), and CrP in neutralized perchloric acid extract were determined according to the standard enzymatic procedures.

Energy charge potential [11] was calculated from the concentration of ATP, ADP, and AMP to estimate the myocardial energy state according to ([ATP] + 0.5 [ADP])/([ATP] + [ADP] + [AMP]). The ratio of ([G6P] + [F6P])/([FDP] was calculated from the concentration of hexose phosphates in order to assume the rate of glycolytic flux through phosphofructokinase (PFK) reaction [8]. The ratio of [lactate]/[pyruvate] was also calculated as an index of the cytoplasmic redox state in the myocardial cell.

Statistical analysis

Data were evaluated using analysis of variance and Student's t-test, and a *P* value of 0.05 or less was considered significant.

Results

Hemodynamics

Hemodynamic data from the dogs, in which the coronary arteries have been ligated for 3 min are illustrated in Fig. 1. Iloprost (0.1, 0.3 or 1 $\mu g \cdot kg^{-1}$) was injected iv 5 min before ligation. These parameters did not change appreciably by the vehicle injection. Immediately after the injection of iloprost at the dose of 0.3 or 1 $\mu g \cdot kg^{-1}$, a rapid and significant decrease in systolic and diastolic blood pressures was observed. Iloprost of 0.1 $\mu g \cdot kg^{-1}$ slightly but significantly decreased diastolic pressure, whereas it did not change systolic pressure significantly. The blood pressure response reached the maximum within 2 min after the injection. Iloprost produced about 2% (0.1 $\mu g \cdot kg^{-1}$), about 18% (0.3 $\mu g \cdot kg^{-1}$) and about 24% (1 $\mu g \cdot kg^{-1}$) decreases in systolic pressure, and about 8% (0.1 $\mu g \cdot kg^{-1}$), about 22% (0.3 $\mu g \cdot kg^{-1}$) and about 40% (1 $\mu g \cdot kg^{-1}$) decreases in diastolic pressure. The decrease in systolic and diastolic blood pressures caused by the iloprost injection was transient. Heart rate transiently increased 1 min after the injection of iloprost of 0.3 and 1 $\mu g \cdot kg^{-1}$. The injection of iloprost at either dose, however, did not change coronary flow significantly. Ligation of the LAD decreased systolic and diastolic pressure slightly, but in some cases significantly. Coronary flow, of course, reduced significantly and became 0 $ml \cdot min^{-1}$ immediately after LAD ligation. The data from the animals in which the coronary arteries have not been ligated, are not shown in Fig. 1, but the hemodynamic responses to iloprost were similar to those in the ischemic group.

Energy metabolism

Changes in the levels of adenine nucleotides in the myocardium are summarized in Table 1. Before ischemia, the level of ATP in the 0.3 $\mu g/kg^{-1}$ of iloprost-treated heart was higher, and the levels of ADP in the 0.1 and 0.3 $\mu g \cdot kg^{-1}$ of iloprost-treated heart were lower than those in the vehicle-treated heart. The other values of the metabolites in the iloprost-treated group did not significantly differ from those in the vehicle-treated group. Three min after ligation of the LAD, the level of ATP decreased, while the levels of ADP and AMP increased. Pretreatment with iloprost (0.3 $\mu g \cdot kg^{-1}$) significantly attenuated the decrease in ATP and total adenine nucleotides, and the increase in AMP caused by ischemia. The decrease in the total adenine nucleotide level caused by ischemia was also attenuated significantly by 1.0 $\mu g \cdot kg^{-1}$ of iloprost.

Changes in energy charge potential during ischemia in each group are shown in Fig. 2. After 3 min of ischemia, energy charge potential decreased from 0.89 ± 0.01 to 0.82 ± 0.01 in the vehicle-treated heart. The difference was so small but significant ($P < 0.01$). Iloprost at either dose significantly attenuated the decrease in energy charge potential caused by ischemia. When the reduction of energy charge potential in the vehicle-treated hearts was taken as 100%, those in the iloprost-treated hearts were 94% at 0.1 $\mu g \cdot kg^{-1}$, 74% at 0.3 $\mu g \cdot kg^{-1}$, and 86% at 1.0 $\mu g \cdot kg^{-1}$. The maximum effect of iloprost was observed at the dose of 0.3 $\mu g \cdot kg^{-1}$.

The level of CrP is shown in Table 2. Iloprost did not affect the level of CrP in the nonischemic myocardium. The CrP level significantly decreased during ischemia in both vehicle- and iloprost-treated hearts. The effects of iloprost did not obtain significance as compared with those in the vehicle-treated hearts.

Carbohydrate metabolism

Changes in the glucose and glycogen levels during ischemia are shown in Table 3. In the nonischemic myocardium, the level of glucose in the 0.1 $\mu g \cdot kg^{-1}$ of iloprost-treated group was significantly lower than that in the vehicle-treated group. Iloprost in the other groups, however, did not affect the levels of glucose and glycogen. In the vehicle-treated myocardium, ischemia significantly decreased the levels of glycogen and glucose. In the presence of iloprost, the levels of glycogen and glucose were not significantly changed by ischemia. The level of glucose in the iloprost (1.0 $\mu g \cdot kg^{-1}$) treated hearts appeared to be decreased by ischemia, and the difference with that in the vehicle-treated ischemic hearts was significant.

Changes in the hexose phosphate levels during ischemia are shown in Table 4. The difference in the levels of G6P, F6P and FDP between the vehicle- and iloprost-treated nonischemic myocardium was not significant, except that the level of G6P in the iloprost (0.1 $\mu g \cdot kg^{-1}$)-treated nonischemic myocardium was significantly lower than that in the vehicle-treated heart. Ischemia significantly increased the levels of G6P and F6P in both vehicle- and iloprost-treated myocardium. The level of FDP did not change in the vehicle-treated myocardium

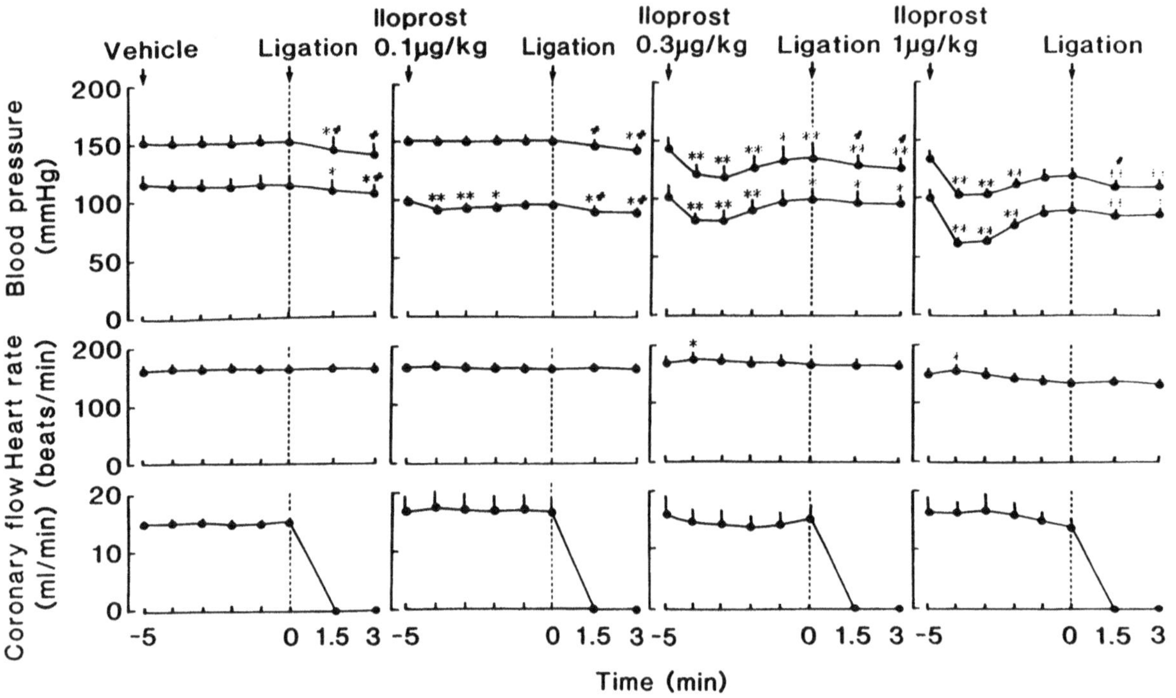

Fig. 1. Hemodynamic changes in the dog whose LAD was ligated for 3 min.

Either vehicle or iloprost (0.1, 0.3 or 1 $\mu g \cdot kg^{-1}$) was injected iv 5 min before the LAD ligation. Values at − 5 min are in the absence of vehicle and iloprost. Data are means ± SE of 6–10 observations in each group. Although the changes in coronary flow after LAD ligation (0 $ml \cdot min^{-1}$) are statistically significant as compared with either value before drug injection, or just before LAD ligation, symbols indicating significance are not shown at the value obtained after LAD ligation to avoid complication. * $P < 0.05$; ** $1P < 0.01$, compared with the value obtained 5 min before injection. # $P < 0.05$, compared with the value obtained just before LAD ligation (0 min).

during ischemia, while it increased in the iloprost-treated ischemic myocardium, and significant differences in the FDP level between nonischemic and ischemic hearts were obtained in 0.1 and 0.3 $\mu g \cdot kg^{-1}$ of iloprost-treated groups. To assume the rate of glycolytic flux through the PFK reaction, the ratio of ([G6P] + [F6P])/[FDP] is

Table 1. Effects of iloprost on changes in adenine nucleotide levels during ischemia

	n	ATP	ADP	AMP	Total adenine nucleotide
		(μmoles·g wet weight^{-1})			
Vehicle					
Nonischemia	6	5.30 ± 0.15	1.00 ± 0.03	0.25 ± 0.02	6.54 ± 0.15
Ischemia	6	3.96 ± 0.14**	1.44 ± 0.07**	0.33 ± 0.01**	5.77 ± 0.18**
Iloprost (0.1 $\mu g \cdot kg^{-1}$)					
Nonischemia	5	5.53 ± 0.15	0.83 ± 0.02##	0.24 ± 0.02	6.60 ± 0.15
Ischemia	10	4.37 ± 0.17**	1.39 ± 0.06**	0.30 ± 0.01*	6.06 ± 0.22
Iloprost (0.3 $\mu g \cdot kg^{-1}$)					
Nonischemia	6	5.83 ± 0.07##	0.88 ± 0.03#	0.23 ± 0.01	6.50 ± 0.06
Ischemia	7	4.66 ± 0.16**, ##	1.37 ± 0.07**	0.29 ± 0.01**, #	6.33 ± 0.15#
Iloprost (1 $\mu g \cdot kg^{-1}$)					
Nonischemia	7	5.40 ± 0.23	0.92 ± 0.05	0.23 ± 0.01	6.58 ± 0.26
Ischemia	6	4.16 ± 0.11**	1.33 ± 0.04**	0.31 ± 0.01**	5.80 ± 0.15#

Myocardial tissue sample was taken just before (nonischemia) or 3 min after coronary artery ligation (ischemia). Either vehicle or iloprost was injected iv 5 min before coronary artery ligation. Data are means ± SE). * $P < 0.05$; ** $P < 0.01$, vs nonischemia. # $P < 0.05$; ## $P < 0.01$, vs vehicle.

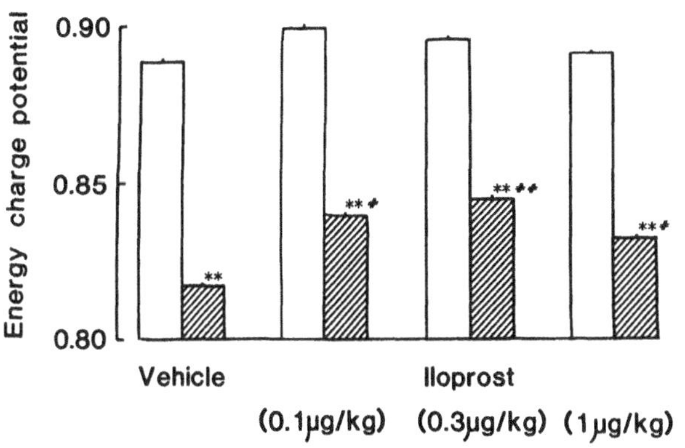

Fig. 2. Effect of iloprost on changes in energy charge potential during ischemia.

Ischemia was induced by LAD ligation for 3 min. Vehicle or iloprost at the dose of 0.1, 0.3 or 1 $\mu g \cdot kg^{-1}$ was injected iv 5 min before LAD ligation. Heart samples were taken just before (nonischemia, □) or 3 min after LAD ligation (ischemia, □). Energy charge potential was calculated from the data shown in Table 1 according to ([ATP] + 0.5 [ADP])/([ATP] + [ADP] + [AMP]). Data are means ± SE (vertical bars) of 6–7 observations in each group. ** $P < 0.01$, compared with 'nonischemia' group. # $P < 0.05$; ## $P < 0.01$, compared with either nonischemic or ischemic value in 'vehicle' group.

calculated and illustrated in Fig. 3. In the nonischemic heart, iloprost did not alter the ([G6P] + [F6P])/[FDP] ratio significantly. After 3 min of ischemia, the ratio of ([G6P] + [F6P])/[FDP] increased from 1.8 to 29.0 in the vehicle-treated group. Similarly, the ratio increased during ischemia in the iloprost-treated groups. However, the increase in the ratio caused by ischemia was significantly attenuated by iloprost at either dose. When the elevation of ([G6P] + [F6P])/[FDP] in the vehicle-treated hearts was taken as 100%, those in the

Table 2. Effect of iloprost on the CrP level in the ischemic myocardium

	n	CrP (μmoles·g wet weight^{-1})
Vehicle		
Nonischemia	6	6.34 ± 0.40
Ischemia	6	1.46 ± 0.26**
Iloprost (0.1 $\mu g \cdot kg^{-1}$)		
Nonischemia	5	7.36 ± 0.23
Ischemia	10	1.38 ± 0.26**
Iloprost (0.3 $\mu g \cdot kg^{-1}$)		
Nonischemia	6	7.70 ± 0.56
Ischemia	7	2.64 ± 0.55**
Iloprost *1 $\mu g \cdot kg^{-1}$)		
Nonischemia	7	7.39 ± 0.35
Ischemia	6	1.28 ± 0.21**

CrP Creatine phosphate. Data are means ± SE. ** $P < 0.01$, vs nonischemia.

iloprost-treated hearts were 38% at 0.1 $\mu g \cdot kg^{-1}$, 29% at 0.3 $\mu g \cdot kg^{-1}$, and 32% at 1.0 $\mu g \cdot kg^{-1}$.

Changes in the levels of lactate and pyruvate during ischemia are shown in Table 5. Before LAD ligation, the levels of lactate and pyruvate were not significantly modified by iloprost. In both vehicle- and iloprost-treated hearts, ischemia significantly increased the myocardial level of lactate, but it did not change the level of pyruvate. Iloprost significantly attenuated the lactate accumulation due to ischemia in the 0.1 and 0.3 $\mu g \cdot kg^{-1}$ of iloprost-treated hearts. In Fig. 4, the ratio of [lactate]/[pyruvate] is shown. The ratio of [lactate]/[pyruvate] significantly increased during ischemia. When the elevation of [lactate]/[pyruvate] in the vehi-

Table 3. Effects of iloprost on glycogen and glycose levels in the ischemic myocardium

	n	Glycogen (μmoles glucose eq·g wet weight^{-1})	Glucose (μmoles·g wet weight^{-1})
Vehicle			
Nonischemia	6	34.97 ± 2.26	6.48 ± 0.45
Ischemia	6	27.68 ± 2.18*	4.87 ± 0.34*
Iloprost (0.1 $\mu g \cdot kg^{-1}$)			
Nonischemia	5	30.93 ± 3.58	4.64 ± 0.68#
Ischemia	10	28.95 ± 3.08	4.81 ± 0.72
Iloprost (0.3 $\mu g \cdot kg^{-1}$)			
Nonischemia	6	34.91 ± 3.86	5.91 ± 0.79#
Ischemia	7	29.05 ± 1.87	5.19 ± 0.69
Iloprost (1 $\mu g \cdot kg^{-1}$)			
Nonischemia	7	29.70 ± 2.64	4.90 ± 0.88
Ischemia	6	29.27 ± 2.71	2.96 ± 0.30##

Data are means ± SE. * $P < 0.05$, vs nonischemia. # $P < 0.05$; ## $P < 0.01$, vs. vehicle.

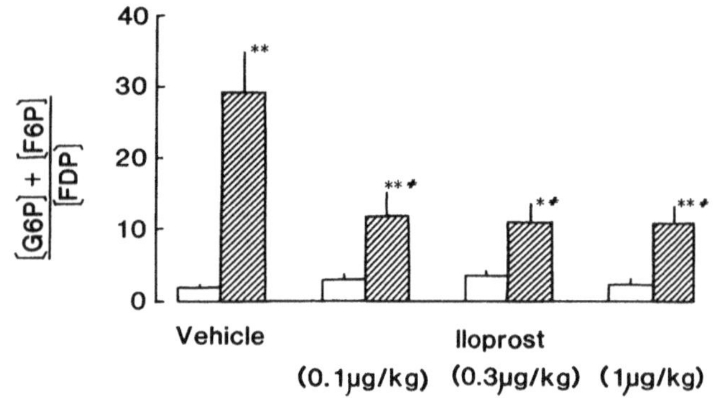

Fig. 3. Effect of iloprost on changes in ([G6P] + [F6P])/[FDP] ratio during ischemia.

□, and □ represent the nonischemic and ischemic values, respectively. The data shown in Table 4 were used for the calculation. Data are means ± SE (vertical bars). * $P < 0.05$; ** $P < 0.01$, compared with 'nonischemia' group. # $P < 0.05$; ## $P < 0.01$, compared with either nonischemic or ischemic value in 'vehicle' group.

cle-treated hearts was taken as 100%, those in the iloprost-treated hearts were 67% at 0.1 $\mu g \cdot kg^{-1}$, 45% at 0.3 $\mu g \cdot kg^{-1}$, and 65% at 1.0 $\mu g \cdot kg^{-1}$. The ratio in the iloprost (0.3 and 1 $\mu g \cdot kg^{-1}$)-treated groups were significantly lower than that in the vehicle-treated heart.

Discussion

It has been reported that iloprost prevents myocardial ischemic damage caused by permanent coronary ligation in the cat [12], by coronary artery microembolization in the pig [13], and by coronary ligation in the rat [14]. In the present study, iloprost attenuated the ischemia-induced metabolic derangement in the myocardium. Iloprost reduced the changes in energy charge potential (Fig. 2), ([G6P] + [F6P])/[FDP] ratio (Fig. 3), and [lactate]/[pyruvate] ratio (Fig. 4) produced by ischemia. These results suggest that iloprost has a protective effect on the ischemic myocardium. Why does iloprost reduce the ischemic effects on the myocardium? Iloprost reduces the arterial blood pressure [12], dilates the coronary arteries [4], increases the collateral blood flow to the ischemic region [15], has a membrane stabilizing effect [12], inhibits neutrophil activation and neutrophil accumulation [16], has antiaggregatory effects on platelets [4], and has antiarrhythmic properties against ischemia and reperfusion arrhythmias in dogs [17].

Iloprost did not increase coronary blood flow significantly (Fig. 1). This may be explained by a decrease of arterial blood pressures. Ribeiro *et al.* [2] have also found the reduction of myocardial blood flow to the normal myocardium by PGI_2, resulting from a combination of factors including reduced coronary perfusion pressure and decreased myocardial metabolic demands with consequent autoregulation. The blood flow measured with an electromagnetic flow probe placed around a main trunk of the LAD does not mean the regional tissue blood flow in the ischemic myocardium, because the collateral vessels can also supply blood to the ischemic myocardium. Therefore it is impossible to exclude the possibility that iloprost may change the blood distribution to the ischemic myocardium. PGI_2 maintains ischemic myocardial tissue blood flow, while it reduces nonischemic flow [2]. The decrease in blood

Table 4. Effects of iloprost on hexose phosphate levels in the ischemic myocardium

	n	G6P	F6P (μmoles·g wet weight^{-1})	FDP
Vehicle				
Nonischemia	6	0.17 ± 0.03	0.04 ± 0.05	0.13 ± 0.02
Ischemia	6	1.93 ± 0.19**	0.45 ± 0.05**	0.11 ± 0.02
Iloprost (0.1 $\mu g \cdot kg^{-1}$)				
Nonischemia	5	0.09 ± 0.02#	0.03 ± 0.01	0.09 ± 0.02
Ischemia	10	1.50 ± 0.16**	0.36 ± 0.04**	0.17 ± 0.02*
Iloprost (0.3 $\mu g \cdot kg^{-1}$)				
Nonischemia	6	0.15 ± 0.02	0.04 ± 0.01	0.08 ± 0.02
Ischemia	7	1.10 ± 0.23**, #	0.24 ± 0.05**, #	0.13 ± 0.02*
Iloprost (1 $\mu g \cdot kg^{-1}$)				
Nonischemia	7	0.13 ± 0.02	0.01 ± 0.01	0.11 ± 0.03
Ischemia	6	1.21 ± 0.21**, #	0.31 ± 0.07**	0.16 ± 0.03

G6P Glucose 6 phosphate; F6P Fructose 6 phosphate; FDP Fructose 1, 6 diphosphate. Data are means ± SE. * $P < 0.05$; ** $P < 0.01$, vs nonischemia. # $P < 0.05$, vs vehicle.

pressure may decrease the myocardial oxygen demand. This may contribute, at least in part, to the protective effect of iloprost on the ischemic myocardial metabolism. Farber *et al.* [18] have stated, however, that the hypotensive effect of iloprost cannot explain its beneficial effect on myocardial function recovery, because the greater protective effect appears at a time when blood pressure has returned to normal. In fact, the blood pressure that had been decreased by iloprost returned to its control level at a time when the coronary artery was ligated in the present study (Fig. 1). In addition, iloprost at the high dose of 1 $\mu g \cdot kg^{-1}$ decreased blood pressure to a greater extent than the drug at the low dose of 0.3 $\mu g \cdot kg^{-1}$ did. Nevertheless, iloprost at 0.3 $\mu g \cdot kg^{-1}$ reduced the ischemia-induced metabolic derangement more than iloprost at 1 $\mu g \cdot kg^{-1}$ did. Our results accord with that of Farber *et al.* [18], and suggest that the reduction of ischemic effects on the myocardium by iloprost is not only due to the decrease in blood pressure.

There is a release of oxygen-derived free radicals during ischemia, leading to membrane damage [19]. Iloprost prevents an ischemia-induced decrease in superoxide dismutase activity [19]. However, it is very difficult to determine whether or not iloprost has a direct action on free radical generating or scavenging system. Whatever the mechanism is, iloprost preserves membrane integrity in the ischemic heart [12, 19]. Schrör *et al.* [12] have demonstrated that prevention of catecholamine overflow with iloprost from damaged adrenergic nerve endings in the ischemic myocardium may indicate its membrane stabilizing activity. They [12] emphasize that the prevention of excessive increases in myocardial metabolism caused by catecholamine may be quite important for the protective effect of iloprost on the ischemic myocardium. In addition, lysosomal membrane is also stabilized by iloprost [3]. This results in the inhibition of release of the lysosomal degradative enzymes, such as cathepsin D, from lysosomes to the cytoplasm [20].

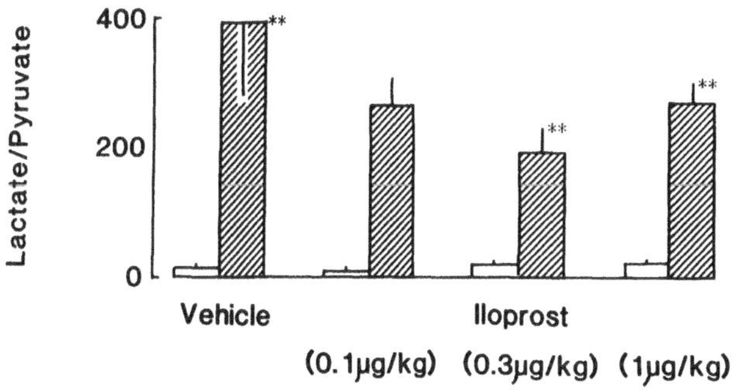

Fig. 4. Effect of iloprost on changes in the ratio of [lactate]/[pyruvate] during ischemia.

□, and □ represent the nonischemic and ischemic values, respectively. The ratio of [lactate]/[pyruvate] was calculated by using the values shown in Table 5. Data are means ± SE (vertical bars). ** $P <$ 0.01, compared with 'nonischemia' group.

The antiaggregatory effect of PGI_2 on platelets [4] may be involved in the mechanism of the favorable effect of iloprost on the ischemic myocardium. Jolly *et al.* [21], however, suggest that the platelet plays a rather passive role in the process of infarction. Because they [21] report the effect of a long period of ischemia on the tissue, we cannot still exclude the possibility of participation of the platelets in the early stage of the ischemic damage.

From the crossover-plot study of the levels of carbo-

Table 5. Effects of iloprost on pyruvate and lactate levels in the ischemic myocardium

	n	Pyruvate (μmoles·g wet weight^{-1})	Lactate
Vehicle			
Nonischemia	6	0.04 ± 0.01	0.56 ± 0.12
Ischemia	6	0.04 ± 0.01	11.49 ± 0.43**
Iloprost (0.1 $\mu g \cdot kg^{-1}$)			
Nonischemia	5	0.06 ± 0.01	0.47 ± 0.17
Ischemia	10	0.05 ± 0.01	9.76 ± 0.51**, #
Iloprost (0.3 $\mu g \cdot kg^{-1}$)			
Nonischemia	6	0.04 ± 0.01	0.69 ± 0.18
Ischemia	7	0.05 ± 0.01	8.91 ± 1.08**, #
Iloprost (1 $\mu g \cdot kg^{-1}$)			
Nonischemia	7	0.04 ± 0.01	0.75 ± 0.05
Ischemia	6	0.04 ± 0.01	10.85 ± 0940**

Data are means ± SE. ** $P < 0.05$, vs nonischemia. # P0.05, vs vehicle.

hydrate metabolites [22], a negative crossover point was obtained between F6P and FDP in the ischemic myocardium, suggesting an inhibition of glycolytic flux at the level of PFK reaction [8, 22, 23]. Opie [23] has demonstrated that glycolysis is accelerated immediately after the onset of ischemia, and then inhibited. Although no direct flux measurement has been made, the negative crossover point obtained in the ischemic myocardium suggests an inhibition of glycolytic flux, because hydrogen ions and lactate accumulated in the ischemic myocardium [22] inhibit the PFK reaction [24]. When glycolysis is inhibited at the site of PFK reaction, the levels of G6P and F6P will increase, whereas the level of FDP will decrease after coronary artery ligation. We calculated the ratio of ([G6P] + [F6P])/[FDP] to assume the extent to which glycolytic flux through the PFK reaction was inhibited by ischemia (Fig. 3). This ratio in the iloprost-treated dog was significantly ($P < 0.05$) lower than that in the vehicle-treated dog. The reduction of glycolytic flux inhibition by iloprost indicates that this drug may lessen the effects of ischemia on the myocardium.

In the present study, an ischemic myocardial sample was obtained from the subendocardial portion of the myocardium, because the portion is more sensitive than the epicardial portion to metabolic changes induced by ischemia [7, 25]. In addition, we chose the very short duration of ischemia of only 3 min. The reason for this is that it is very important for us to know what will happen in the myocardial cells within a few min after the onset of ischemia, because anginal attack in a patient with angina pectoris may occur within a few minutes after the onset of ischemia, and because the most potent influence of acute ischemia (less than 7 min after the LAD ligation) on the myocardial metabolism is observed 3 min after coronary ligation in the dog experiments [7].

In conclusion, iloprost inhibited the activation of anaerobic metabolism in the ischemic myocardium. It is likely that iloprost makes the metabolic state of the myocardium aerobic, even when the heart is in anaerobic condition.

References

1. Rösen R, Rösen P, Ohlendorf R, Schrör K: Prostacyclin prevents ischemia-induced increase of lactate and cyclic AMP in the ischemic myocardium. Eur J Pharmacol 69: 489–491, 1981
2. Ribeiro LGT, Brandon TA, Hopkins DG, Reduto LA, Taylor AA, Miller RR: Prostacyclin in experimental myocardial ischemia: Effects on hemodynamics, regional myocardial blood flow, infarct size and mortality. Am J Cardiol 47: 835–840, 1981
3. Lefer AM, Ogretree ML, Smith JB, Silver MJ, Nicolaou KC, Barnette WE, Gasic GP: Prostacyclin: A potentially valuable agent for preserving myocardial tissue in acute myocardial ischemia. Science 200: 52–54, 1978
4. Schrör K, Darus H, Matzky R, Ohlendorf R: The antiplatelet and cardiovascular actions of a new carbacyclin derivative (ZK 36 374) – equipotent to PGI_2 *in vivo*. Naunyn-Schmiedeberg's Arch Pharmacol 316: 252–255, 1981
5. Aksulu HA, Ercan ZS, Turker RK: Further studies on the antiarrhythmic effect of iloprost. Arch Int Pharmacodyn Ther 227: 223–234, 1985
6. Bing RJ: Cardiac metabolism. Physiol Rev 45: 171–213, 1965
7. Ichihara K, Abiko Y: Difference between endocardial and epicardial utilization of glycogen in the ischemic heart. Am J Physiol 229: 1585–1589, 1975
8. Weishaar R, Ashikawa K, Bing RJ: Effect of diltiazem, a calcium antagonist, on myocardial ischemia. Am J Cardiol 43: 1137–1143, 1979
9. Abiko Y, Ichihara K, Izumi T: Effects of antianginal drugs on ischemic myocardial metabolism. In: MM Winbury and Y Abiko (eds). Ischemic Myocardium and Antianginal Drugs. Raven Press, New York, 1979, pp 155–169
10. Ichihara K, Abiko Y: Inhibition of endo- and epicardial glycogenolysis by propranolol in ischemic hearts. Am J Physiol 232: H349–H353, 1977
11. Atkinson DE, Walton GM: Adenosine triphosphate conservation in metabolic regulation. Rat liver citrate cleavage enzyme. J Biol Chem 242: 3239–3241, 1967
12. Schrör K, Darius H, Matzky R, Ohlendorf R: Beneficial effects of a new carbacyclin derivative, ZK 36 374, in acute myocardial ischemia. J Pharmacol Exp Ther 219: 243–249, 1981
13. De Langen CDJ, Gilst WH, Wesseling H: Sustained protection by iloprost of the porcine heart in the acute and chronic phases of myocardial infarction. J Cardiovasc Pharmacol 7: 924–928, 1985
14. Osborne JA, Darius H, Lefer AM: Beneficial effects of the prostacyclin analogue iloprost in acute myocardial ischemia in rats. Fed Proc 45: 788, 1986
15. Smith III EF, Gallenkämper W, Beckmann R, Thomsen T, Mannesmann G, Schrör K: Early and late administration of a PGI_2-analogue ZK 36 374 (iloprost): Effects of myocardial preservation, collateral blood flow and infarct size. Cardiovasc Res 18: 163–173, 1984
16. Simpson PJ, Mickelson J, Fantone JC, Gallagher KP, Lucchesi BR: Iloprost inhibits neutrophil function *in vivo* and *in vitro* and limits experimental infarct size in canine heart. Circ Res 60: 666–673, 1987
17. Coker SJ, Parratt JR: Prostacyclin-antiarrhythmic or arrhythmogenic? Comparison of the effects of intravenous and intracoronary prostacyclin and ZK 36 374 during coronary artery occlusion and reperfusion in anaesthetised greyhounds. J Cardiovasc Pharmacol 5: 557–567, 1983
18. Farber NE, Pieper GM, Thomas JP, Gross GJ: Beneficial effects of iloprost in the stunned canine myocardium. Circ Res 62: 204–215, 1988
19. Thiemermann CE, Steinhagen-Thiessen E, Schrör K: Inhibition of oxygen centered free radical formation by the stable prostacy

clin-mimetic iloprost (ZK 36374) in acute myocardial ischemia. J Cardiovasc Pharmacol 6: 365–366, 1984

20. Ichihara K, Haneda T, Onodera S, Abiko Y: Inhibition of ischemia-induced subcellular redistribution of lysosomal enzymes in the perfused rat heart by the calcium entry blocker, diltiazem. J Pharmacol Exp Ther 242: 1109–1113, 1987
21. Jolly SR, Schumacher WA, Kunkel SL, Abrams GD, Liddicoat J, Lucchesi BR: Platelet depletion in experimental myocardial infarction. Basic Res Cardiol 80: 269–279, 1985
22. Ichihara K, Abiko Y: Crossover plot study of glycolytic intermediates in the ischemic canine heart. Jap Heart J 23: 817–828, 1982
23. Opie LH: Effects of regional ischemia on metabolism of glucose and fatty acids. Relative rates of aerobic and anaerobic energy production during myocardial infarction and comparison with effects of anoxia. Circ Res 38 (Suppl): I 52-I 74, 1976
24. Ui M: A role of phosphofructokinase in pH-dependent regulation of glycolysis. Biochim Biophys Acta 124: 310–320, 1966
25. Winbury MM: Experimental coronary disease – Models and methods of drug evaluation. In: GVH Born, O Eichler, A Farah, H Herken, AD Welch (eds). Handbook of Experimental Pharmacology New Series. Springer-Verlag, Berlin, 1975, pp 1–69.

Molecular and Cellular Biochemistry **119**: 143–150, 1993.

Cardioprotective actions of wild garlic (Allium ursinum) in ischemia and reperfusion

B. Rietz, H. Isensee, H. Strobach,[1] S. Makdessi and R. Jacob
Physiologisches Institut II, Universität Tübingen/FRG; [1] Pharmakologisches Institut, Universität Düsseldorf, Germany

Abstract

The susceptibility to ventricular arrhythmias under the conditions of cardiac ischemia and reperfusion was investigated in the Langendorff heart preparation of rats fed for eight weeks a standard chow enriched with 2% of pulverized wild garlic leaves. The isolated hearts were perfused with a modified Krebs-Henseleit solution. The incidence of ventricular fibrillation (VF) during 20 min occlusion of thedescending branch of the left coronary artery (LAD) was significantly reduced in the wild garlic group as compared to untreated controls (20% vs 88%). The same holds for the size of the ischemic zone (33.6% vs 40.9% of heart weight). In the reperfusion experiments (5 min after 10 min ischemia), ventricular tachycardia (VT) occurred in 70% of the wild garlic group vs 100% in untreated controls and VF in 50% vs 90%. The time until occurrence of extrasytoles, VT or VR was prolonged. No significant alterations in cardiac fatty acid composition could be observed. Although the prostacyclin production was slightly increased in hearts of the wild garlic group, inhibition of cyclooxygenase by acetylsalicylic acid (ASA; aspirin) could not completely prevent the cardioprotective effects suggesting that the prostaglandin system does not play a decisive role in the cardioprotective action of wild garlic. Furthermore, a moderate angiotensin converting enzyme (ACE) inhibiting action of wild garlic was found *in vitro* as well as *in vivo* that could contribute to the cardioprotective and blood pressure lowering action of wild garlic. Whether a free radical scavenging activity of wild garlic is involved in its cardioprotective effects remains to be established. (Mol Cell Biochem **119:** 143–150, 1993)

Key words: wild garlic, ischemia, reperfusion, ventricular arrhythmias, Langendorff preparation

Introduction

Ventricular arrhythmias arising predominantly from ischemic heart disease are a major cause of death in the western industrialized countries. However, the prevention and the pharmacological management of life-threatening arrhythmias remain a great challenge. It is, therefore, important to search for alternative dietary and phytotherapeutic strategies. Previously, we investigated the action of various diets on the susceptibility of the isolated rat heart (Langendorff preparation) to ventricular arrhythmias under the conditions of acute ischemia and reperfusion [1]. In the framework of these studies it was found that feeding of garlic powder (allium sativum) as well as of pulverized dried wild garlic (allium ursinum) leaves for several weeks has significant protective effects with respect to the size of the ischemic zone and the incidence of arrhythmias [2].

The aim of the present study was to quantify such cardioprotective actions for wild garlic. Furthermore, an attempt was made to define the underlying mechanisms. To obtain indications of a conceivable role of

Address for offprints: R. Jacob, Physiologisches Institut II, Universität Tübingen, Germany prostaglandins [3, 4], the enzyme cyclooxy-genase was inhibited by acetylsalicylic acid (aspirin; ASA). Since inhibition of the angiotensin converting enzyme (ACE) has been shown to be beneficial in ischemic hearts [5, 6] and *in vitro* investigations have revealed an ACE inhibiting component in wild garlic [7], *in vivo* studies on this action were performed additionally. Finally, we examined whether wild garlic can induce changes in the fatty acid composition of cardiac membranes. Considering our studies on oil diets, a higher polyunsaturated fatty acid content of membrane phospholipids could be assumed which protects against ischemia- and reperfusion-induced arrhythmias [8, 9].

Material and methods

Animals

Two-months-old male Wistar rats (SAVO Ivanovas, Kisslegg/FRG) were used in the studies of the anti-arrhythmic effects of wild garlic (study A). Six-week-old spontaneously hypertensive rats (SHR) were used for monitoring blood pressure lowering effects and the ACE-inhibition (study B). The rats were housed (two per cage) at 25° C and a 12 h light-dark cycle. Food and water were ad libitum.

Diet

The control rats were fed a commercial rat chow (ssniff R, ssniff Specialdiäten GmbH, Soest/FRG). The experimental rats were fed the standard chow enriched with 2% pulverized dried leaves of wild garlic (Teutopharma, Telgte/FRG) for 8 weeks (study A) or 12 weeks (study B).

The standard rat chow consisted of carbohydrate 50.0%, protein 21.0%, fat 3.0%, crude fibre 4.8%.

The fatty acid composition of the standard chow was 14:0, 1.8%; 16:0, 18.9%; 18:0, 4.4%; 18:1 (n = 7), 0.5%; 18:1 (n = 9), 21.3%; 18:2 (n = 6), 48.4%; 18:3 (n = 3), 2.8%.

(Study A)
Heart perfusion

Rats were anaesthetised with thiopental-sodium (2.5 mg/kg i.p.) and heparin (200 IU) was injected into the vena cava. The hearts were excised and temporarily placed in ice-cold perfusion medium. Each heart was cannulated via the aorta and perfused according to Langendorff [10] at a constant flow of 6.0 ml/min in the occlusion experiments and 7.0 ml/min in the reperfusion experiments. The respective pressures were 5.2 kPa and 6.1 kPa. In the occlusion experiment the potassium ion concentration was reduced from the Krebs-Henseleit value (4.7 mM) to 3.0 mM because pilot studies had revealed a higher incidence of severe arrhythmias under these conditions. In the reperfusion experiments the potassium concentration was 4.7 mM. -Acetylsalicytlic acid was added to the perfusate at a concentration of 10 μM.

Occlusion experiments

Hearts were perfused for an initial 10 min period. At the end of this time, the left descending coronary artery (LAD) was ligated for 20 min. The electrocardiogram (ECG) was continuously recorded using two electrodes attached directly to the heart.

Reperfusion experiments

After 10 min equilibration, the LAD was occluded for 10 min which was followed by reperfusion. Rhythm disturbances were monitored for a 5 min period. The conditions of the reperfusion experiments were based on a study of Manning and Hearse [11] who demonstrated that reperfusion after a 10- to 15 min period of ischemia resulted in the highest incidence of rhythm disturbances and most of the reperfusion-induced arrhythmias occurred with the first minutes of reperfusion.

Evaluation of rhythm disturbances

The ECGs were analysed to determine (a) the duration during which each heart remained in normal sinus rhythm without any extrasystole during the occlusion or reperfusion phase, (b) the incidence, latency and duration of ventricular tachycardia (VT) (defined as four or more following beats of the same morphology without preceding P-wave) and (c) the incidence and latency of ventricular fibrillation (VF) (and also whether fibrillation was spontaneously reversible or whether hearts remained in irreversible fibrillation). VF was identified by its chaotic morphology of the ECG [12].

Determination of the size of ischemic zone

After the occlusion time of 20 min, a 5.0 ml bolus of 0.4% (w/v) malachite green dye was injected into the aorta to distinguish between the perfused green region from the underperfused flesh-coloured region due to dye exclusion. The underperfused region was cut out, blotted dry and weighed to determine the ischemic zone as a percentage of total heart weight.

Determination of prostacyclin as 6-oxo-$PGF_{1\alpha}$

Samples of coronary effluent were obtained at time 0, i.e., before the beginning of the occlusion and after three and eight min of occlusion. One ml fractions of the perfusate were collected in tubes and immediately frozen at −65° C. 6-oxo-$PGF_{1\alpha}$, the stable degradation product of prostacyclin, was determined by a radioimmunoassay (RIA).

Lipid extraction and analysis

At the end of the reperfusion experiments, the Langendorff hearts were rinsed in ice-cold homogenizing medium (250 mM sucrose, 1 mM EDTA, pH 7.4). The left ventricle was minced with scissors in 10 vol/g of homogenizing medium. The homogenization was carried out using a Polytron at a setting of 7 for 15s total, in three bursts. The homogenate was centrifuged at 2000 xg for 30 min. The pellet was homogenized using a teflon-glass homogenizer. The lipids were extracted according to Folch [13], using BHT as antioxidant and after formation of the fatty acid methyl esters [14], the fatty acid composition of total myocardium was determined by gas chromatography (VEGA 6000, Carlo Erba, Milan/Italy; CP-Sil 88, 100-230° C). The fatty acid composition of the basal diet was determined in the same way.

(Study B)
Blood pressure measurement

During the 12 week feeding period the systolic blood pressure of 5 rats of the wild garlic group and of 5 control animals was determined on the tail artery using a piezoresistive pressure transducer.

Plasma sampling

After the feeding period the rats were sacrificed, the heparinized blood was centrifuged at 10.000 g for 20 min at 4° C, the plasma was separated and kept on ice. The ACE activity was measured immediately.

ACE determination

ACE was measured by a modified fluorometric method according to Depierre and Roth [15] using Z-Phe-His-Leu as substrate. Substrate solution (50 μl, 10 mM) was diluted with 400 μl of cold phosphate buffer (70 mM,

Table 1. Effects of a diet containing 2% wild garlic powder on ischemia-induced arrhythmias (Langendorff heart preparation) without and with inhibition of cyclooxygenase by acetylsalicylic acid (ASA)

	Time until occurrence of extra-systoles (min)	Ventricular tachycardia			Ventricular fibrillation		Ischemic zone (%)
		time until occurrence (min)	incidence (%)	duration (s)	time until occurrence (min)	incidence (%)	
Control (n = 17)	3.4 ± 1.0	9.5 ± 1.7	35.5	20.0 ± 5.0	11.5 ± 1.8	88.0	40.9 ± 2.3
Wild garlic (n = 12)	7.1 ± 1.2*	16.0 ± 1.1*	30.0	15.0 ± 4.1	16.0 ± 1.0*	20.0*	33.6 ± 4.4*
Control + ASA (n = 10)	2.2 ± 0.7	8.7 ± 1.2	33.3	13.3 ± 2.4	9.1 ± 1.1	100.0	46.4 ± 4.0
Wild garlic + ASA (N = 12)	5.0 ± 1.4	9.5 ± 2.8°	41.7	24.0 ± 8.0	10.4 ± 2.2°	58.0°	37.5 ± 3.7

* signifies statistical significance, $p < 0.05$; control vs wild garlic.
° signifies statistical significance, $p < 0.05$; wild garlic vs wild garlic + ASA.

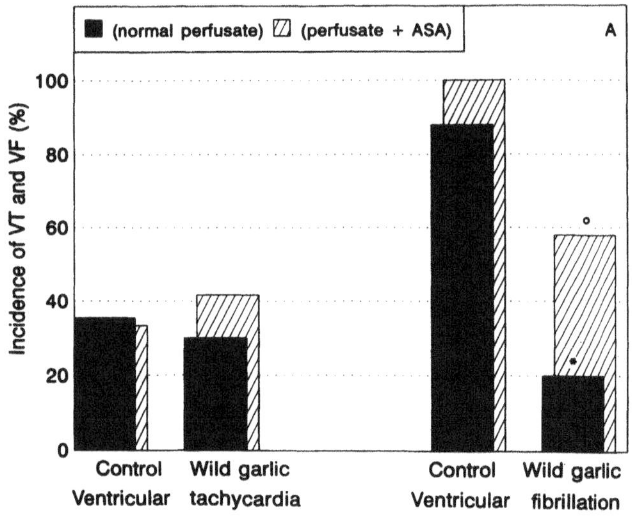

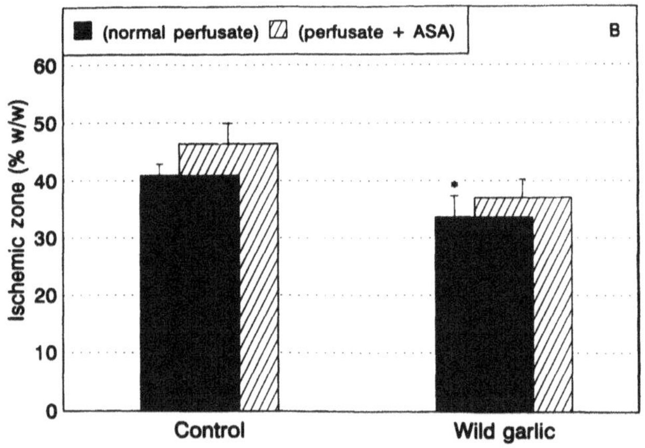

Fig. 1. Effects of diet containing 2% wild garlic on the incidence of ventricular tachycardia (VT) and ventricular fibrillation (VF) (Fig. 1A) and the size of the ischemic zone (Fig. 1B) without and with inhibition of cyclooxygenase by acetylsalicylic acid (ASA). * signifies statistical significance ($p < 0.05$) control vs wild garlic; o wild garlic vs wild garlic + ASA.

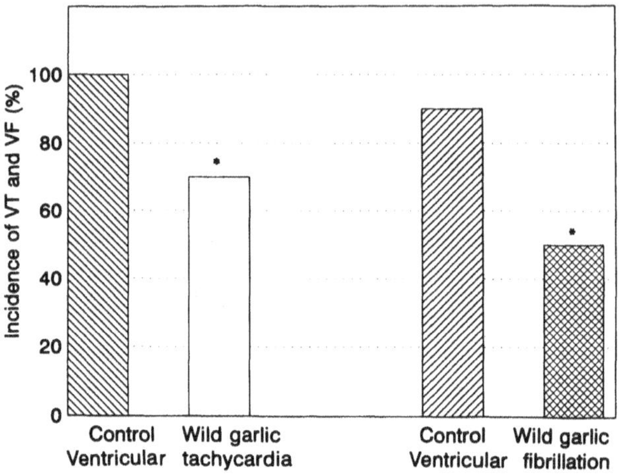

Fig. 2. Effects of a diet containing 2% wild garlic on reperfusion induced ventricular tachycardia (VT) and ventricular fibrillation (VF). * signifies statistical significance ($p < 0.05$).

pH 8.0, containing 300 mM sodium chloride). When the lyophilized extract of wild garlic was added, an aliquot of buffer was replaced by an equal volume of the wild garlic solution (*in vitro* study). The reaction was started by adding 50 μl of plasma to the samples and incubating at 37° C. After 30 min the reaction was terminated by transferring 200 μl aliquots from the incubation solution into 0.5 ml of 0.2 N NaOH. All subsequent steps in the assay were continued in the dark. Orthophthaldialdehyd solution (50 μl, 2%) was added to the samples. The reaction was stopped 30 min later with 0.5 ml of 1.6 N HCl; precipitates were spun down by a 3000 g centrifugation for 6 min and fluorescence ($\lambda ex = 360$ nm, $\lambda em = 500$ nm) was measured within 60 min [15]. Zero blank values were substracted from the corresponding test values. All measurements were performed at least fivefold. The results were expressed as nmol His-Leu/ml plasma/min.

Statistical analysis

Comparisons between the wild garlic and the control group were made by Student's t-test. Differences in the percent incidence of arrhythmias were assessed by using chi-square analysis. Means ± SD were calculated, $p < 0.05$ was considered statistically significant.

Results

Effects of wild garlic on ischemia-induced arrhythmias and size of the ischemic zone

Table 1 summarizes the effects of wild garlic on arrhythmias and ischemic zone, induced by a 20 min period of ischemia. The time until occurrence of extrasystoles was increased significantly from 3.4 min (control group) to 7.1 min (wild garlic group). Although the time until occurrence of ventricular tachycardia was likewise prolonged there was no significant difference in the incidence of VT.

The most interesting findings, shown in Table 1, concern the incidence of VF and the size of the ischemic zone. The incidence of VF was reduced from 88% (control group) to 20% (wild garlic group) (Fig. 1A). If VF

occurred under these conditions, it never reverted to sinus rhythm.
The ischemic zones of the hearts of the wild garlic group were significantly smaller than those of the control hearts (33.6% vs 40.9% of total cardiac tissue, Fig. 1B).

Effects of wild garlic on reperfusion-induced arrhythmias

The effects of wild garlic on the susceptibility to arrhythmias during coronary reperfusion after 10 min of occlusion are shown in Table 2 and Fig. 2. When the ligation of the LAD was opened after 10 min, extrasystoles appeared immediately both in the control and the wild garlic group. Rats fed the diet containing wild garlic powder showed significant reductions in both the incidence of ventricular tachycardia (70% vs 100%) and ventricular fibrillation (50% vs 90%). In contrast to the phase of occlusion reversion to sinus rhythm was observed in 10% of the control group as well as the wild garlic group.

Effects of wild garlic on prostaglandin synthesis

Coronary perfusate sampling was performed for measuring 6-oxo-$PGF_{1\alpha}$-generation. A slightly increased release was observed in the wild garlic group. The difference was significant at the beginning of the occlusion. Figure 3 shows the time course of changes in the level of 6-oxo-$PGF_{1\alpha}$ in the wild garlic group compared to the control group.

Influence of cyclooxygenase inhibition on the antiarrhythmic effects of wild garlic

To investigate the conceivable role of prostaglandins in preventing severe arrhythmias during ischemia in the wild garlic group, we inhibited their formation by blocking cyclooxygenase by ASA. In comparison to the experiments without inhibition of cyclooxygenase, the susceptibility to arrhythmias was increased in both groups (Table 1; Fig. 1A and 1B). However, cyclooxygenase inhibition was not able to prevent the cardioprotective effects completely neither with respect to the incidence of VF nor to the size of the ischemic zone.

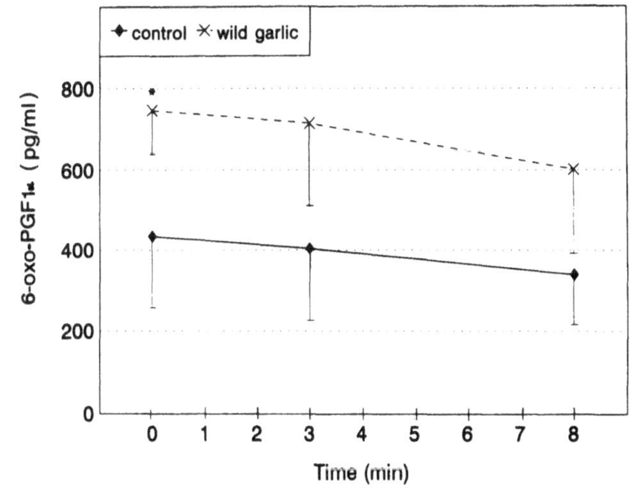

Fig. 3. Concentration of prostacyclin in the perfusate at the time zero, i.e., before the occlusion and after 3 and 8 min (n = 36). * signifies statistical significance ($p < 0.05$).

Inhibition of plasma angiotensin converting enzyme (ACE) and effects on systolic blood pressure

In order to examine whether wild garlic shows any inhibitory effect on ACE, the activity of this enzyme in plasma was determined *in vitro,* i.e., the lyophilized extract of wild garlic was added to the reaction medium. In the *in vivo* experiments plasma of rats fed the wild garlic diet was tested. The *in vitro* inhibition of ACE by a wild garlic extract is shown in Fig. 4. The inhibitory

Table 2. Effects of a diet containing 2% wild garlic powder on reperfusion-induced arrhythmias (Langendorff heart preparation)

	Ventricular tachycardia			Ventricular fibrillation	
	time until occurrence (min)	incidence (%)	duration (s)	time until occurrence (min)	incidence (%)
Control (n = 10)	0.25 ± 0.05	100	16.5 ± 2.3	0.6 ± 0.17	90
Wild garlic (n = 10	1.0 ± 0.3	70*	21.4 ± 3.9	1.45 ± 0.4	50*

* signifies statistical significance, $p < 0.05$.

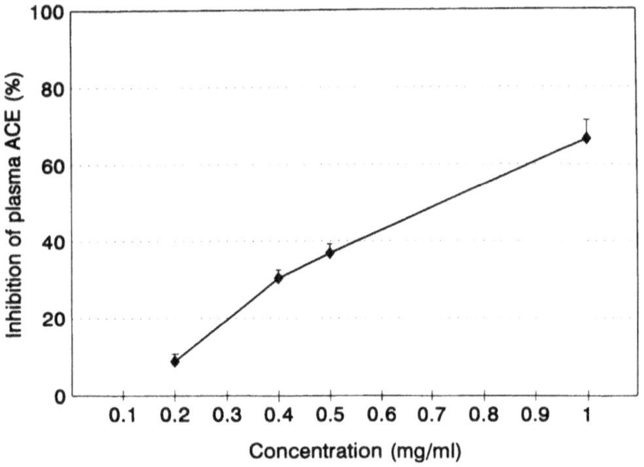

Fig. 4. In vitro inhibition of plasma angiotensin converting enzyme (ACE) by an extract of wild garlic (n = 36).

potency increased proportionally to the concentration of wild garlic. It can be seen, however, that high concentrations of wild garlic were required.
The plasma ACE activity in the wild garlic group was significantly lower than in the control group. The activity in plasma (expressed as nmol His-Leu) released per 1 ml plasma in 1 min, was 26.4 ± 1.0 (n = 5) in the wild garlic group as compared to 30.0 ± 1.2 (n = 5) (= − 12%).

During the 12 weeks feeding the systolic blood pressure of five SHR was determined. A tendency towards lower values could be seen in the wild garlic group (average − 9 mmHg) which was, however, not statistically significant.

Fatty acid composition of heart muscle

To test whether the antiarrhythmic effects of wild garlic arise from an altered fatty acid composition of heart muscle, this composition was analyzed both in the diet group and control group. However, significant changes could not be observed (Table 3).

Discussion

Considering the well-known difficulties in the prevention and pharmacological treatment of life-threatening cardiac arrhythmias, it seems appropriate to systematically investigate also dietary and phytotherapeutic interventions. A number of studies demonstrated cardioprotective actions of diets enriched with polyunsaturated fatty acids [8, 16, 17]. Recently, we could confirm antiarrhythmic effects of various oil diets, particularly fish oil and linseed oil, in the isolated heart preparation of the rat [9]. Similar results were obtained for the first time after long-term feeding of garlic powder [2]. The results of the present study demonstrate that a dietary supplement of wild garlic can significantly reduce the size of the ischemic zone as well as ischemia- and reperfusion-induced arrhythmias. It remains to be shown to what extent such effects can be obtained under more physiological conditions involving, for example, the blood perfused heart or the *in vivo* heart. The Langendorff preparation has to be regarded as a model which nonetheless is suitable for clarifying certain aspects. In the present study the perfusion pressure and potassium concentration of the perfusion fluid were adjusted to result in a high incidence of ventricular flutter and fibrillation during ischemia and reperfusion. The results of our investigations cannot identify the constituents of wild garlic which might be decisive for the cardioprotective action. We used the pulverized dried leaves that contain various substances. Besides aminoacids, proteins and adenosine various sulfur-containing compounds are present. There are the two presursor aminoacids alliin and methylcysteinesulfoxide which are converted into allicin, methyl-allyl-thiosulfinate and dimethyl-thiosulfinate by the enzyme allii-

Table 3. Effects of a diet containing 2% wild garlic on the fatty acid composition of the heart

Fatty acid	Control (%)	Wild garlic (%)
14:0	0.6 ± 0.1	0.3 ± 0.04
16:0	16.5 ± 0.9	14.6 ± 0.5
16:1, n-7	1.0 ± 0.32	0.7 ± 0.3
18:0	18.5 ± 0.8	20.6 ± 1.0
18:1, n-9	6.4 ± 0.9	5.9 ± 1.8
18:2, n-6	23.6 ± 1.5	23.0 ± 1.8
20:3, n-6	0.2 ± 0.04	0.2 = 0.1
20:4, n-6	18.0 ± 0.5	19.1 ± 3.9
18:3, n-3	0.3 ± 0.05	0.3 ± 0.1
20:5, n-3	traces	traces
22:5, n-3	2.0 ± 0.4	1.2 ± 0.3
22:6, n-3	7.5 ± 1.5	8.6 ± 2.0
saturated	35.6	35.5
monounsaturated	11.3	10.5
polyunsaturated	54.5	54.7
total n-6	41.8	42.3
total n-3	9.8	10.1
n-3/n-6	0.23	0.24

nase, whenever the cell integrity is destroyed. Since these substances are unstable, further metabolites involving mono-, di- and trisulfides, thioles and thiosulfinates are produced. These sulfur-containing compounds are supposed to be the active ingredients. Compared to garlic, there are only minor differences in the quantity of these compounds [18, 19].

As regards the cardioprotective mechanisms of these compounds, no definite conclusions can be drawn. Certainly, the size of the ischemic zone plays a role. An important protective mediator against arrhythmias is prostacyclin which is produced in vascular walls by cyclooxygenase [20]. Whereas Wagner *et al.* [19] demonstrated as *in vitro*-inhibition of cyclooxygenase by an extract of garlic and wild garlic, we found in the Langendorff preparation an enhanced generation of prostacyclin. The inhibition of cyclooxygenase by ASA in the perfusate evoked a higher incidence of ventricular fibrillation in the hearts of both the wild garlic group and the control hearts. Cyclooxygenase inhibition did, however, not abolish the cardioprotective effect of wild garlic completely, and thus it seems that despite a moderate increase in synthesis, prostacyclin does not play a decisive role in the antiarrhythmic action of wild garlic. Inhibition of ACE results in well documented beneficial effects in ischemic hearts [6]. ACE converts angiotensin I into angiotensin II which, diluted in the perfusate, aggravates the arrhythmogenic effects of ischemia, increases infarction zones, and catabolizes bradykinin. The latter is known to reduce arrhythmias and the size of the ischemic zones in isolated ischemic hearts [21].

The *in vitro*-inhibition of ACE by an extract of wild garlic as found in our experiments is consistent with the results of Elbl [7], who showed a dose-dependent inhibition of ACE. The proposed active substance was γL-glutamyl-(+)-S-allyl-L-cysteinesulfoxide, which fits in a molecular model of the ACE active site. After 12 weeks feeding the ACE-activity in the plasma of the wild garlic group was reduced compared to the control group. This inhibition could contribute to the cardioprotective effects of wild garlic, although blood pressure was only slightly influenced.

After feeding various oil diets, a close correlation was observed between the changes in cardiac membrane fatty acid composition and the antiarrhythmic action [9]. Since wild garlic like garlic [2] did not induce significant changes in the myocardial fatty acids, the cardioprotective effect of these agents appear not to depend on the membrane fatty acid composition.

Oxygen free radical generation has been proposed to be a major mechanism particularly in the pathogenesis of reperfusion injury [22]. In this context studies on antioxidant and free radical scavenging activity of garlic and alliin [23] are of great interest. As the compounds of garlic and wild garlic are very lipophilic, they might be incorporated into the myocardial membranes acting as radical scavengers. This presumption still needs to be examined.

Acknowledgement

We thank Dr. Pandalis, Teutopharma, Glandorf/Germany, for supplying the wild garlic powder and Mr. H. Sweidan for gas chromatographic analysis.

References

1. Isensee H, Rietz B, Jacob R: Antiarrhythmic effects of diets rich in polyunsaturated fatty acids (PUFA) or garlic. Internat. Congress on Phytotherapie, München, 1992
2. Isensee H, Rietz B, Jacob R: Antiarrhythmic actions of garlic (Allium sativum). Drug Res. in press, 1992
3. Johnston KM, MacLeod BA, Walker MJA: Effects of aspirin and prostacyclin on arrhythmias resulting from coronary artery ligation and infarct size. Br J Pharmac 78: 29–37, 1983
4. Coker SJ, Parratt JR: The effects of prostaglandins E2, F2, prostacyclin, flurbiprofen and aspirin on arrhythmias resulting from coronary artery ligation in anaesthetized rats. Br J Pharmac 74: 155–160, 1981
5. Linz W, Schölkens BA, Han YF: Beneficial effects of converting enzyme inhibitor, ramipril, in ischemic rat hearts. J of Card Pharm 8 (suppl 10): S91–S99, 1986
6. Hock CE, Ribeiro LGT, Lefer AM: Preservation of ischemic myocardium by a new converting enzyme inhibitor, enalaprilic acid, in acute myocardial infarction. Am Heart J 109: 222–228, 1985
7. Elbl G: Chemisch-biologische Untersuchungen pflanzlicher Inhibitoren des Angiotensin I Converting Enzyms (ACE) insbesondere die der Arzneipflanzen Lespedeza capitate (MICHX) und Allium ursinum (L). Inaugural-Dissertation, München 1991
8. McLennan PL, Abeywardena MY, Charnock JS: Dietary fish oil prevents ventricular fibrillation following coronary artery occlusion and reperfusion. Am Heart J 116: 709–17, 1988
9. Isensee H, Al Makdessi S, Jacob R: Effects of diets rich in Omega-3 and Omega-6 fatty acids on cardiac membrane phospholipids and susceptibility to arrhythmias in the rat. Internat. Congress on Omega-3 Fatty Acids. Metabolism and Biological Effects. Oslo/Norway, 1992
10. Langendorff O: Untersuchungen am überlebenden Säugetierherzen. Pfluegers Arch 61: 291–332, 1895
11. Manning SA, Hearse DJ: Reperfusion-induced arrhythmias: Mechanisms and prevention. J Moll Cell Cardiol 16: 497–518, 1984

12. Walker MJA, Curtis MJ, Hearse DJ *et al.*: The Lambeth Convention: guidelines for the study of arrhythmias in ischemia, infarction and reperfusion. Cardiovasc Res 22: 447–455, 1988
13. Folch J, Lees M, Sloane-Stanley GH: A simple method for the isolation and purification of total lipides from animal tissues. J Biol Chem 226: 497–509, 1975
14. Morrison WR, Smith LM: Preparations of fatty acid methylesters and dimethylacetals from lipids with fluoride-methanol. J Lipid Res 5: 262–270, 1964
15. Depierre D, Roth H: Fluorimetric determination of dipeptidyl carboxypeptidase. Enzyme 19: 65–70, 1975
16. Leprán I, Nemecz G, Koltai M, Szekeres L: Effect of a linoleic acid-rich diet on the acute phase of coronary occlusion in conscious rats: Influence of indometacin and aspirin. J Cardiovasc Pharmac 3: 847–853, 1981
17. Riemersma RA, Sargent CA: Dietary fish oil and ischaemic arrhythmias. J Int Med 225 (suppl 1): 111–6, 1989
18. Wagner H, Sendl A: Bärlauch und Knoblauch. Deutsche Apothekerzeitung 33: 1809–1815, 1991
19. Sendl A, Wagner H: Comparative chemical and pharmacological investigations of extracts of Allium ursinum (wild garlic) and Allium sativum (garlic). Planta med 56: 588–589, 1990
20. Bönner G, Rahn KH: Prostacyclin und Hypertonie. Springer-Verlag, Belin, Heidelberg, New York, London, Paris, Tokyo, Hong Kong, 1989
21. Linz W, Martorana PA, Schölkens A: Local inhibition of bradykinin degradation in ischemic hearts. J Cardiovasc Pharmacol 15 (suppl 6): S99–S109, 1990
22. Meerson FZ, Kagan VE, Kozlov YuP, Belkina LM, Arkhipenko YuV: The role of lipid peroxidation in the pathogenesis of ischemic heart damage and the antioxidant protection of the heart. Basic Res Cardiol 77: 465–485, 1982
23. Kourounakis PN, Rekka EA: Effect on active oxygen species of Alliin and Allium sativum (garlic) powder. Res Comm Chem Path Pharmacol 72: 249–252, 1991.

Molecular and Cellular Biochemistry **119**: 151–157, 1993.

Na^+/H^+ exchanger and reperfusion-induced ventricular arrhythmias in isolated perfused heart: possible role of amiloride

Seibu Mochizuki, Shingo Seki, Masa-aki Ejima, Tatsuyuki Onodera, Masayuki Taniguchi and Shin-ichiro Ishikawa
Department of Internal Medicine, Aoto Hospital, The Jikei University School of Medicine, Japan

Abstract

The roles of the Na^+/H^+ exchange system in the development and cessation of reperfusion induced ventricular arrhythmias were studied in the isolated perfused rat heart. The hearts were perfused in the working heart mode with modified Krebs Henseleit bicarbonate (KHB) buffer and whole heart ischemia was induced by a one-way ball valve with 330 beat/min pacing. Ischemia was continued for 15 min followed by 20 min of aerobic reperfusion (control). Amiloride (1.0 mM), an inhibitor of the Na^+/H^+ exchange system, was added to the KHB buffer only during reperfusion (group B) or only during ischemic periods (group C). Electrocardiographic and hemodynamic parameters were monitored throughout the perfusion. Coronary effluent was collected through pulmonary artery cannulation and PO_2, PCO_2, HCO_3^- and pH were measured by blood-gas analyzer.

The incidence of reperfusion induced ventricular arrhythmias was 100%, 100% and 0% in control, group B and group C, respectively. The mean onset time of termination of reperfusion arrhythmias was significantly shorter in group B than in control. PCO_2 increased from 39.0 ± 0.9 to 89.3 ± 6.0 mmHg at the end of ischemia in control and from 40.6 ± 0.4 to 60.5 ± 5.8 in group C, the difference between groups was statistically significant. HCO_3^- level decreased from 21.8 ± 0.1 to 18.3 ± 0.5 mmol/l in control, however, this decrease was significantly inhibited in group C (from 22.0 ± 0.5 to 20.3 ± 0.2). The increase in PCO_2 and the decrease in HCO_3^- in group B were similar over time to those observed in control. The decrease in pH produced by ischemia was marked in control (from 7.35 ± 0.01 to 6.92 ± 0.04) and group B (from 7.34 ± 0.01 to 6.94 ± 0.02), whereas a decrease in pH was significantly prevented in group C (from 7.34 ± 0.01 to 7.15 ± 0.04). There were no significant differences in PCO_2, HCO_3^- or pH among the three groups during reperfusion.

These experiments provide evidence that amiloride significantly prevented the incidence of reperfusion arrhythmias when added only during ischemia and significantly terminated reperfusion arrhythmias when added only during reperfusion. Amiloride may prevent a decrease in pH, due to alterations in PCO_2 and/or HCO_3^-. These changes in PCO_2 and HCO_3^- might be indirectly influenced by inhibition of the Na^+/H^+ exchange system via Cl^-/HCO_3^- exchange. The mechanism by which amiloride terminates reperfusion arrhythmias seems to involve electrophysiological effects which were not directly addressed in this experiment. (Mol Cell Biochem **119**: 151–157, 1993).

Key words: Na^+/H^+ exchanger, amiloride, reperfusion-induced arrhythmias

Address for offprints: S. Mochizuki, Department of Internal Medicine, Aoto Hospital, The Jikei University School of Medicine, 6-41-2, Aoto, Katsushika-ku, Tokyo / 125, Japan

Introduction

It is generally believed that intracellular Ca^{2+} ($[Ca^{2+}]_i$) plays a key role in the regulation of myocardial cell viability [1–4]. Two mechanisms have been suggested for Ca^{2+} extrusion by the sarcolemma: 1) a Na^+/Ca^{2+} exchange system which uses the energy of the Na^+ gradient to extrude Ca^{2+} [5–8]; and 2) a Ca^{2+} ATPase in which the energy for Ca^{2+} extrusion is provided by ATP hydrolysis [9]. Thus, inhibition of either of these two systems is likely to lead to an increase in $[Ca^{2+}]_i$; such inhibition occurs with decreased intracellular pH (pH_i) resulting from hypoxia and/or ischemia [10–12]. Therefore, alterations in intracellular H^+ concentrations ($[H^+]_i$) may play a primary role in cell damage. To maintain the pH_i level, a proton expelling system is required.

Various studies have shown the existence of an electroneutral Na^+-H^+ exchange system in cardiac cells that is important in the regulation of intracellular Na^+ ($[Na^+]_i$) and $[H^+]_i$ [13–22]. This system is a critical determinant of ion regulation, particularly during reperfusion [23, 24]. However, the mechanisms underlying reperfusion injury are not well known, although many authors have demonstrated the important role of the Na^+/H^+ exchanger in the development of reperfusion injury or arrhythmias [25–27]. The main purpose of this work was to establish whether, in the working heart preparation, the Na^+-H^+ exchange system plays a role in the genesis and termination of reperfusion arrhythmias.

Methods

Heart perfusion

Male Sprague Dawley rats (300–350 g) were anesthetized with pentobarbital sodium (50 mg/kg IP), and the hearts were removed. The hearts were then perfused by the Langendorff technique and the pulmonary vein and pulmonary artery were cannulated. After perfusion, a working heart mode was established as previously described with non-recirculating buffer, using a 7 mmHg preload and a 60 mmHg afterload [28, 29]. The control perfusion fluid was modified Krebs-Henseleit bicarbonate buffer containing in mmolar: NaCl 118; KCl 4.7; $NaHCO_3$ 25; $CaCl_2$ 2.5; $MgSO_4 \cdot H_2O$ 1.2; EDTA 0.5; KH_2PO_4 1.2; and glucose 11; the fluid was continuously equibrated with 95% O_2: 5% CO_2 (pH 7.4, 37°C). Subsequent to the 5 min control perfusion, whole heart ischemia was induced by the use of a one-way ball valve which prevented retrograde perfusion during diastole [30] and the hearts were perfused with electrical pacing (300 beats/min) for 15 min. After reperfusion, the pacing and the one-way ball valve were stopped for an additional 20 min (control group: n = 7). Amiloride (1 mM) was added to the perfusate only during reperfusion (n = 7) or only during ischemia (n = 7). Amiloride was purchased from Sigma.

Hemodynamic measurement and electrocardiogram

Aortic flow and coronary flow were measured with an electromagnetic blood flowmeter (Nihon-Koden, MFV 2100), which was set at the aortic outflow tract and pulmonary artery, respectively. Left ventricular (LV) pressure and LV max dP/dt (LV dP/dt) were measured with a polygraph system (Fukuda Denshi Co. Tokyo MIC 8600) by insertion of the catheter through the pulmonary vein and left atrium into the left ventricular cavity. The electrocardiogram was monitored by the use of electrodes (Carbonrode® Fukuda Denshi Co. Tokyo) which were attached to the surface of the heart.

Coronary effluent was collected at 5 min intervals and used for the measurement of venous PO_2, PCO_2, HCO_3^- and pH by a blood-gas analyzer (Corning 175, USA).

Statistical analysis

Student's *t*-test was used to determine the differences among the 3 groups, and the Fisher exact probability test was used to analyze the incidence of ventricular arrhythmias during the reperfusion periods. All values were expressed as mean ± SEM.

Results

1. Mean changes in coronary flow during preischemic, ischemic and reperfused periods (Fig. 1).

   Coronary flow decreased during ischemia to the same extent among the 3 groups. The recovery rate

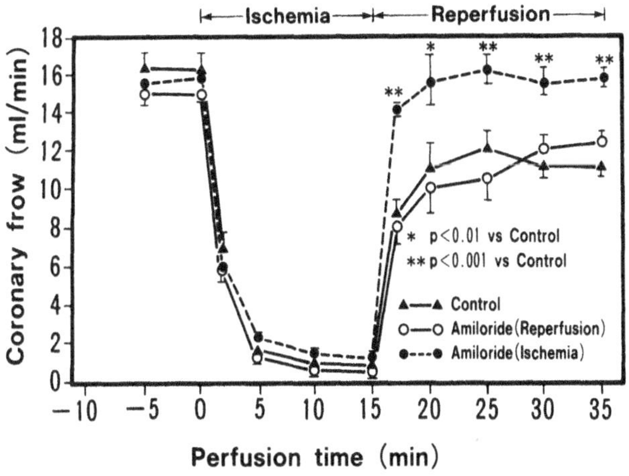

Fig. 1. Variations in coronary flow as a function of time. At time 0, whole heart ischemia was induced for 15 min, followed by reperfusion for 20 min. Closed triangles with solid lines represent the control group, open circles with solid lines represent the hearts to which 1.0 mM amiloride was added only during ischemia. Closed circles with dotted lines represent the hearts to which 1.0 mM amiloride was added only during ischemia.

Each point represents the mean ± standard error of the mean for 7 hearts, in this and all subsequent figures.

of coronary flow during reperfusion in the amiloride-treated hearts during ischemia was significantly higher than in either control or the amiloride-treated hearts during reperfusion.

2. The incidence and duration of ventricular arrhythmias (Fig. 2)

The incidence of ventricular arrhythmias during reperfusion was 100% (7/7), 100% (7/7) and 0% (0/7) in control and in amiloride-treated hearts during reperfusion and during ischemic periods, respectively. Mean duration of ventricular arrhythmias was significantly less in amiloride-treated hearts during reperfusion than in control.

3. Mean changes in R-R intervals calculated from the electrocardiogram as a function of time (Fig. 3).

The R-R intervals did not change throughout the perfusion in control, whereas they lengthened with time in amiloride-treated hearts during reperfusion.

4. Mean changes in PCO_2 during preischemic, ischemic and reperfused periods (Fig. 4).

An increase in mean PCO_2 at the end of ischemia was significantly suppressed in the amiloride-treated hearts during ischemia when compared with the control and the amiloride-treated hearts during reperfusion. During reperfusion, however, there were no significant differences among the 3 groups.

5. Mean changes in HCO_3^- during preischemic, ischemic and reperfused periods (Fig. 5).

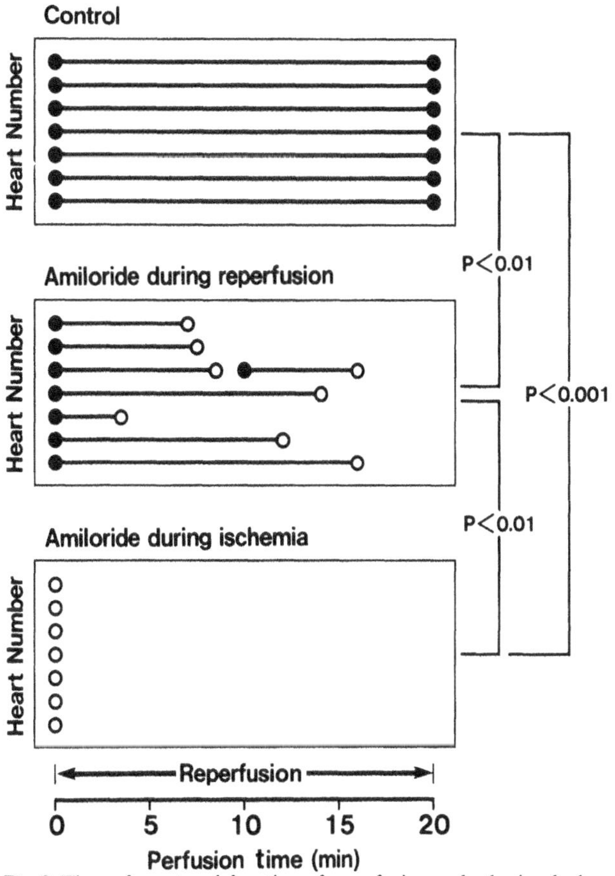

Fig. 2. Time of onset and duration of reperfusion arrhythmias during reperfusion period.

The time of onset is indicated by a closed circle at the left end of each line, and the time of termination of reperfusion arrhythmias is shown by a closed circle at the right end of each line. The distance between the left and right closed circles represents the duration of reperfusion arrhythmias. Open circles represent the individual heart in which arrhythmias did not occur during reperfusion.

A decrease in mean HCO_3^- at the end of ischemia was significantly inhibited in the amiloride-treated hearts during ischemia when compared with the control and the amiloride-treated hearts during reperfusion. During reperfusion, however, there were no significant differences among the 3 groups.

6. Mean changes in pH during preischemic, ischemic and reperfused periods (Fig. 6).

There were no significant differences in mean pH in the coronary effluent during the preischemic period. However, pH decreased to 6.92 ± 0.04 and 6.96 ± 0.05 in the control and amiloride-treated hearts during reperfusion, respectively, at the end of ischemia, whereas the decrease in pH in the amiloride-treated hearts during ischemia was significantly less (7.12 ± 0.04) than in the control.

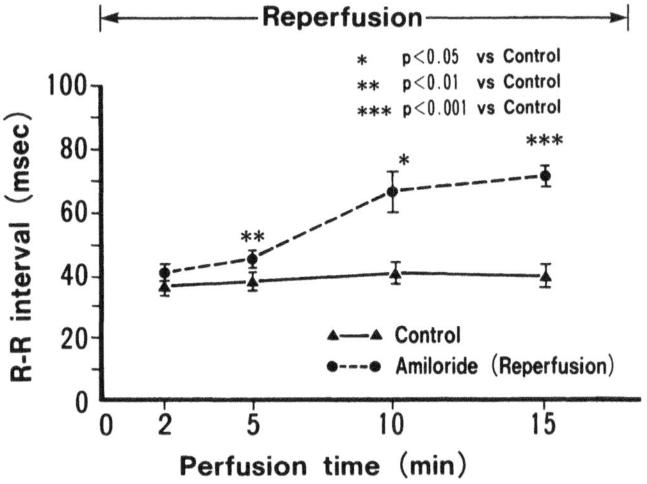

Fig. 3. Effect of amiloride on the R-R intervals in the electrocardiogram in control (closed triangles with solid lines) and amiloride-treated hearts (closed circles with dotted lines) only during reperfusion.

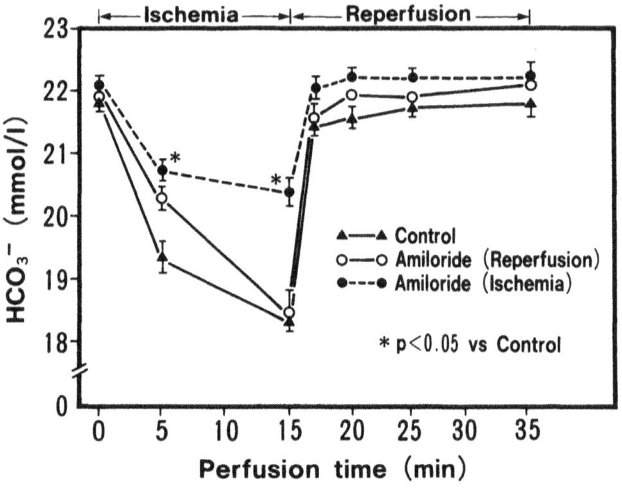

Fig. 5. Variations in bicarbonate in the coronary effluent during preischemic, ischemic and reperfusion periods.

These data were obtained from the hearts described in Fig. 1 and 2.

Discussion

The contractility of the heart muscle is dependent upon the relative intracellular concentrations or activities of Na^+, H^+ and Ca^{2+} [31–33]. Each ion has a potent influence on cell volume regulation as well as on myocardial contractility [20, 34]. In addition, there are several complex interactions among Na^+, H^+ and Ca^{2+} in heart muscle.

The Na^+/H^+ exchange system is not a primary regulator of pH_i under physiological conditions, however this system becomes a major pH_i regulator under conditions of acidosis (pH 6.6–6.7) [14]. Therefore, intracellular acidification stimulates Na^+ influx via the Na^+/H^+ exchange system, leading to an increase in $[Na^+]_i$ which, in turn, promotes Ca^{2+} accumulation via the Na^+/Ca^{2+} exchange system [19]. An increase in $[Na^+]_i$ is another critical determinant in the regulation of cell integrity [35–37]. An increase in $[Na^+]_i$ occurs in hypoxia, metabolic inhibition [38] and/or ischemic myocardium [39, 40], due to the operation of the Na^+/H^+ exchange system. Thus, any intervention which blocks the Na^+/H^+ exchange system leads to a reduction in $[Na^+]_i$. In fact, several authors have reported that amiloride

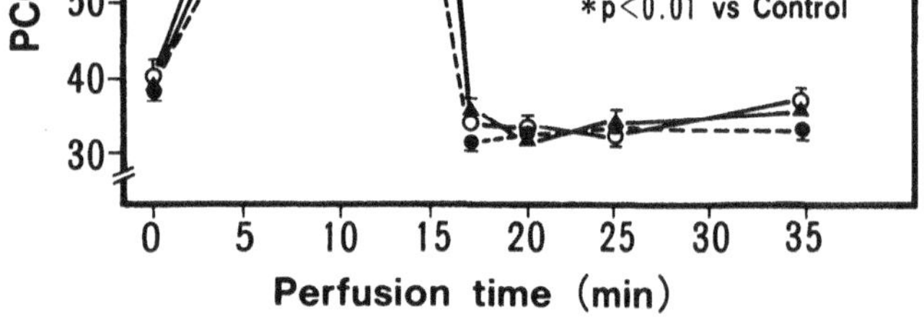

Fig. 4. Variations in CO_2 tension in coronary effluent during preischemic, ischemic and reperfused periods.

Each sample was collected at the time indicated.

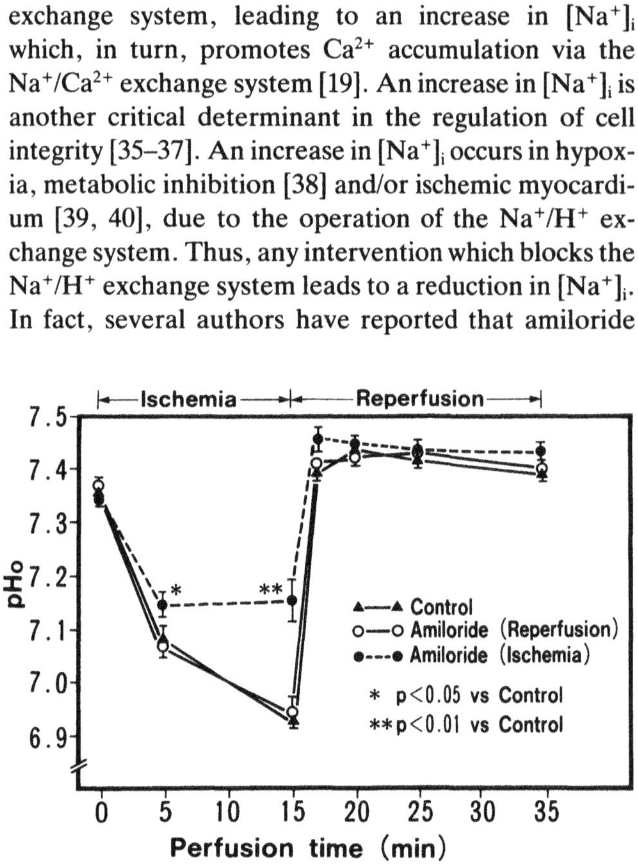

Fig. 6. Variations in pH in the coronary effluent during preischemic, ischemic and reperfused periods.

and its derivatives prevented any increase in intracellular Na^+ activity (a^iNa) during ischemia [41] or intracellular acidosis [42]. However, it is controversial whether the inhibition of the Na^+/H^+ exchange system alters $[pH]_i$ during either ischemia or reperfusion. Theoretically, if the Na^+/H^+ exchange system is blocked by amiloride, $[H^+]_i$ remains inside the cell because of the H^+ extrusion system is inhibited. Murphy *et al.* [41] reported that pH_i as well as ATP levels were the same in the presence and absence of amiloride during 20 min of ischemia, although the improved recovery of contractile function correlates with low $[Ca^{2+}]_i$ and $[Na^+]_i$ in the presence of amiloride. In the present study, however, a fall in extracellular pH (pH_o) during ischemia was prevented in hearts treated with amiloride only during ischemic perfusion. This protective effect of amiloride might be associated with a low incidence of reperfusion arrhythmias. We have estimated that the fall in pH_o reflects the fall in pH_i, based on the fact that pH_o and pH_i are related [43]. Accordingly, it is more likely that the beneficial effect of amiloride is due to its effect on pH_i, which is associated with alterations in PCO_2 and HCO_3^- during ischemia, rather than a direct effect of amiloride on the Na^+/H^+ exchange system.

The mechanism by which the increase in PCO_2 during ischemia is attenuated in hearts treated with amiloride is unknown, however it is speculated that amiloride might have preserved the mitochondrial function as reported by Duan *et al.* [44], which may lead to modifications in CO_2 production. The retention of carbon dioxide and subsequent intracellular respiratory acidosis is one of the major causes of depressed myocardial contraction [45]. Carbon dioxide will freely diffuse from the intracellular to the extracellular fluid, so that CO_2 in the coronary effluent will rapidly follow changes in intracellular CO_2 [45].

Amiloride significantly prevented a fall in extracellular HCO_3^- during ischemia when compared with the control group. Less is known about the mechanism underlying the preservation of HCO_3^- seen during ischemia. When pH_i is lowered, H^+ is extruded and HCO_3^- is transported into the cell; therefore, HCO_3^- in the extracellular space declines during ischemia. When amiloride is present during ischemia, this decrease in HCO_3^- is inhibited, suggesting that amiloride may indirectly contribute to the Cl^-/HCO_3^- exchange system, although Cl^-/HCO_3^- exchange accounts for approximately 20% of the pH_i regulating system [46, 47]. In addition, the Cl^-/HCO_3^- exchange system is reported to prevent the cell from becoming too alkaline by enhancing HCO_3^- efflux and Cl^- influx [48]. It will be of interest in the future to determine whether an inhibitor such as 4-acetamido-4'-isothiocyanostilbene-2, 2'-disulfonic acid (SITS) interferes with the action of amiloride.

During reperfusion, the Na^+/H^+ exchange system becomes more operational since the washout of interstitial H^+ is promoted by the reperfusion. Indeed, many investigators have shown that the stimulation of the Na^+/H^+ exchange system, particularly in the early phase of reperfusion, may be responsible for an accumulation of $[Na^+]_i$ which subsequently leads to an increase in $[Ca^{2+}]_i$ [1, 16, 50]. Therefore, interactions which lessen $[Na^+]_i$ at the onset of reperfusion are believed to attenuate injury following reperfusion of ischemic myocardium. Amiloride infused during reperfusion may, therefore, reduce reperfusion injury.

Karmazyn [27] and Meng *et al.* [50] have shown that post ischemic ventricular recovery was enhanced with amiloride. The improved contractile recovery was accompanied by a substantial reduction in the release of creatine kinase, although amiloride had no apparent effect when it was added only at the onset of reperfusion. Weiss *et al.* [50] have shown that amiloride improved recovery of developed pressure and pH_i 45 min after reoxygenation. They also demonstrated that amiloride limits cellular Ca^{2+} gain during reoxygenation. However, a direct role of the Na^+/H^+ exchange system in reperfusion injury has not been fully established.

In the present study, the Na^+/H^+ exchange system might be involved in the development of reperfusion arrhythmias, however, it is difficult to explain why the reperfusion arrhythmias were terminated by amiloride since there was no significant difference in pH_o during reperfusion between the heart with and that without amiloride. Electrophysiologycal effects of amiloride may contribute to this mechanism under these circumstances.

Amiloride has been reported to lengthen the action potential duration [51, 52], and to suppress the slow inward current and the outward current in voltage clamp experiments with rabbit sino-atrial node [53]. Duff *et al.* [54, 55] have reported that amiloride has antiarrhythmic activity in that it suppresses sustained ventricular tachyarrhythmias in the post infraction model, and this antiarrhythmic activity relates to selective blockade of the inwardly rectifying potassium current (IK_1). Thus, we suggest that this potent electrophysiological effect of amiloride, which was added to the buffer, might be the primary mechanism during reperfusion in this experiment. In fact, prolongation of R-R intervals was observed in amiloride-treated hearts during reperfusion, which may relate to the beneficial effect of the drug on termination of reperfusion arrhythmias.

One unresolved issue in this experiment is whether amiloride inhibits the Na^+/Ca^{2+} exchange system during either ischemia or reperfusion. Amiloride has proven to be extremely useful for defining the physiological regulation of the ion transport system [14], however it may block a number of other transport systems. Amiloride is a weak inhibitor of the Na^+/Ca^{2+} exchange system with an IC_{50} of 1 mM [56], and the Na^+/Ca^{2+} exchange in sarcolemmal vesicle isolated from beef heart is inhibited at concentrations of amiloride ranging between 0.1 and 0.8 mM [14].

The Na^+/Ca^{2+} exchange system plays an important role in the regulation of $[Ca^{2+}]_i$ and pH_i, and this system is inhibited during acidosis or hypoxia and metabolic inhibition results in intracellular Ca^{2+} overload [57]. Thus, inhibition of the Na^+/Ca^{2+} exchange system might be disadvantageous in the regulation of Na^+ and Ca^{2+} as well as H^+. However, at the concentration of amiloride used in this study, the Na^+/H^+ exchange system is reportedly blocked almost completely, whereas Na^+/Ca^{2+} exchange is still 70% or more active [16]. Thus, the Na^+/Ca^{2+} exchange system may have been operative during either ischemia or reperfusion in this experiment, which may have induced myocardial cells to expel Ca^{2+}.

In conclusion, the inhibitory effect of amiloride on the incidence of reperfusion arrhythmias most likely results from prevention of a decrease in pH_i as a consequence of alterations in PCO_2 and HCO_3^-. The Na^+/H^+ exchanger and Cl^-/HCO_3^- exchanger are suspected of being linked to the maintenance of $[pH]_o$ and $[pH]_i$. This relative preservation of pH homeostasis might serve as an anti-arrhythmogenic property. The decrease in the duration of reperfusion arrhythmias, which was observed in the amiloride-treated hearts only during reperfusion, may be attributed to an electrophysiological effect of amiloride.

Acknowledgements

This study was supported by a grant from the Ministry of Education, Science and Culture, Japan, grant number 04770564.

We thank Drs. Tomoko Kinoshita, Fumio Tanaka and Mitsuru Okumura for their assistance. The expert assistance of Miss Yumiko Ikimi in the preparation of the paper is gratefully acknowledged.

References

1. Allen DG, Eisner DA, Lab MJ, Orchard CH: The effects of low sodium solutions on intracellular calcium concentration and tension in ferret ventricular muscle. J Physiol 345: 391–407, 1983
2. Carafoli E: The homeostasis of calcium in heart muscle. J Mol Cell Cardiol 17: 203–212, 1985
3. Steenbergen C, Murphy E, Watts JA, London RE: Correlation between cytosolic free calcium, contracture, ATP and irreversible ischemic injury in perfused rat heart. Circ Res 66: 135–146, 1990
4. Nayler WG, Panagiotopoulos S, Elz JS, Daly MJ: Calcium-mediated damage during post-ischemic reperfusion. J Mol Cell Cardiol 20 (Suppl 2): 41–54, 1988
5. Reuter H, Seitz N: The dependence of calcium efflux from cardiac muscle on temperature and external ion composition. J Physiol 195: 451–470, 1968
6. Eisner DA, Lederer WJ: Na-Ca exchange: Stoichiometry and electrogenicity. Am J Physiol 248: C189–C202, 1985
7. Chapman RA, Coray A, McGuigan JAS: Sodium/Calcium exchange in mammalian ventricular muscle: A study with sodium – sensitive micro-electrodes. J Physiol 343: 253–276, 1983
8. Kirby MS, Orchard C, Boyett MR: The control of calcium influx by cytoplasmic calcium in mammalian heart muscle. Mol Cell Biochem 89: 109–113, 1989
9. Barry WH, Rasmussen CAF Jr, Ishida H, Bridge JHB: External Na-independent Ca extrusion in cultured ventricular cells. Magnitude and functional significance. J Gen Physiol 88: 393–411, 1986
10. Bersohn MM, Philipson KD, Fukushima JY: Sodium-calcium exchange and sarcolemmal enzymes in ischaemic rabbit heart. Am J Physiol 248: C288–C295, 1982
11. Daly MJ, Elz JS, Nayler WG: Sarcolemmal enzymes and Na^+-Ca^{2+} exchange in hypoxic, ischaemic and reperfused rat hearts. Am J Physiol 247: H237–H243, 1984
12. Mochizuki S, MacLeod KT: The effects of hypoxia on cytoplasmic Ca^{2+} during low Na exposure in isolated cardiac myocytes from guinea-pig. J Physiol 446: 334, 1991
13. Piwnica-Worms D, Jacob R, Horres CR, Lieberman M: Na/H exchange in cultured chick heart cells. J Gen Physiol 85: 43–64, 1985
14. Frelin C, Vigne P, Lazdunski M: The role of the Na^+/H^+ exchange system in the regulation of the internal pH in cultured cardiac cells. Eur J Biochem 149: 1–4, 1985
15. Pierce GN, Philipson KD: Na^+-H^+ exchange in cardiac sarcolemmal vesicles. Biochim Biophys Acta 818: 109–116, 1985
16. Luzdunski M, Frelin C, Vigne P: The sodium/hydrogen exchange in cardiac cells: Its biochemical and pharmacological properties and its role in regulating internal concentrations of sodium and internal pH. J Mol Cell Cardiol 17: 1029–1042, 1985
17. Grinstein S, Rothstein A: Mechanisms of regulation of the Na^+/H^+ exchanger. J Membrane Biol 90: 1–12, 1986
18. Kaila K, Vaughan-Jones RD: Influence of sodium-hydrogen exchange on intracellular pH, sodium and tension in sheep cardiac Purkinje fibres. J Physiol 390: 93–118, 1987
19. Kim D, Cragoe EJ Jr, Smith TW: Relations among sodium pump inhibition, Na-Ca and Na-H exchange activities, and Ca-H interaction in cultured chick heart cells. Cirs Res 60: 185–193, 1987
20. Cala PM, Anderson SE, Cragoe EJ Jr: Na/H exchange-depend-

ent cell volume and pH regulation and disturbances. Comp Biochem Physiol 90A: 551–555, 1988
21. Meno H, Jarmakani JM, Philipson KD: Developmental changes of sarcolemmal Na^+/H^+ exchange. J Mol Cell Cardiol 21: 1179–1185, 1989
22. Blatter LA, McGuigan JAS: Intracelluar pH regulation in ferret ventricular muscle. The role of Na-H exchange and the influence of metabolic substrates. Circ Res 68: 150–161, 1991
23. Meng HP, Lonsberry BB, Pierce GN: Influence of perfusate pH on the postischemic recovery of cardiac contractile function: Involvement of sodium-hydrogen exchange. J Pharmacol Exp Ther 258: 772–777, 1991
24. Khandoudi N, Bernard M, Cozzone P, Feuvray D: Intracellular pH and role of Na^+/H^+ exchange during ischaemia and reperfusion of normal and diabetic rat hearts. Cardiovasc Res 24: 873–878, 1990
25. Karmazyn M: Ischemic and reperfusion injury in the heart. Cellular mechanisms and pharmacological interventions. Can J Physiol Pharmacol 69: 719–730, 1990
26. Dennis SD, Coetzee WA, Cragoe EJ Jr, Opie LH: Effects of proton buffering and of amiloride derivatives on reperfusion arrhythmias in isolated rat hearts. Circ Res 66: 1156–1159, 1990
27. Karmazyn M: Amiloride enhances postischemic ventricular recovery: Possible role of Na^+-H^+ exchange. Am J Physiol 255: H608–H615, 1988
28. Neely JR, Liebermeister H, Battersby EJ, Morgan HE: Effect of pressure development on oxygen consumption by isolated rat heart. Am J Physiol 212: 804–814, 1967
29. Mochizuki S, Okumura M, Tanaka F, Sato T, Kagami A, Tada N, Nagano M: Ischemia-reperfusion arrhythmias and lipids: Effect of human high- and low-density lipoproteins on reperfusion arrhythmias. Cardiovasc Drug Ther 5: 269–276, 1991
30. Neely JR, Rovetto MJ, Whitmer JT, Morgan HE: Effects of ischemia on function and metabolism of isolated working rat heart. Am J Physiol 225: 651–658, 1973
31. MacLeod KT: Regulation and interaction of intracellular calcium, sodium and hydrogen ions in cardiac muscle. Cardioscience 2: 71–85, 1991
32. Kim D, Smith TW: Cellular mechanisms underlying calcium-proton interactions in cultured chick ventricular cells. J Physiol 398: 391–410, 1988
33. Kaila K, Vaughan-Jones RD, Bountra C: Regulation of intracellular pH in sheep cardiac Purkinje fibre: Interactions among Na^+, H^+ and Ca^{2+}. Can J Physiol Pharmacol 65: 963–969, 1986
34. Jennings RB, Reimer KA: The cell biology of acute myocardial ischemia. Annu Rev Med 42: 225–246, 1991
35. Renlund DG, Gerstenblith G, Lakatta EG, Jacobus WE, Kallman CH, Weisfeldt ML: Perfusate sodium during ischemia medifies post-ischemic functional and metabolic recovery in the rabbit heart. J Mol Cell Cardiol 16: 795–801, 1984
36. Tani M, Neely JR: Role of intracellular Na^+ in Ca^{2+} overload and depressed recovery of ventricular function of reperfused ischemic rat hearts. Possible involvement of H^+-Na^+ and Na^+-Ca^{2+} exchange. Circ Res 65: 1045–1056, 1989
37. Haigney MCP, Miyata H, Lakatta EG, Stern MD, Silverman HS: Dependence of hypoxic cellular calcium loading on Na^+-Ca^{2+} exchange. Circ Res 71: 547–557, 1992
38. MacLeod KT: Effects of hypoxia and metabolic inhibition on the intracellular sodium activity of mammalian ventricular muscle. J Physiol 416: 455–468, 1989
39. Poole-Wilson PA, Tones MA: Sodium exchange during hypoxia and on reoxygenation in the isolated rabbit heart. J Mol Cell Cardiol 20 (Suppl 2): 15–22, 1988
40. Anderson SE, Murphy E, Steenbergen C, London RE, Cala PM: Na-H exchange in myocardium: Effects of hypoxia and acidification on Na and Ca. Am J Physiol 259: C940–C948, 1990
41. Murphy E, Perlman M, London RE, Steenbergen C: Amiloride delays the ischemia – induced rise in cytosolic free calcium. Circ Res 68: 1250–1258, 1991
42. Bountra C, Vaughan-Jones RD: Effect of intracellular and extracellular pH on contraction in isolated mammalian cardiac muscle. J Physiol 418: 163–187, 1989
43. Poole-Wilson PA, Cameron IR: Intracellular pH and K^+ of cardiac and skeletal muscle in acidosis and alkalosis. Am J Physiol 228: 1305–1310, 1975
44. Duan J, Karmazyn M: Protective effects of amiloride on the ischemic reperfused rat heart. Relation to mitochondrial function. Eur J Pharmacol 210: 149–157, 1992
45. Cobbe SM, Poole-Wilson PA: The time of onset and severity of acidosis in myocardial ischaemia. J Mol Cell Cardiol 12: 745–760, 1980
46. Vanheel B, de Hemptinne A, Leusen I: Analysis of Cl^-/HCO_3^- exchange during recovery from intracellular acidosis in cardiac Purkinje strands. Am J Physiol 246: C391–C400, 1984
47. Mahnensmith RL, Aronson PS: The plasma membrane sodium-hydrogen exchanger and its role in physiological and pathophysiological processes. Circ Res 57: 773–788, 1985
48. Vanghan-Jones RD: Chloride activity and its control in skeletal and cardiac muscle. Phil Trans R Soc Lond B 299: 537–548, 1982
49. Meng HP, Pierce GN: Protective effects of 5-(N, N-dimethyl) amiloride on ischemia-reperfusion injury in hearts. Am J Physiol 258: H1615–H1619, 1990
50. Weiss RG, Lakatta EG, Gerstenblith G: Effects of amiloride on metabolism and contractility during reoxygenation in perfused rat hearts. Circ Res 66: 1012–1022, 1990
51. Kennedy RH, Berlin JR, Ng YC, Akera T, Brody TM: Amiloride: Effects on myocardial force of contraction, sodium pump and Na^+/Ca^{2+} exchange. J Mol Cell Cardiol 18: 177–188, 1986
52. Marchese AC, Hill JA Jr, Xie PD, Strauss HC: Electrophysiologic effects of amiloride in canine Purkinje fibers: Evidence for a delayed effect on repolarization. J Pharmacol Exp Ther 232: 485–491, 1984
53. Satoh H, Hashimoto K: An electrophysiological study of amiloride on sino-atrial node cells and ventricular muscle of rabbit and dog. Arch Pharmacol 333: 83–90, 1986
54. Duff HJ, Brown E, Cragoe EJ, Rahmberg M: Antiarrhythmic activity of amiloride: Mechanisms. J Cardiovasc Pharmacol 17: 879–888, 1991
55. Duff HJ, Lester WM, Rahmberg M: Amiloride. Antiarrhytmic and electrophysiological activity in the dog. Circulation 78: 1469–1477, 1988
56. Kleyman TR, Cragoe EJ Jr: The mechanism of action of amiloride. Seminars in Nephrology 8: 242–248, 1988
57. DiPolo R, Beaugé L: The effect of pH on Ca^{2+} extrusion mechanism in dialyzed squid axons. Biochim Biophys Acta 688: 237–245, 1982

Molecular and Cellular Biochemistry **119**: 159–169, 1993.

Effects of a new class I antiarrhythmic drug bidisomide on canine ventricular arrhythmia models

Wu Zhenjiu*, Takeo Awaji, Akira Hirasawa, Shigeru Motomura and Keitaro Hashimoto

Department of Pharmacology, Yamanashi Medical College, Tamaho-cho, Nakakoma-gun, Yamanashi, Japan 409–38 (Present address: Department of Pharmacology, Faculty of Basic Medicine, Yanbian Medical College, Jilin Province, China)*

Abstract

The antiarrhythmic and direct cardiovascular effects of a new antiarrhythmic agent, bidisomide, α-{2-[acetyl(1-methylethyl)amino]ethyl}-α-(2-chlorophenyl)-1-piperidinebutanamide, were investigated. To determine the antiarrhythmic effects, spontaneously occurring adrenaline-, digitalis- and two-stage coronary ligation-induced arrhythmias were used. Bidisomide suppressed these three arrhythmia models. The antiarrhythmic plasma concentration, IC_{50}, of bidisomide for digitalis-induced arrhythmia was 22.1 μg/ml, and those calculated for intravenous bidisomide in 24 h and 48 h coronary ligation-arrhythmias were 15.1 and 11.6 μg/ml and that calculated for oral bidisomide in 24 h coronary ligation-arrhythmia was 5.4 μg/ml and that for adrenaline induced arrhythmia was 58.7 μg/ml. In the blood perfused sinoatrial node and papillary muscle preparations, bidisomide decreased the sinoatrial rate and contractile force and increased the intraventricular conduction time and coronary blood flow. These results indicate that bidisomide is similar to other class I antiarrhythmic agents such as pirmenol and KW-3401 in its antiarrhythmic profile and is expected to become a clinically useful antiarrhythmic drug. (Mol Cell Biochem **119:** 159–169, 1993)

Key words: bidisomide, ventricular arrhythmias, contractile force, intraventricular conduction

Introduction

Bidisomide, α-{2-[acetyl(1-methylethyl)amino]ethyl}-α-(2-chlorophenyl)-1-piperidinebutanamide (Fig. 1), is a newly synthesized antiarrhythmic agent effective on canine ventricular and atrial arrhythmia models [1, 2] and is reported to decrease Vmax of cardiac action potential with slow kinetic properties and shorten action potential duration (APD) [3]. These electrophysiological effects are characteristics of class I antiarrhythmic drugs and sub-classified as class Ib according to their effects on the action potential duration, but bidisomide is different from lidocaine or mexiletine whose kinetics of interaction with Na channels are slow. Bidisomide is also reported to be devoid of serious cardiovascular side effects [4]. We have been reporting effects of various antiarrhythmic drugs using canine ventricular arrhythmia models, and classified antiarrhythmic drugs based on their pharmacological effectiveness [5–8]. Bidisomide was used for the same experimental arrhyth-

Address for offprints: K. Hashimoto, Department of Pharmacology, Yamanashi Medical College, Tamaho-cho, Nakakoma-gun, Yamanashi, Japan 409–38

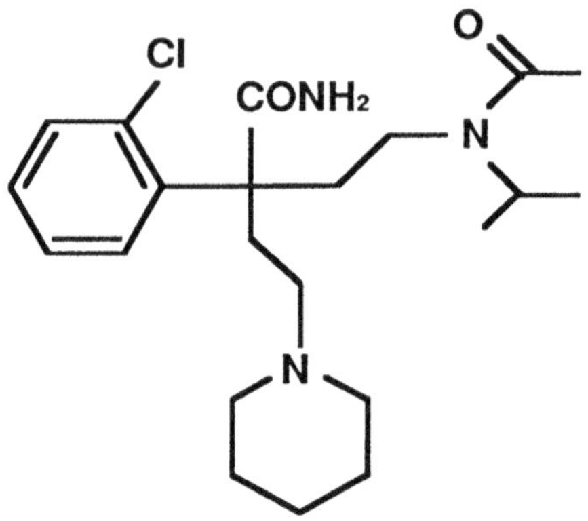

Fig. 1. Structure of bidisomide.

mia models to examine to which class bidisomide belongs compared to other antiarrhythmic drugs and to evaluate its direct effects on cardiac functions of the cross circulated blood-perfused canine sinoatrial node (SA) and papillary muscle (PM) preparations.

Material and methods

Production of two-stage coronary ligation-induced arrhythmia

Twelve female beagle dogs, weighing 9–13 kg, were anesthetized initially with intravenous thriopental sodium, 30 mg/kg, and intubated. As reported earlier [5], the chest was opened and a two-stage coronary ligation of the left anterior descending artery (LAD) was performed under halothane anesthesia.

Experiments were done without anesthesia 24 h and 48 h after coronary ligation. The lead II ECG, atrial electrogram from implanted electrodes sutured on the left atrial appendage, and the instantaneous and mean blood pressure were recorded continuously using a telemetry system (Nihon Kohden, Tokyo, Japan). A constant rate infusion of bidisomide (5 mg/kg) was performed for 10 min using a syringe pump (Terumo, Tokyo, Japan), and arterial blood samples were drawn from one lumen of the arterial double lumen cannula at 0, 1, 3, 5, 7, 9, 15, 30 and 60 min after the start of the infusion.

For oral administration studies, 35 mg/kg bidisomide in a gelatin capsule was administered to 6 dogs, 24 h after coronary ligation. Arterial blood samples were drawn at 0, 0.5, 1, 1.5, 2, 2.5, 3, 3.5, 4, 5, 6 and 24 h after drug administration.

Production of digitalis-induced arrhythmia

Six mongrel dogs of either sex, weighing 9–13 kg, were anesthetized with intravenous pentobarbital sodium, 30 mg/kg. As reported earlier [6], 40 μg/kg ouabain was injected intravenously and then followed by an additional 10 μg/kg every 20 min until stable ventricular tachycardia was produced. A constant rate infusion of bidisomide (5 mg/kg) was performed for 10 min using a syringe pump, and arterial blood samples were drawn from one lumen of the arterial double lumen cannula at 0, 1, 3, 5, 7, 9, 15, 30 and 60 min after the start of the infusion.

The lead II ECG, atrial electrogram from catheter tip electrodes in the right atrium, and the instantaneous and mean blood pressure were continuously recorded.

Production of adrenaline-induced arrhythmia

Six mongrel dogs of either sex, weighing 8–10.5 kg, were anesthetized initially with thiopental sodium. As reported earlier [7], after intubation, 1.0% halothane, vaporized with 100% oxygen, was administered with a volume-limited ventilator (20 ml/kg, 15 strokes/min). Adrenaline was infused through the left femoral vein at a rate of 2.5 μg/kg/min using a syringe pump. If multifocal ventricular tachycardia was not induced, a higher infusion rate was employed. A constant rate infusion of bidisomide (20 mg/kg) was started 3 min after adrenaline infusion and continued for 10 min using another syringe pump, and arterial blood samples were drawn from one lumen of the arterial double lumen cannula 1 min before and 1, 3, 5, 7, 9, 13 and 15 min after the start of the infusion.

The lead II ECG, atrial electrogram from catheter tip electrodes in the right atrium and blood pressure were continuously recorded.

Determination of bidisomide plasma levels

The venous blood samples were collected into heparinized syringes at the predetermined time and centrifuged at 3000 rpm for 5 min. The plasma layer was frozen until subsequent determination of bidisomide levels using a high-performance liquid chromatographic method [1].

Evaluation of the antiarrhythmic effects

The severity of ventricular arrhythmia was expressed by the arrhythmic ratio: the number of ventricular ectopic beats divided by the total heart rate. The total heart rate is the number of all beats counted from the 5 sec strip of ECG (i.e., the number of ventricular ectopic beats plus the number of conducted beats), and the ventricular beats were judged by the different shape of the ventricular complex from the normal QRS complex. The arrhythmic ratio before drug injection was almost 1 as shown in the control values of the figures and there were no spontaneous improvements in these ratios. If the values after drug administration were decreased significantly from the 0 time value, the drug was judged as having significant effects. The antiarrhythmic plasma concentrations of drugs were determined as follows: a simple linear regression was calculated between the data of plasma concentrations of a drug and the data of arrhythmic ratio when the plasma was obtained using a computer program Stat View 512+ (Brain power Inc., Calabasas, CA, U.S.A.), and the plasma concentration that decreased the arrhythmic ratio to 50% of that at 0 concentration was calculated as 50% inhibitory concentration, IC_{50}. The probability and the range of 95% confidence limit were also calculated by the computer program.

Cross circulated blood-perfused cardiac preparations

Two mongrel dogs of either sex were used. One dog, weighing 10–15 kg was used as a donor dog and another recipient dog, weighing approximately 10 kg, was used to obtain isolated preparations. The recipient dog was anesthetized with pentobarbital sodium (30 mg/kg i.v.), given heparin calcium (500 units/kg i.v.) and exsanguinated. The heart was excised and plunged into a cold Tyrode's solution kept at about 4° C, then the sinoatrial node and the papillary muscle preparations were prepared.

As reported earlier [9], the SA node preparation consisted of the entire right atrium, and the sinus node artery was cannulated *via* the right coronary artery (RCA). The sinoatrial rate (SAR) of the preparation was measured with a tachometer (NEC San-ei Instrument, 1321, Tokyo, Japan) triggered by an electrogram recorded from bipolar atrial electrodes sutured close to the SA node.

The PM preparation consisted of an anterior papillary muscle attached to the intraventricular septum and the anterior septal artery (ASA) was directly cannulated. This preparation was electrically driven at a fixed rate of 2 Hz using a stimulator and an isolation unit (Dia Medical, DHM-226-3 and DPS-110, Tokyo, Japan) with rectangular pulses of 1–3 V and 5 msec duration *via* bipolar silver-silver chloride stimulating electrodes sutured onto the His bundle of the ventricular septum. Developed tension (DT) of the PM was measured isometrically using a force displacement transducer (Dia Medical, DRM-T200) and an amplifier (Dia Medical, DRM-T20) under a resting tension at which maximum tension developed (approximately 2 g). The intraventricular conduction time was measured as the time between the stimulus artifact and distal electrogram on the tip of the PM.

Donor dogs were anesthetized initially with pentobarbital sodium (30 mg/kg i.v.), and given an additional 4–5 mg/kg every hour and 500 units/kg of heparin calcium was injected, followed by an additional 200 units/kg every hour. After intubation, dogs were artificially ventilated with room air. The left carotid artery and the external jugular vein were cannulated for drainage of the arterial blood and return of the blood to the venous system. The systemic blood pressure from the left femoral artery and the heart rate triggered by the R wave of standard lead II ECG were continuously monitored.

For cross circulation, the SA node and PM preparations were placed in double-wall glass jackets maintained at 38° C. Then both preparations were simultaneously cross-circulated *via* each cannulated artery with heparinized arterial blood of the donor dog under a constant perfusion pressure of 100 mmHg using a Starling pneumatic resistance placed parallel to the perfusion system. Drained venous blood from the preparations and excess blood passing through the pneumatic resistance were collected in a blood reservoir and returned to the donor dog. The rate of blood flow *via* the ASA and PSA was measured by an electromagnetic flow meter (Nihon Kohden MFV-1100).

Analysis of results

Data are expressed as the mean ± S.D. in the experiments for two-stage ligation-, digitalis- and adrenaline-induced VTs. Analysis of variance (ANOVA) and Dunnett test for paired data were used for the determination of statistically significant differences between the mean values obtained at various times after treatment and the

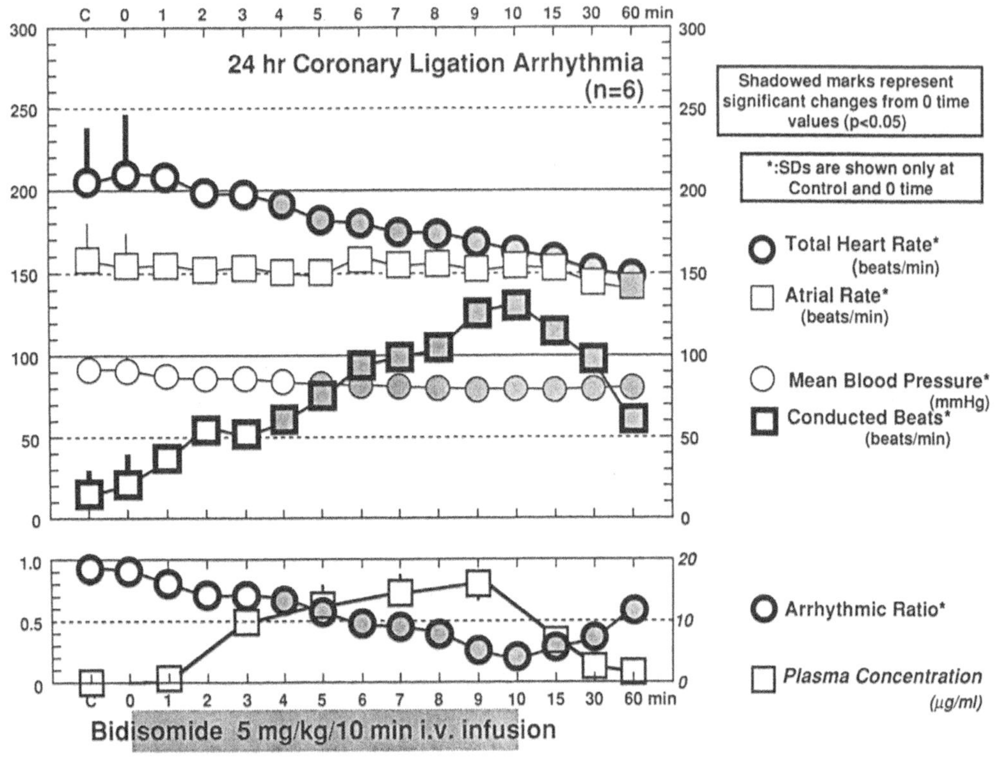

Fig. 2. Effects of intravenous infusion of bidisomide (5 mg/kg/10 min) on the 24 h coronary ligation arrhythmia.

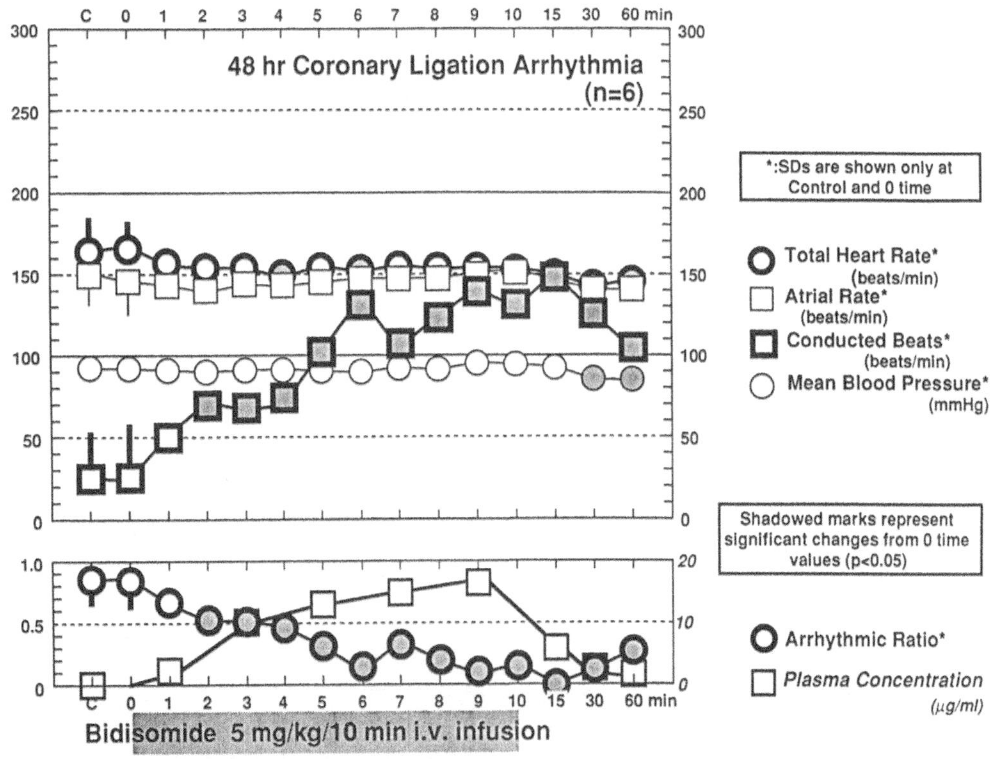

Fig. 3. Effects of intravenous infusion of bidisomide (5 mg/kg/10 min) on the 48 h coronary ligation arrhythmia.

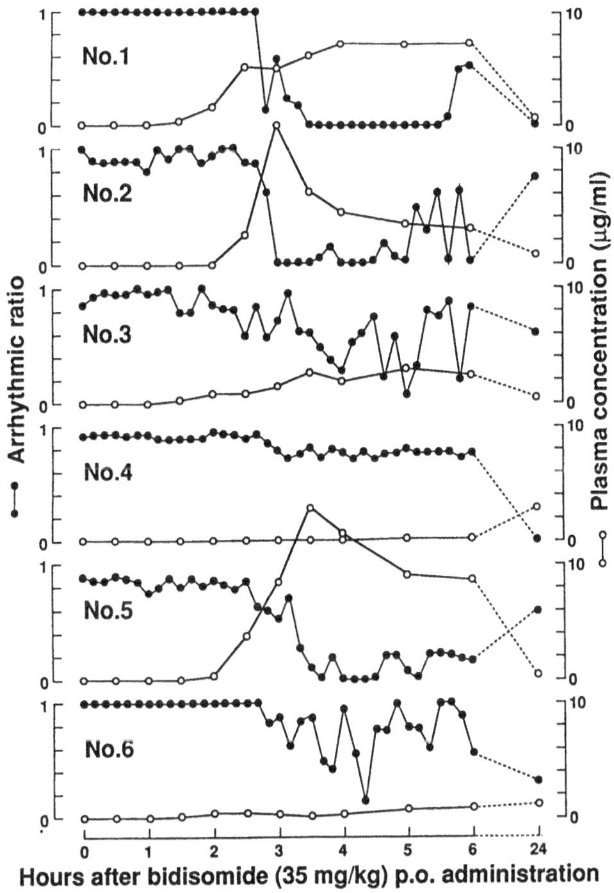

Fig. 4. Effects of oral bidisomide (35 mg/kg) on the 24 h coronary ligation arrhythmia. Only the arrhythmic ratio and plasma concentration of bidisomide are shown.

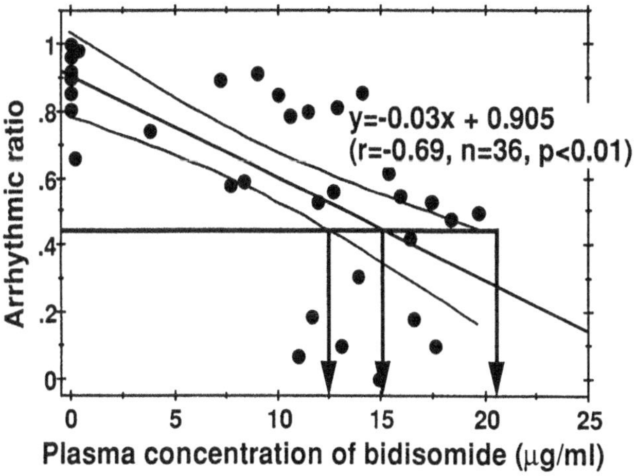

Fig. 5. Correlation between the plasma bidisomide concentration and the arrhythmic ratio of the beagles with 24 h coronary ligation arrhythmias given intravenous infusion of bidisomide (5 mg/kg/10 min). Effective antiarrhythmic concentration was determined as that decreasing the control arrhythmic ratio 50%, i.e., 15.1 µg/ml.

pretreatment, 0 time, values. In cross circulated blood-perfused experiments, data are expressed as the mean ± S.E. The doses of bidisomide that produced 50% change in the parameters of the cross circulated blood-perfused cardiac preparations were determined from dose-response curves. The doses that produced a 15% or 50% decrease were determined from dose-response curves fitted by the least squared method. A p value of 0.05 or less was considered statistically significant.

Results

Two-stage coronary ligation-induced arrhythmia

After 1–2 days of coronary ligation, all dogs showed continuously occurring multifocal ventricular ectopic beats. The arrhythmic ratios after 24 and 48 h of ligation were 0.93 ± 0.07 (mean ± S.D., n = 6) and 0.84 ± 0.17 (n = 6), respectively (Fig. 2 and 3), showing that the 48 h arrhythmia was less severe. In the preliminary experiments using 1–5 mg/kg/10 min i.v. infusion, the 5 mg/kg/10 min infusion rate was chosen as that producing antiarrhythmic effects without minimum cardiovascular depression. For 24 h arrhythmia, bidisomide decreased the total heart rate, increased the number of conducted beats and decreased the arrhythmic ratio after 4 min of infusion as shown in Fig. 2. The blood pressure did not change significantly. No dog vomited nor did they show other central nervous system side effects.

For 48 h arrhythmia, bidisomide at the same infusion rate showed stronger antiarrhythmic effects starting 2 min after the start of infusion (Fig. 3), but effects on the total heart rate, atrial rate and blood pressure were minimal. No dog vomited nor did they show other central nervous system side effects.

Oral administration at 35 mg/kg bidisomide also suppressed 24 h coronary ligation arrhythmia as shown in Fig. 4. Since the absorption of bidisomide varies in each beagle, individual data on the arrhythmic ratio and the plasma concentration are shown. During the observation period of 24 h, there were no deleterious cardiovascular or central nervous system side effects, but 1 out of 6 dogs vomited 3 h after administration.

The linear regression between the plasma concentration of intravenous bidisomide and the arrhythmic ratio data for the first 10 min in the 24 h experiments ($r = -0.69$, $n = 36$, $p < 0.01$) (Fig. 5) and in the 48 h experiments ($r = -0.58$, $n = 36$, $p < 0.01$) was obtained and the calculated antiarrhythmic plasma con-

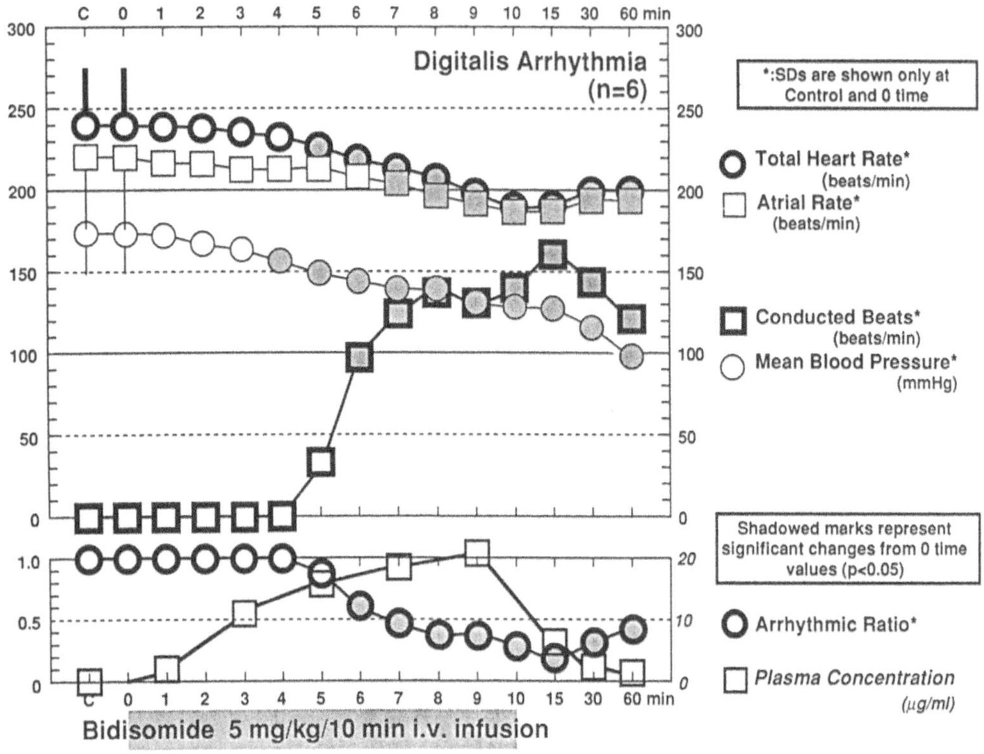

Fig. 6. Effects of intravenous infusion of bidisomide (5 mg/kg/10 min) on the digitalis arrhythmia.

centrations for 24 and 48 h coronary ligation arrhythmias were 15.1 (95% confident limit, 12.5–20.5) µg/ml and 11.6 (9.2–17.6) µg/ml, respectively. The linear regression in the oral bidisomide experiments gave the calculated antiarrhythmic plasma concentration of 5.4 µg/ml (4.3–6.8 µg/ml, $r = -0.71$, $n = 72$, $p < 0.01$) which was smaller than those obtained in intravenous bidisomide experiments.

Digitalis-induced arrhythmia

After intravenous injection of a total dose of about 70–90 µg/kg ouabain, almost all the beats became of ventricular origin. The same infusion rate as that used in the coronary ligation arrhythmia was used. As shown in Fig. 6, bidisomide significantly decreased the total heart rate, atrial rate and blood pressure and increased the number of conducted beats 6 min after the start of infusion. The arrhythmic ratio significantly decreased after 6 min and stayed low even after 60 min. The linear regression between the plasma concentration of bidisomide and the arrhythmic ratio data for the first 10 min ($r = -0.55$, $n = 36$, $p < 0.01$) resulted in the calculated antiarrhythmic plasma concentrations for digitalis arrhythmias as 22.1 (16.9–37.6) µg/ml.

Adrenaline-induced arrhythmia

As reported previously [7], adrenaline infusion for 3 min at a rate of 2.5 µg/kg/min induced tachycardia with almost all the beats consisting of ventricular ectopic beats. The same infusion rate as that used in the coronary ligation and digitalis arrhythmias was ineffective, thus 4 times higher 20 mg/kg/10 min was used. As shown in Fig. 7, bidisomide significantly increased the number of conducted beats 4 min after the start of infusion and decreased the arrhythmic ratio, total heart rate, atrial rate and blood pressure. The linear regression between the plasma concentration of bidisomide and the arrhythmic ratio data for the first 10 min ($r = -0.87$, $n = 36$, $p < 0.01$) resulted in the calculated antiarrhythmic plasma concentrations for adrenaline arrhythmias as 58.7 (53.7–69.5) µg/ml.

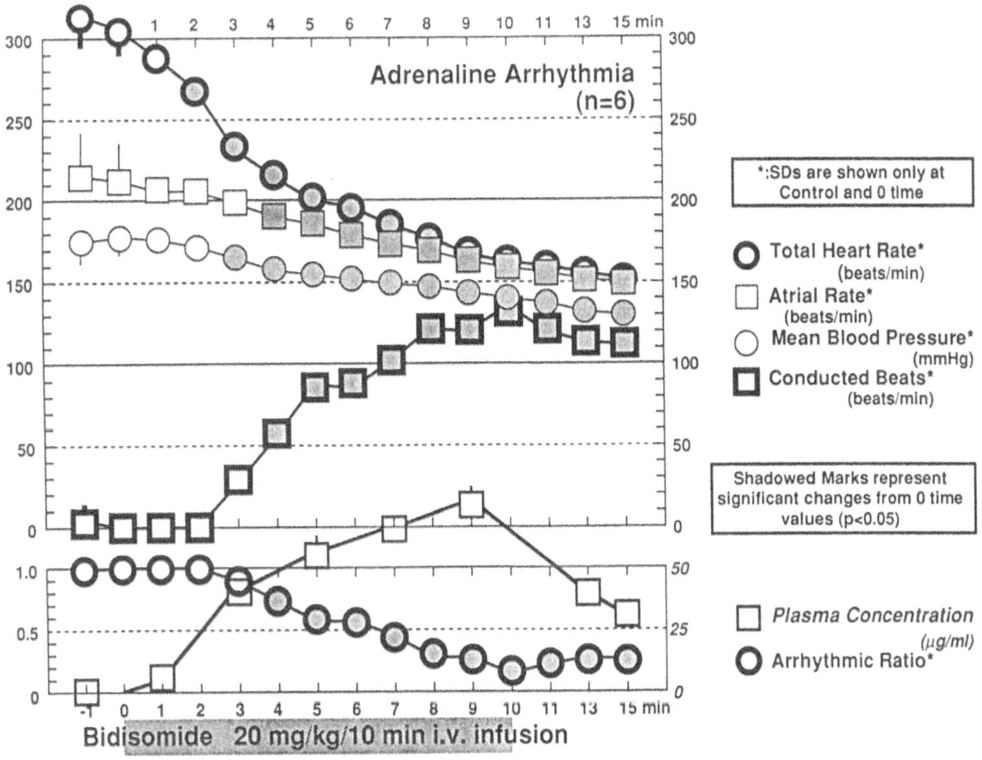

Fig. 7. Effects of higher speed of intravenous infusion of bidisomide (20 mg/kg/10 min) on the adrenaline arrhythmia.

Cross circulated blood-perfused cardiac preparations

In 6 SA node preparations, the basal sinus rate (SAR) was 79 ± 5 beats/min (n = 6). Single bolus injection of bidisomide (1 μg–10 mg) into the RCA produced a dose-dependent decrease in the SAR (Fig. 8). The dose that produced a 15% decrease in SAR was calculated as 1798 μg.

In 6 paced PM preparations, the basal developed tension (DT), basal intraventricular conduction time and basal ASA flow were 6.6 ± 11.3 g, 60 ± 7 msec and 10 ± 4 ml/min, respectively. Bidisomide (1 μg–10 mg) injected into the ASA induced a negative inotropic effect in a dose dependent manner (Fig. 8). The dose that produced a 50% decrease in DT was calculated as 3389 μg. The blood flow of the ASA tended to increase, and the intraventricular conduction time increased significantly from 10 μg, which was far less than the negative inotropic dose (Fig. 8). The doses that produced a 50% increase in the intraventricular conduction time and in the coronary flow were 3744 and 2325 μg, respectively.

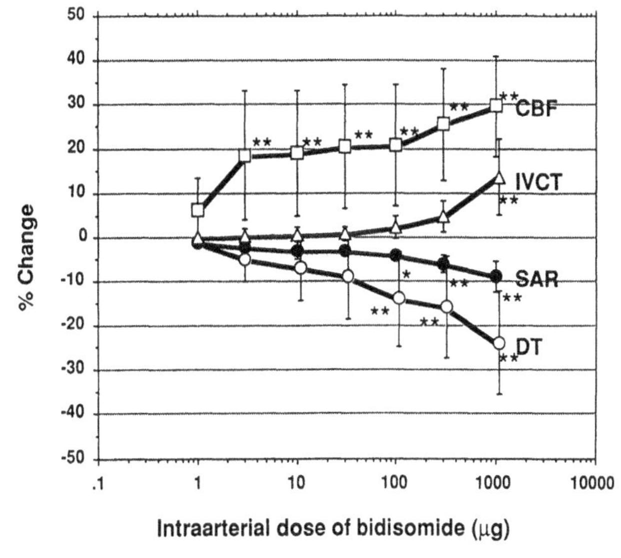

Fig. 8. Dose-response relationships of intraarterial doses of bidisomide and the sinoatrial rate of the isolated blood perfused canine sinoatrial node preparation and the developed tension, intraventricular conduction time and coronary blood flow of the isolated blood perfused canine papillary muscle preparation.

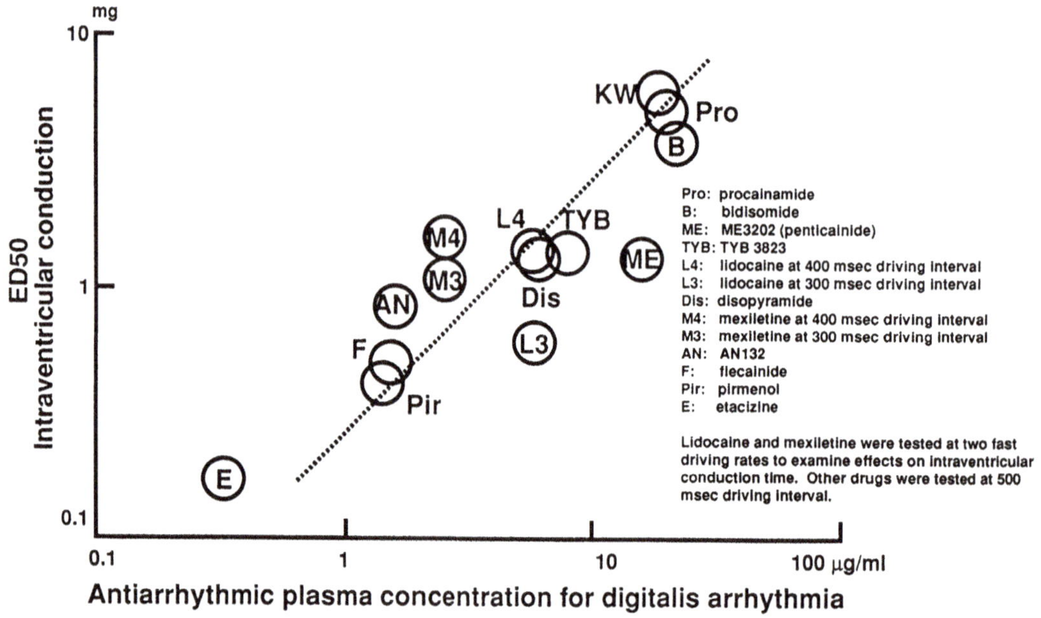

Fig. 9. Correlation between the ED50 (IVCT) and antiarrhythmic plasma concentrations of class I antiarrhythmic drugs on digitalis arrhythmias.

Discussion

The results of the present investigation indicate that bidisomide is an effective antiarrhythmic agent on various canine ventricular arrhythmia models. In 24 h and 48 h coronary artery ligation induced ventricular arrhythmias, both the intravenous and oral administrations of bidisomide suppressed the arrhythmias concentration dependently without much cardiovascular depressant effect, but with some central nervous system effects, such as vomiting and sedation. The antiarrhythmic concentrations of intravenous bidisomide for 24 h coronary ligation arrhythmia, 15.1 μg/ml were higher than those for 48 h coronary ligation, 11.6 μg/ml, but were lower than those of digitalis and adrenaline ar-

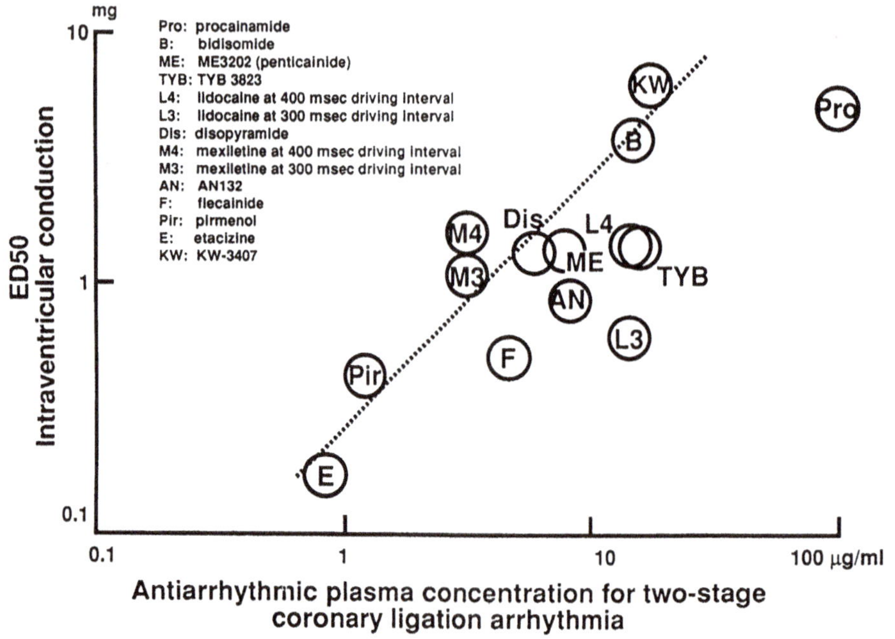

Fig. 10. Correlation between the ED50 (IVCT) and antiarrhythmic plasma concentrations of class I antiarrhythmic drugs on 24 h coronary ligation arrhythmias.

Table 1. Antiarrhythmic profiles of class I drugs on canine ventricular arrhythmias

Drugs	Canine antiarrhythmic plasma concentration (IC_{50}, μg/ml)			Triggered activity arrhythmia		Canine Na-channel block		Action potential duration	Canine Ca-channel block
	Dig Arrhy	Cor Arrhy	Adr Arrhy	Digitalis	Adrenaline	conc. (μg/ml)	kinetics		
AFD-19	1.1	3.3	4.8		(+)	0.9^{GP}	Slow	(−)	
AHR 10718	3.8	9.0	18.9			2		Shorten	
AN-132	1.6	8.3	10.6			5	Slow		
Aprindine	1.1	5.2	1.0			0.5	Slow	Shorten	
Bidisomide	22.1	15.1	58.7			4	Slow	Shorten	
Carocainide	(+)	(−)	(−)				Slow		
Cibenzoline	0.92	3.7	5.0			0.5–1	Slow	(−)	(+)
Disopyramide	6.9	5.9	(−/W)	(+)	(−/W)	5	Slow	Lengthen	(+)
E-0747	2.2	3.6	7.7			1	Slow	Shorten	
Etacizin	0.30	0.83	(−/W)			0.1	Slow	(−)	
Flecainide	1.5	4.6	2.0	(+)	(±)	3	Slow	(−)	
KT-362	6.7	8.1	1.5			1.5–5	Slow	(−)	(+)
KW-3407	18.3	18.1	21.4			5	Slow	Lengthen	?
Lidocaine	5.7	14.7	12.5	(+)	(±)	10	Fast	Shorten	(−)
Mexiletine	2.6	3.3	4.1	(+)	(±)	2	Fast	Shorten	
Nicainoprol	5.4	21.7	4.9			4		Shorten	
NS-2	3.7	4.3	(−)		(+)	0.8^{GP}	Slow	(−)	(−)
OPC88117	8.8	23.0	1.2	(+)	(+)	29	Fast	Lengthen	(+)
Penticainide	15.9	7.9	12.3			6	Slow	Shorten	
Phenytoin	12.7	17.8	11.2			5–20		Shorten	(+)
Pirmenol	1.4	1.2	3.5			0.4–4		Lengthen	
Pilsicainide	3.1	5.8	(−/W)			3	Slow	Shorten	
Procainamide	19.9	98.9	(−/W)	(+)	(−/W)	30	Slow	Lengthen	(−)
Propafenone	2.2	4.5	1.1			0.25–2	Slow	Shorten	(+, β-Block)
SA3212	8.8	52.8	1.8			2	Fast	Shorten	(+)
Tocainide	5.5	9.7	38.7			60	Fast	Shorten	
TYB-3823	8.1	16.2	5.0			0.7^{GP}	Slow	Shorten	(+)
YUTAC	0.043	0.48	0.57	(+)		$0.3–1^{GP}$	Slow	(−)	(+)

Dig Arrhy = digitalis induced arrhythmia, Cor Arrhy = coronary ligation induced arrhythmia, Adr Arrhy = adrenaline induced arrhythmia, Dir = direct effect, Ind = indirect effect via β adrenergic receptor (+) = effective, (−) = no effect, (−/W) = no effect or worsening, (±) = weak effect, conc. = concentration, ^{GP} = guinea pig data.

rhythmias. Contrary to our previous experience that the calculated antiarrhythmic plasma concentrations in intravenous and oral administration experiments are almost the same [10–14], oral bidisomide experiments gave a much lower antiarrhythmic plasma concentration of 5.4 μg/ml. Previous canine data show similar concentration ranges of less than 10 μg/ml for suppressing experimental arrhythmias [1]. Bidisomide may need time to equilibrate between the plasma concentrations and tissue concentrations, since after stopping intrave-

Table 2. Effects of class I antiarrhythmic drugs on blood perfused canine heart preparation

Drugs	ED15 (SAR)	ED50 (DT)	ED50 (IVCT)	ED50 (CBF)	ED50 (AH)	ED50 (HV)
Bidisomide	1798	3389	3744	2325		
Disopyramide	454	800		833	1410	1154
Etacizine	290	280		201	49	201
Flecainide	221	361		260	229	304
KW-3407	729	> 10000		3000	1300	6000
Pirmenol	689	922		504	234	416
TYB 3823	780	1380	1250	5200		

nous infusion of bidisomide, the antiarrhythmic effect persisted even though the blood plasma concentrations decreased to low levels in all four intravenous bidisomide experiments (Fig. 2, 3, 6, 7).

In the digitalis induced arrhythmia, bidisomide was effective and decreased the arrhythmic ratio in a concentration dependent manner. The antiarrhythmic plasma concentration was 22.1 μg/ml and was about 5 times higher than the concentrations *in vitro* [3] to suppress $\dot{V}$max of the action potentials of canine ventricular muscle, namely about 4 μg/ml (10^{-5} M). This drug concentration in crystalline solution is that of a free drug and naturally the concentration of the free drug in plasma is lower than the plasma concentration by virtue of the protein binding, thus the present result confirmed our previous results that regardless of the precise electrophysiological mechanism of generation of coronary ligation and digitalis arrhythmias, the antiarrhythmic effect on these arrhythmias is due to suppression of Na channels, probably suppressing automaticity in the ventricular tissues or decreasing conduction velocity [6, 8]. Figure 9 and 10 show the relationships between the *in vitro* Na channel inhibitory action assessed by the intracoronary inhibitory doses of class I antiarrhythmic drugs and the effective plasma concentrations of class I antiarrhythmic drugs of digitalis and 24 h coronary ligation arrhythmia experiments. The bidisomide plots were added to those reported previously [15]. There was a fairly good linear relationship between the *in vitro* Na channel inhibitory dose and effective concentrations for coronary ligation and digitalis arrhythmias, confirming again the importance of Na channel blockade in the mechanism of antiarrhythmic actions [5, 6, 8].

In the adrenaline-induced arrhythmia, bidisomide was also effective, but required higher doses and the antiarrhythmic plasma concentration, 59 μg/ml was more than 3 times higher than those obtained in other arrhythmias. We have shown that class I Na channel blockers are divided into two groups; one effective on the adrenaline arrhythmia and the other ineffective, and that all β blockers and all Ca channel blockers we tested are effective on this arrhythmia [7, 8]. It is not yet clear whether the effectiveness of class I drugs on adrenaline arrhythmia is solely due to their additional Ca channel blocking or β receptor blocking action or due to Na channel blocking action. However the present results indicate that Na channel blockade may be the antiarrhythmic mechanism even for adrenaline arrhythmia, since bidisomide has been reported not to have either Ca channel blocking effect or β blocking effect [3]. Compared to other class I antiarrhythmic drugs, the pattern of efficacy of bidisomide, i.e., effective on three arrhythmias, but least effective on adrenaline arrhythmia, resembles those of E-0747, pirmenol, tocainide and KW-3401, when compared with other class I antiarrhythmic drugs that have already been reported (Table 1) [8].

In the cross circulated blood-perfused preparations, bidisomide decreased intraventricular conduction, ventricular contractile force and sinoatrial automaticity dose-dependently. As already reported in isolated cardiac preparations in crystalline solution that bidisomide has a negative inotropic effect and coronary vasodilator effects [4], the negative inotropic and coronary dilator effects were weak in our preparations. The dose to increase intraventricular conduction time was also very high. So among the antiarrhythmic drugs tested using the blood perfused canine heart preparations (Table 2) [9, 16, 17], it is similar to etacizine, flecainide or TYB-3823 where the dose relating to the antiarrhythmic efficacy {ED_{50} (IVCT) or ED_{50} (HV)} is close to the negative inotropic dose, indicating that bidisomide can be expected to show antiarrhythmic effects without prominent negative inotropic effects.

In conclusion, the results of the present investigation indicate that bidisomide is an antiarrhythmic agent; 1) possessing efficacy in three canine arrhythmia models; digitalis, coronary ligation and adrenaline-induced arrhythmias especially for adrenaline arrhythmia, 2) having minimal vascular and central nervous system side effects, 3) having only a weak direct negative inotropic effect, and 4) having a coronary dilating effect. Based on these findings, bidisomide may deserve further study for use as an antiarrhythmic drug clinically.

Acknowledgements

The authors thank Searle Yakuhin K.K., Osaka, Japan for supplying bidisomide and for the determination of bidisomide plasma concentration. We also thank Miss Mie Yamada for preparing the manuscript. This study was supported by a Grant-in-Aids for Scientific Research (03253103) of the Japanese Ministry of Education, Science and Culture.

References

1. Garthwaite S, Hatley F, Frederick L, Ruby J, Cook C: Effect of

SC-40230, a new class I antiarrhythmic agent, on canine ventricular tachycardias. Drug Dev Res 17: 119–133, 1989
2. Spinelli W, Hoffman BF: Mechanisms of termination of reentrant atrial arrhythmias by class I and class III antiarrhythmic agents. Circ Res 65: 1565–1579, 1989
3. Martin CL, Borowicz LE, Schniepp HC, Sanguinetti MC: Voltage- and rate-dependent reduction of $\dot{V}max$ of cardiac action potentials by SC-40230. Drug Dev Res 17: 51–61, 1989
4. Frederick LG, McDonald SJ, Garthwaite SM: Cardiovascular profile of a new anti-arrhythmic agent, SC-40230. Cardiovasc Res 23: 897–903, 1989
5. Hashimoto K, Satoh H, Shibuya T, Imai S: Canine-effective plasma concentrations of antiarrhythmic drugs on the two-stage coronary ligation arrhythmia. J Pharmacol Exp Ther 223: 801–810, 1982
6. Hashimoto K, Ishii M, Komori S, Mitsuhashi H: Canine digitalis arrhythmia as a model for detecting Na-channel blocking antiarrhythmic drugs: A comparative study using other canine arrhythmia models and the new antiarrhythmic drugs, propafenone, tocainide and SUN 1165. Heart and Vessels 1: 29–35, 1985
7. Shibuya T, Hashimoto K, Imai S: Effective plasma concentrations of antiarrhythmic drugs against sustained halothane-adrenaline arrhythmia in dogs. J Cardiovasc Pharmacol 5: 538–545, 1983
8. Hashimoto K, Haruno A, Matsuzaki T, Sugiyama A, Akiyama K: Effects of antiarrhythmic drugs on canine ventricular arrhythmia models: Which electrophysiological characteristics of drugs are related to their effectiveness? Cardiovasc Drugs Ther 5: 805–818, 1991
9. Sugiyama A, Motomura S, Tamura K, Hashimoto K: Comparison of cardiovascular effects of pirmenol with those of disopyramide in isolated canine heart preparations cross-circulated with a donor dog. Jap J Pharmacol 53: 97–110, 1990
10. Hashimoto K, Komori S, Ishii M, Kamiya J: Effective plasma concentrations of aprindine in canine ventricular arrhythmias. J Cardiovasc Pharmacol 6: 12–19, 1984
11. Hashimoto K, Ishii M, Komori S: Antiarrhythmic plasma concentrations of mexiletine on canine ventricular arrhythmias. J Cardiovasc Pharmacol 6: 213–219, 1984
12. Hashimoto K, Akiyama K, Mitsuhashi H: Antiarrhythmic plasma concentrations of cibenzoline on canine ventricular arrhythmias. J Cardiovasc Pharmacol 9: 148–153, 1987
13. Hashimoto K, Akiyama K, Mitsuhashi H: Antiarrhythmic effect of a new class I antiarrhythmic drug, nicainoprol, on canine ventricular arrhythmias. Jap J Pharmacol 49: 245–254, 1989
14. Hashimoto K, Watanabe K, Mochizuki S, Tomiyama A: Effects of KT-362, a new Na and Ca influx and Ca release inhibitor, on canine ventricular arrhythmias. Jap J Pharmacol 51: 475–482, 1989
15. Hashimoto K: Pharmacology of TYB-3823, a new class I antiarrhythmic drug. Cardiovasc Drug Rev 9: 413–423, 1991
16. Hashimoto K, Sugiyama A, Haruno A, Matsuzaki T, Hirasawa A: Effects of a new antiarrhythmic drug TYB-3823 on canine ventricular arrhythmia models. J Cardiovasc Pharmacol 17: 336–342, 1991
17. Sugiyama A, Motomura S, Hashimoto K: Comparison of cardiovascular effects of a novel class Ic antiarrhythmic agent, NIK-244, with those of flecainide in isolated canine heart preparations cross-circulated with a donor dog. Jap J Pharmacol 56: 1–12, 1991

Molecular and Cellular Biochemistry **119**: 171–178, 1993.

Antimicrobial activity of lipoprotein particles containing apolipoprotein AI

Norio Tada, Takuya Sakamoto, Akihiko Kagami, Keiko Mochizuki and Kosei Kurosaka[1]

Department of Internal Medicine and [1] Department of Laboratory Medicine, Aoto Hospital, The Jikei University School of Medicine, Japan

Abstract

Human plasma *in vitro* inhibits the growth of coagulase negative staphylococci, *S. epidermidis,* which may be pathogenic in the immunocompromised host. To determine the antimicrobial components, serum was fractionated by column chromatography, which revealed that elution areas where lipoproteins can be yielded had high antimicrobial activity against *S. epidermidis*. Therefore, lipoprotein fractions, including very low density lipoprotein (VLDL), low density lipoprotein (LDL) and high density lipoprotein (HDL), were separated by ultracentrifugation and incubated with *S. epidermidis*. All 3 lipoprotein fractions suppressed bacterial growth within the first 3 h but VLDL enhanced bacterial growth after 9 h of incubation compared with the control. HDL, however, inhibited bacterial growth throughout 21 h of incubation.

To confirm these results, serum from healthy volunteers was separated by ion exchange column chromatography and again by HPLC to purify the antimicrobial fraction. In the protein analysis with gradient polyacrylamide-SDS gel, apolipoprotein AI (apo AI), which is a major apolipoprotein of HDL, was detected in the antimicrobial fraction. Therefore, this fraction was loaded onto an immunoaffinity column coupled with the anti-apo AI monoclonal antibody (Mab). Unbound fraction had no antimicrobial activity, but anti-*S. epidermidis* activity was recovered from the bound fraction which consisted mainly of apo AI, AII and apo C in protein composition.

These results indicated that the antimicrobial activity was associated with the apo AI-containing lipoprotein particles (HDL). This property of HDL may directly affect bacterial growth and promote the self-defense mechanisms of normal and immunocompromised individuals. (Mol Cell Biochem **119:** 171–178, 1993)

Key words: Antimicrobial activity, apolipoprotein AI, high density lipoprotein, immunoaffinity column, *Staphylococcus epidermidis*

Introduction

It is well recognized that ready susceptibility to pathogens and opportunistic infections are often encountered in immunocompromised hosts. There are many reports of opportunistic pathogens and great progress has been made in the field of clinical microbiology. However, only a few reports have been published on the internal defense system other than studies of phagocytic cell function and the immune mechanism including lymphocytes, immunoglobulin and complement [1]. It has been, however, reported that human serum has some

Address for offprints: N. Tada, Department of Internal Medicine, Aoto Hospital, The Jikei University School of Medicine, 6-41-2 Aoto, Katsushika, Tokyo, 125 Japan

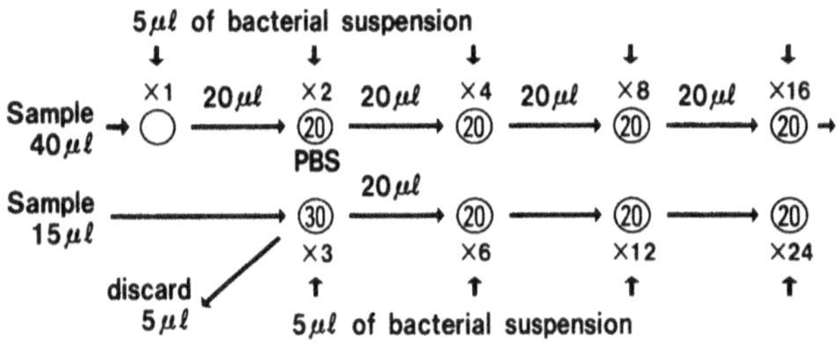

Fig. 1. Assay procedure for the measurement of antimicrobial activity (microwell plate method). The sample volume in each well is 20 micro l. After the dilution of samples by serial two-fold dilution technique, 5 micro l of bacterial suspension was added into the well in the microwell plate. Then, the plate was incubated overnight at 37° C. PBS: phosphate buffered saline.

inhibitory activity against microorganisms [2, 3] and it is well known that addition of serum into the assay system increases the antimicrobial activity of antibiotics [4]. Such serum inhibitory components, which constitute the last defense mechanism against pathogens of low virulence especially in the compromised host, are assumed to include immunoglobulin, complement, lysozymes [5] and transferrin [6, 7]. However, it is not known whether other serum components also participate in this mechanism. In the process of separating the inhibitory component from human serum, we found that the serum fraction containing apolipoprotein Al (apo Al) showed antimicrobial activity. This paper describes the preparation of the apo Al-containing serum fraction and the activity of this fraction against *S. epidermidis*.

Material and methods

Antimicrobial activity determination with microwell plate method

An ordinary assay for antimicrobial activity was performed as described by Machida *et al.* [7]. Briefly, 20 micro l of phosphate buffered saline (PBS, 0.15 M NaCl in 2 mM phosphate, pH 7.4) was poured into each well of a microplate (6 × 10 well, Nunc, Denmark, 60 micro l of well volume) and 40 micro l of each sample, which had been sterilized by passage through a 0.45 micro m Millipore filter, was added to the first well in every two rows. Then, serial two-fold dilutions of the sample were performed as shown in Fig. 1. For the preparation of the bacterial suspension, *Staphylococcus epidermidis* (H-155 strain) was incubated with tryptosoy broth (TSB) overnight at 37° C, then diluted 1: 100 in TSB and incubated another 3 h at 37° C. The bacterial count as this time was around 10^8 cfu/ml. The bacterial suspension was diluted again 1: 10^4 by 5-fold concentrated bactocasiton broth with 5% glucose to yield a bacterial concentration of 10^4 cfu/ml. Five micro l of this bacterial suspension was added to each microplate well (50 cfu/well). The microwell plates were incubated overnight at 37° C and examined for bacterial growth in the wells.

Determination of antimicrobial activity of lipoproteins (test tube method)

The lipoproteins, VLDL (d< 1.006 g/ml), LDL (1.006 < d < 1.063 g/ml) and HDL (1.063 < d < 1.21 g/ml), from 3 normal subjects were separated sequentially by ultracentrifugation according to Havel's method [8] and sterilized by passage through a 0.45 micro m Millipore filter. Lipoprotein samples from each subject were adjusted to the same protein concentration with PBS. Then, 1 ml of lipoprotein solution (VLDL, LDL and HDL) was placed in a sterile test tube and 10 micro l of bacterial suspension, containing 5 × 10^3 cfu/ml of *S. epidermidis,* was added and the test tube was incubated at 37° C. Fifty micro l of medium in the test tube was collected before and at 3, 9, and 21 h after the start of incubation and bacteria in the medium were counted.

Separation and semipurification of serum antimicrobial component

Human serum (about 3 ml) was loaded on an Ultrogel column (AcA54, 2.5 × 100 cm, IBF, France). The eluant was 0.3 M NaCl in 2.5 mM Tris-HCl, pH 7.8 and the antimicrobial activity of each column fraction was measured by the microwell plate method.

In addition, human serum was fractionated with ion exchange chromatography (DE52 Cellulose, 10 g–18 ml, Whatman U.S.A.) using an eluant of 2.5 mM Tris-HCl, pH 7.8, with a 0/1 M NaCl gradient. Fractions with antimicrobial activity separated by this ion exchange column were collected and loaded on a high performance liquid chromatography column (HPLC, TSK G3000SW, Toso, Japan) with an eluant of 0.2 M Na_2HPO_4-HCl, pH 6.8, and 0.5 ml/min as an eluting speed. For further purification of the serum

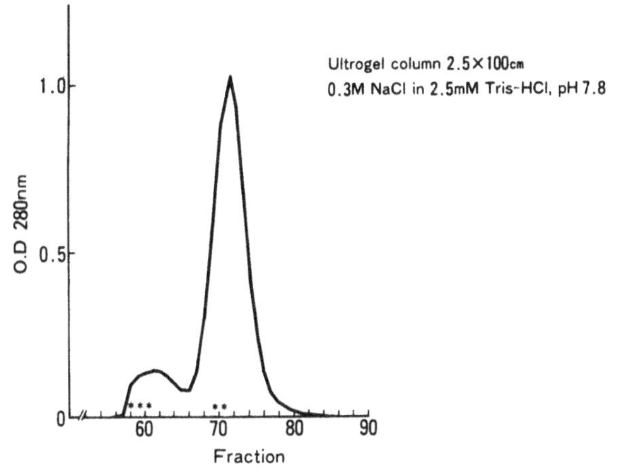

Fig. 2. The elution profile of human serum by Ultrogel column (AcA54) and antimicrobial activity. (*) denotes the fractions which inhibit the growth of *S. epidermidis* assayed with microwell plate method. Serum antimicrobial activity was distributed in the fraction from which serum lipoproteins were also recovered.

antimicrobial component, an immunoaffinity column coupled with anti-apo Al Mab (apo Al immunoaffinity column) was employed.

Apolipoprotein Al immunoaffinity chromatography

The monoclonal antibody against apo Al (H-12) was a gift of Dr. Katsuyuki Nakajima, Japan Immunoresearch Laboratories, Takasaki Japan. This Mab, raised in a Balb/c mouse immunized with apo Al purified by Sephacryl S-200 (Pharmacia, Sweden) and again by 10% polyacrylamide-SDS gel, was obtained by standard techniques [9], and produced routinely by injecting 5×10^6 hybridoma cells into pristane-primed adult Balb/c mice. The immunoglobulin fraction of anti-apo Al Mab was conjugated to CNBr-activated Sepharose 4B (Pharmacia) at a ratio of 5 mg/ml. 0.2 M glycine was used to block the remaining active sites on the gels. The column was conditioned with calf serum for 1 h at room temperature to reduce nonspecific binding and washed with 1.0 M acetic acid in a 0.5 M NaCl solution. This procedure was repeated twice before the column was used. The affinity column capacity for apo Al was 1.47 mg/ml.

An antimicrobial fraction semipurified by an ion exchange Cellulose column and further purified by HPLC was loaded on the apo Al immunoaffinity column (10 ml of column volume) equilibrated with 0.15 M NaCl in 0.01 M phosphate, pH 7.2 (first eluant). Then the column was eluted with the first eluant and the unbound fractions were collected. After the column was washed with 20 ml of the first eluant, the column was eluted with 0.5 M NaCl in 1 M acetic acid (second eluant) and the bound fractions were collected. For neutralization, a half volume of 3.5 M Tris-HCl was added to the bound fractions and the total aliquot was rapidly dialyzed against the first eluant.

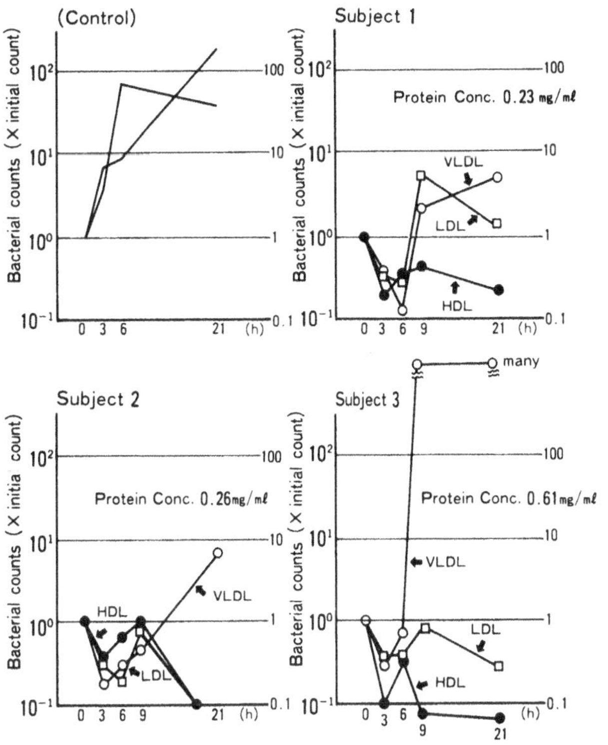

Fig. 3. Antimicrobial activity of lipoproteins. Lipoproteins (VLDL, LDL and HDL) from 3 different healthy subjects were incubated with *S. epidermidis* (H155) and bacteria were counted in the medium at 0, 3, 6, 9, and 21 h after incubation began, as described in Methods. The protein concentrations of lipoproteins were adjusted to 0.23, 0.26, and 0.61 mg/ml in subject 1, 2, and 3, respectively. In all experiments, HDL consistently suppressed bacterial growth in the medium.

Additional methods

Protein was estimated by Lowry's method [10]. The protein compositions of the unbound and bound fractions to the apo Al immunoaffinity column were determined by a 4/20% gradient polyacrylamide-SDS slab gel.

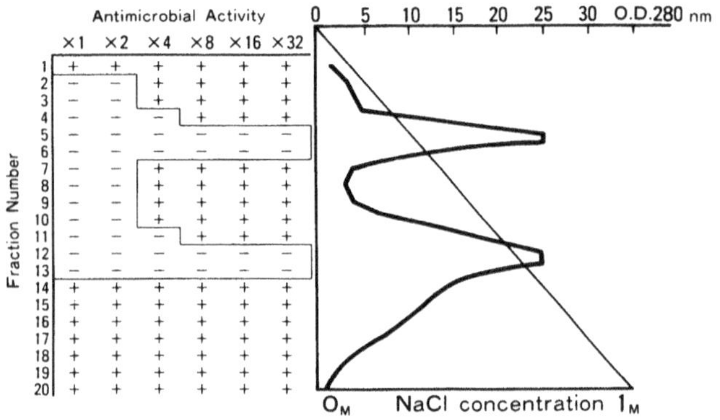

Fig. 4. The elution profile of healthy human serum by ion exchange Cellulose column (DE52) and distribution of antimicrobial activity. The antimicrobial activity in the each fraction was measured by microwell plate method. (+) shows the presence and (−) shows the absence of bacterial growth in the well.

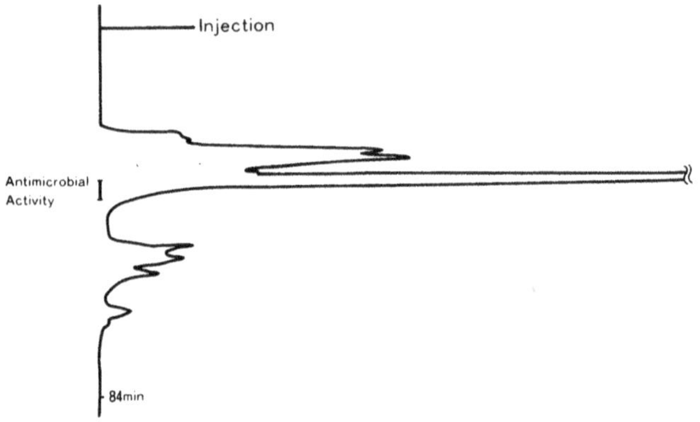

Fig. 5. The elution profile of the semipurified serum antimicrobial component (second peak derived from the ion exchange column) by high performance liquid chromatography with a TSK G3000SW column and the distribution of the antimicrobial activity. —— denotes the area where antimicrobial activity was distributed.

Results

The contribution of human lipoproteins to serum antimicrobial activity

Figure 2 shows the elution profile of healthy human serum fractionated with an Ultrogel column (AcA 54) and the distribution of the antimicrobial activity (∗). Serum antimicrobial activity, determined by the microwell plate method, was detected in the fraction around 60 and 70 of fraction number in which serum lipoproteins were also fractionated. To investigate whether lipoproteins had antimicrobial activity, lipoprotein fractions were separated from serum drawn from 3 healthy subjects after overnight fasting, and were incubated with *S. epidermidis* for 21 h (Fig. 3). Antimicrobial activity was assayed by the test tube method as described in Methods. As a control, two test tubes containing 1 ml of PBS with bacteria but without lipoproteins were prepared. The protein concentration of lipoproteins (VLDL, LDL and HDL) were adjusted as 0.23, 0.26 and 0.61 mg/ml, for subject 1, 2 and 3, respectively. Changes in bacterial count are expressed as multiples of the initial count before incubation. As shown in Fig. 3, all tested lipoproteins reduced bacterial growth during the first 3 h, but VLDL enhanced bacterial growth at 9 h of incubation and this increment of bacterial growth with VLDL exceeded the control in subject 3. However, in all experiments with lipoproteins from 3 different subjects, HDL consistently suppressed bacterial growth in the medium at all times.

Semipurification of antimicrobial component from human serum

Freshly isolated human serum was loaded on an ion exchange Cellulose column (DE52). The elution profile of the column and the antimicrobial activity measured by the microwell plate method are shown in Fig. 4. Column fractions in which antimicrobial activity was detected (No. 11 to 13) were collected and the antimicrobial component was further purified by HPLC with a TSK G3000SW column (Fig. 5), then the eluted area in which antimicrobial activity was distributed was collected.

Isolation of antimicrobial component by apo AI immunoaffinity column

Three original samples (N31, N32 and N33) of semipurified serum antimicrobial component following HPLC were separated by an apo AI immunoaffinity column. Figure 6 shows the elution profile of the sample by immunoaffinity column. The protein concentration and composition of the original samples, and those of unbound and bound components to the immunoaffinity column are shown in Table 1 and Fig. 7, respectively. The unbound components were composed mainly of albumin and bound components were composed of apo AI and apo C groups.

Figure 8 shows the results of antimicrobial activity assays of 3 original samples (N31, N32 and N33) and

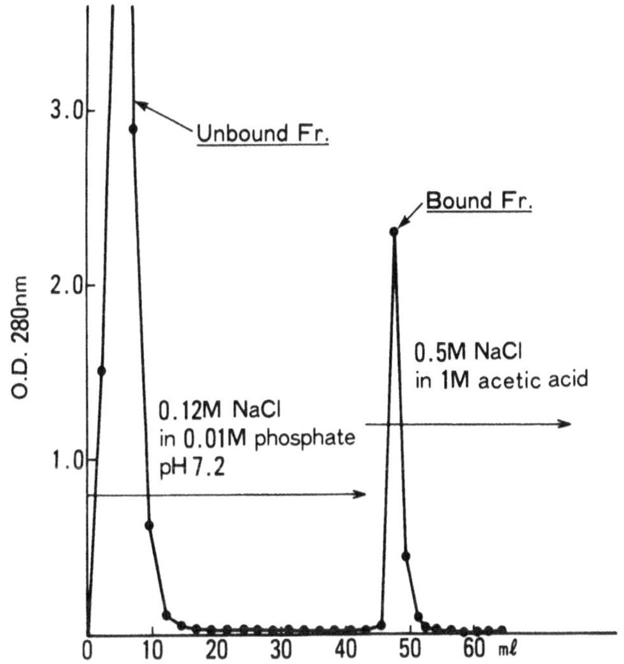

Fig. 6. Elution profile of sample by apo AI immunoaffinity column. Each sample from HPLC was loaded on the apo AI immunoaffinity column and eluted with 0.12 M NaCl in 0.01 M phosphate, pH 7.2 and then with 0.5 M NaCl in 1 M acetic acid.

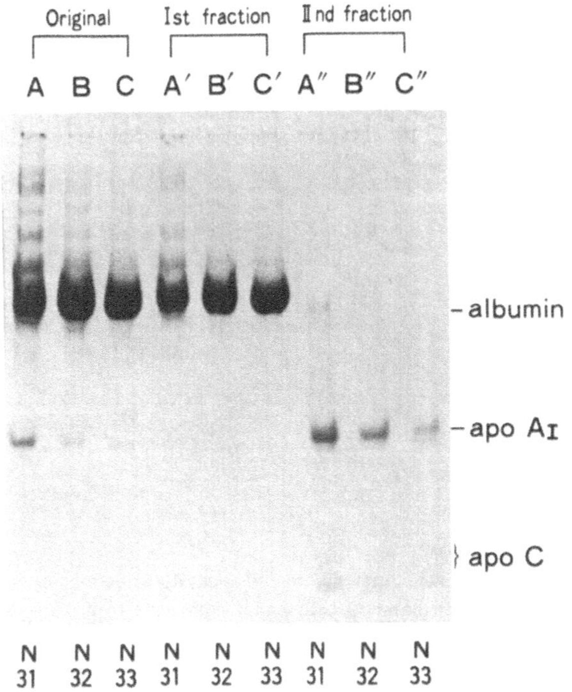

Fig. 7. Protein composition of fractions separated by apo AI immunoaffinity column on 4/20% gradient polyacrylamide-SDS slab gel. A, B, and C represent samples of N31, N32 and N33, respectively. Unbound component (1st fraction) from the affinity column was composed mainly of albumin. Bound component (IInd fraction) was composed of apo AI, AII and apo C group, presumably formed apo AI-containing particles.

each fractionated sample (unbound and bound) in microwell plate. Bound components which contained apo AI showed strong antimicrobial activity but unbound components had no antimicrobial activity. Antimicrobial activities of the first and second eluants of the immunoaffinity column, determined as a control, were also negative.

Discussion

The present study indicated that the fraction containing apo AI, separated by anti-apo AI Mab-coupled immunoaffinity column (apo AI immunoaffinity column), had antimicrobial activity against *S. epidermidis,* H155 strain. Other proteins, mainly apo AII and apo C, were also present in this fraction, indicating that this serum antimicrobial component is the HDL particles. Evidence supporting the participation of HDL was also derived from experiments in which we compared the ability of lipoproteins (VLDL, LDL and HDL) separated by ultracentrifuge to inhibit bacterial growth of *S. epidermidis.* VLDL added to the incubation actually enhanced bacterial growth in the medium (Fig. 3).

HDL is one of the major lipoproteins in humans and other vertebrates. HDL floats in the ultracentrifuge in the density range of 1.063 to 1.21 and has alpha 1 mobility on electrophoresis [11]. The protein composition includes about 85% apo AI and AII and 10% apo C while apo E, B, and D exist as minor components of HDL. Recently, HDL has been regarded as important because of its anti-atherogenic effects, which include the removal of cholesterol from atherosclerotic lesions; thus, HDL-cholesterol (HDL-C) is considered a negative risk factor for coronary heart disease [12–14].

It is also well known that lipoprotein metabolism becomes deranged in infectious disease [15, 16]. For instance, Kerttula *et al.* [17] reported that VLDL and LDL increased and HDL decreased during an attack of bacterial or viral meningitis, but lipoprotein levels normalized in the recovery phase of the infection. In addition, they found a striking increase in HDL-C level in the plasma after complete recovery from the illness. Sammalkorpri *et al.* [18] also described a significant decrease in HDL-C, especially cholesterol in HDL2, in

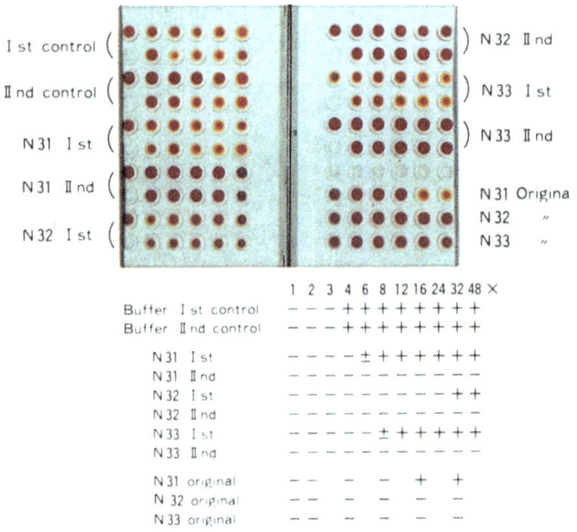

Fig. 8. Antimicrobial activity of two separated fractions by apo AI immunoaffinity column. Upper panel represents results on the microwell plate with the detection of bacterial growth by the addition of phenol red. Yellow color denotes the presence and red color the absence of bacterial growth. Lower panel shows the same results schematically. Note that each bound component (IInd fraction) of the 3 samples (N31, N32 and N33) shows strong antimicrobial activity. Buffer 1st control: first eluant of the column. Buffer IInd control: second eluant of the column.

the early stage of infection. Lanza-Jacoby *et al.* [19] reported that microorganisms inhibited the activity of lipoprotein lipase (LPL) which hydrolyzes triglyceride in VLDL particles and enhances the conversion of VLDL into LDL. In this stage of metabolism, nascent HDL is produced from the surface components of VLDL. Therefore, inhibition of LPL activity results in decreased HDL production. During infection, high catecholamine levels due to stress, serum amyloid protein AA, cytokines (IL-l, tumor necrosis factor, interferon etc.) from lymphocytes [20, 21], activated macrophages and poor nutrition may influence lipoprotein metabolism and reduce plasma HDL levels, it is of interest that microorganisms themselves reduce HDL by inhibiting LPL by the mechanisms listed above and promote subject susceptibility to infection.

There are several substances in the serum which inhibit bacterial growth [7]. Serum complement is one such substance but the antimicrobial component separated in this study differed from complement because its antimicrobial activity lasted more than one week and was heat-resistant up to 80° C (data not shown). Immunoglobulins are another major serum antimicrobial component but only a negligible amount of immunoglobulin was detected in the fraction containing apo Al, as determined by polyacrylamide-SDS gel electrophoresis (Fig. 7).

Rifkin [22] recently reported that HDL showed cytotoxic action against *Trypanosoma brusei*. Also, several papers reported an antiviral effect of apo Al itself against Herpes simplex virus [23], HIV [24] or xenotropic murine virus [25]. What, then, is the mechanism by which HDL induced antimicrobial activity? The answer is not well documented. The ability of HDL to remove cholesterol from the cell membrane may play a role in changing membrane integrity and function, resulting in the inhibition of bacterial or viral growth. Rifkin suggested that HDL-induced cell lysis differed according to species, *i.e.*, baboon and human HDL had high cytotoxic activity but rat and rabbit HDL did not; this author presented evidence that the phospholipid composition, especially a high content of phosphatidyl inositol, but not apo Al might be important for trypanocidal activity [26, 27]. In our preliminary study, apo Al purified by Sephacryl S-200 and by apo Al immunoaffinity column also had antimicrobial activity by itself. Therefore, the mechanism of HDL antimicrobial activity may differ from the trypanocidal action reported for HDL.

Indeed, there are several mechanisms by which HDL participates in the defense system of the body. For instance, Ulevitch *et al.* [28] reported that HDL particles bound to bacterial lipopolysaccharide and reduced the virulence of endotoxin. Alternatively, it has been suggested that inhibition of bacterial growth may be due to the insertion of a particular HDL content, lipids or proteins into the bacterial cell membrane. In more recent studies, defensins, which are cationic, cystine-rich peptides [29–31] from the cytoplasmic granules of neutrophils and macrophages, have been reported to possess antimicrobial activity. Actually, several peptides, including magainin-2 [32, 33], tachyplesin-1 [34] and basic amphipathic fluorescent peptides [35] have been reported to show antimicrobial activity. The cytotoxic

Table 1. Protein concentrations of the fractions separated by apo AI immunoaffinity column

Samples	mg/ml		
	N31	N32	N33
Original sample	19.2	33.7	28.5
Unbound component	4.58	6.90	3.74
Bound component	1.11	1.13	1.27

action of these peptides targets plasma membranes of bacteria. Certain lipids, presumably types of fatty acids, glycosphingolipids or phospholipids, were also reported to have antimicrobial activity [26, 36, 37]. These peptides and lipids may circulatc in thc blood with HDL and move into the bacterial membranes in an exchange process to inhibit bacterial activity, although it was recently discovered that defensins bind to activated alpha 2-macroglobulin [38].

As an another consideration, apo Al, which is an amphipathic protein composed of 243 amino acids and having 11 alpha-helices and 2 beta-turns in its constitution, may have antimicrobial activity as detected in our preliminary experiments. Agawa and his colleagues [39] demonstrated that a synthesized basic amphipathic alpha-helical model peptide showed antimicrobial activity against both Gram-positive and -negative bacteria due to interactions with the structure of ion channels of the bacterial membrane. Additional research is needed to elucidate the mechanism of the antimicrobial activity of HDL. We are currently investigating the clinical significance of this HDL antimicrobial property in elderly men and subjects susceptible to infection.

Acknowledgements

We would like to thank Mrs. Minako Hasegawa, Dr. Sadayori Hoshina and Dr. Kunihiro Ryuuno for expert technical helps and discussions and Miss Noriko Sato for preparation of the manuscript.

References

1. Drutz DJ, Mills J: Immunity and infection. In: DP Sites *et al.* (eds) Basic and Clinical Immunology. 5th ed. Lange Med Publ, California, 1984, pp 197–222
2. Hirsch JG: Comparative bactericidal activities of blood serum and plasma serum. J Exp Med 112: 15–21, 1960
3. Gold MJ, Calmon J, Wendeler M, Levison ME, Johnson CC: Synergistic bactericidal activity of rat serum with vancomycin against Enterococci. J Infect Dis 163: 1358–1361, 1991
4. Hancock RE, Mutharia LM, Chan L, Darveau RP, Speert DP, Pier GB: *Pseudomonas aeruginosa* isolated from patients with cystic fibrosis: A class of serum-sensitive nontypable strains deficient in lipopolysaccharide O side chains. Infect Immun 42: 170–177, 1983
5. Sophianopoulos AJ, Rhodes CK, Holcomb DN, Van Holde KE: Physical studies of lysozyme. 1. Characterization. J Biol Chem 237: 1107–1112, 1962
6. Rogers HJ, Bullen JJ, Gushnie GH: Iron compounds and resistance to infection: Further experiments with *Clostridium welchii* type A *in vivo* and *in vitro*. Immunology 19: 521–538, 1970
7. Machida K, Moriwaki M, Hoshina S, Sakurai S, Kurosaka K, Ikawa S: Antimicrobial component in human sera against *Staphylococcus epidermidis*. Jikeikai Med J 35: 343–349, 1988
8. Havel RJ, Eder HA, Bragdon JH: The distribution and chemical composition of ultracentrifugally separated lipoproteins in human serum. J Clin Invest 34: 1345–1353, 1955
9. Galfre G, Milstein C: Preparation of monoclonal antibodies: Strategies and procedures. Methods Enzymol 73: 3–46, 1981
10. Lowry OH, Rosebough NJ, Farr AL, Randall RJ: Protein measurement with the Folin phenol reagent. J Biol Chem 193: 265–275, 1951
11. Lewis B: The Hyperlipidemias: Clinical and Laboratory Practice. Blackwell, Oxford, 1976, pp 34–36
12. Miller GJ, Miller NE: Plasma high-density-lipoprotein concentration in the development of ischaemic heart disease. Lancet i: 16–19, 1975
13. Miller NE, Weinstein DB, Carew ET, Koschinsky T, Steinberg D: Interaction between high density and low density lipoproteins during uptake and degradation by cultured human fibroblast. J Clin Invest 60: 78–88, 1977
14. Steinberg D: Metabolism of lipoproteins and their role in the pathogenesis of atherosclerosis. Atherosclerosis Rev 18: 1–23, 1988
15. Gallin JI, Kaye D, O'Leary WM: Serum lipids and infection. New Engl J Med 281: 1081–1086, 1969
16. Alvarez C, Ramos A: Lipids, lipoproteins and apoproteins in serum during infection. Clin Chem 32: 142–145, 1986
17. Kerttula Y, Weber TH: Serum lipids in viral and bacterial meningitis. Scand J Infect Dis 18: 211–215, 1986
18. Sammalkorpi K, Valtonen V, Kerttula Y, Nikkilä E, Taskinen M-R: Changes in serum lipoprotein pattern induced by acute infections. Metabolism 37: 859–865, 1988
19. Lanza-Jacoby S, Lansey SC, Cleary MP, Rosato FE: Alterations in lipogenic enzymes and lipoprotein lipase activity during gram-negative sepsis in the rat. Arch Surg 117: 144–147, 1982
20. Beutler BA, Cerami A: Recombinant interleukin 1 suppresses lipoprotein lipase activity in 3T3-Ll cells. J Immunol 135: 3969–3971, 1985
21. Ehnholm C, Aho K, Huttunen JK, Kostiainen E, Mattila K, Pikkarainen J, Cantell K: Effect of interferon on plasma lipoproteins and on the activity of postheparin plasma lipases. Arteriosclerosis 2: 68–73, 1982
22. Rifkin MR: Identification of the trypanocidal factor in normal human serum: High density lipoprotein. Proc Natl Acad Sci USA 75: 3450–3454, 1978
23. Srinivas RV, Birkedal B, Owens BJ, Anantharamaiah GM, Segrest JP, Compans RW: Antiviral effects of apolipoprotein A-l and its synthetic amphipathic peptide analogs. Virology 176: 48–57, 1990
24. Owens BJ, Anantharamaiah GM, Kahlon JB, Srinivas RV, Compas RW, Segrest JP: Apolipoprotein A-l and its amphipathic helix peptide analogues inhibit human immunodeficiency virus-induced syncytium formation. J Clin Invest 86: 1142–1150, 1990
25. Kane JP, Hardman DA, Dimpfl JC, Levy JA: Apolipoprotein is responsible for neutralization of xenotropic tyce C virus by mouse serum. Proc Natl Acad Sci USA 76: 5957–5961, 1979
26. Rifkin MR: Role of phospholipids in the cytotoxic action of high

density lipoprotein on trypanosomes. J Lipid Res 32: 639–647, 1991

27. Rifkin MR: *Trypanosoma brucei*: Cytotoxicity of host high density lipoprotein is not mediated by apolipoprotein A-1. Exp Parasitol 72: 216–218, 1991
28. Ulevitch RJ, Johnston AR, Weinstein DB: New function for high density lipoproteins: Their participation in intravascular reactions of bacterial lipopolysaccharides. J Clin Invest 64: 1516–1524, 1979
29. Schafer WM, Martin LE, Spitznagel JK: Cationic antimicrobial proteins isolated from human neutrophil granulocytes in the presence of diisopropyl fluorophosphate. Infect Immun 45: 29–35, 1984
30. Kagan BL, Selsted ME, Ganz T, Lehrer RI: Antimicrobial defensin peptides form voltage-dependent ion-permeable channels in planar bilayer membranes. Proc Natl Acad Sci USA 87: 210–214, 1990
31. Borenstein LA, Selested ME, Lehrer RI, Miller JN: Antimicrobial activity of rabbit leukocyte defensins against *Treponema pallidum* subsp. *pallidum.* Infect Immun 59: 1359–1367, 1991
32. Hook WA, Tsuji S, Siraganian RP: Magainin-2 release histamine from rat mast cells. Proc Soc Exp Biol Med 193: 50–55, 1990
33. Darveau RP, Cunningham MD, Seachord CL, Cassiano-Clough L, Cosand WL, Blake J, Watkins CS: Beta-lactam antibiotics potentiate magainin 2 antimicrobial activity *in vivo* and *in vitro.* Antimicrob Agents Chemother 35: 1153–1159, 1991
34. Matsuzaki K, Fukui M, Fujii N, Miyajima K: Interaction of an antimicrobial peptide, tachyplesin I with lipid membranes. Biochim Biophys Acta 1070: 259–264, 1991
35. Lee S, Yoshida M, Mihara H, Aoyagi H, Kato T, Yamasaki N: The spectroscopic analysis for binding of amphipathic and antimicrobial model peptides containing pyrenylalanine and tryptophan to lipid bilayer. Biochim Biophys Acta 984: 174–182, 1989
36. Miller SJ, Aly R, Shinefield HR, Elias PM: *In vitro* and *in vivo* antistaphylococcal activity of human *stratum corneum* lipids. Arch Dermatol 124: 209–215, 1988
37. Bibel DJ, Miller SJ, Brown BE, Pandey BB, Elias PM, Shinefield HR, Aly R: Antimicrobial activity of *stratum corneum* lipids from normal and essential fatty acid-deficient mice. J Invest Dermatol 92: 632–638, 1989
38. Panyutich A, Ganz T: Activated alpha 2-macroglobulin is a principal defensin-binding protein. Am J Respir Cell Mol Biol 5: 101–106, 1991
39. Agawa Y, Lee S, Ono S, Aoyagi H, Ohno M, Taniguchi T, Anzai K, Kirino Y: Interaction with phospholipid bilayers, ion channel formation, and antimicrobial activity of basic amphipathic alpha-helical model peptides of various chain lengths. J Biol Chem 266: 20218–20222, 1991

Molecular and Cellular Biochemistry **119**: 179–184, 1993.

Molecular characteristics in Japanese patients with lipidosis: Novel mutations in metachromatic leukodystrophy and Gaucher disease

Yoshikatsu Eto, Hiroshi Kawame, Yoriyasu Hasegawa, Toya Ohashi, Hiroyuki Ida and Toshiharu Tokoro

Division of Medical Genetics and Endocrinology, Department of Pediatrics, Tokyo Jikei University School of Medicine, Tokyo, Japan

Abstract

The characterization of mutations in Japanese patients with lipidosis, particularly in metachromatic leukodystrophy (MLD) and Gaucher disease has been studied in detail. Metachromatic leukodystrophy is characterized by an accumulation of sulfatide in nervous tissues and kidney due to a deficiency of arylsulfatase A (ASA). We analyzed the presence of three known mutant arylsulfatase A alleles in Japanese patients with MLD. Among 10 patients of Japanese patients with MLD, we found that allele 445A mutation has moderately high incidence and also homozygosity of this mutation results in the late infantile form. Allele 2381T was not found in Japanese patients. Furthermore, we found novel mutation which is G- to A mutation at the 1070 nucleotide of the ASA gene (designated 1070 A) in Japanese patients with juvenile onset. This mutation results in a amino acid substitution of Gly245 by Arg and found in heterozygote form. Our studies of molecular analysis in 10 Japanese patients with MLD indicate that Japanese MLD patients have unique characteristics of ASA mutations compared with those of Caucasian patients. On the other hand, Gaucher disease is the most prevalent sphingolipidosis, characterized by an accumulation of glucocerebroside in macrophage derived cells due to a deficiency of lysosomal hydrolase glucocerebrosidase. To study the molecular basis of Gaucher disease in Japanese patients, we analyzed the presence of the two known mutations (6433C and 3548A) in the glucocerebrosidase gene of 15 patients with Gaucher disease. We found that the 6433C and 3548A mutations occur in all subtypes of Japanese patients with Gaucher disease. Most frequent mutations among them was the 6433C mutation, 40% of 30 chromosomes, whereas the novel mutation of the 3548A found in Japanese patients with neuronopathic Gaucher disease was found in 20% (6 out of 30 chromosomes). The characteristics of these mutations in Japanese patients with Gaucher disease is different from those of Caucasian populations reported previously. (Mol Cell Biochem **119:** 179–184, 1993)

Key words: gene analysis, arylsulfatase A gene, metachromatic leukodystrophy, glucocerebrosidase gene, Gaucher disease

Address for offprints: Y. Eto, Division of Medical Genetics and Endocrinology, Department of Pediatrics, Tokyo Jikei University School of Medicine, Tokyo, Japan

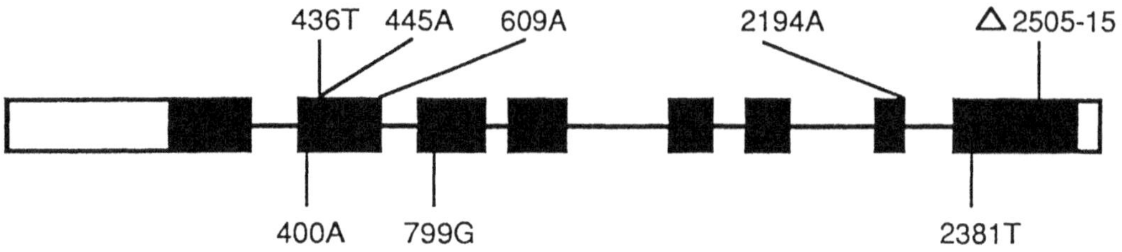

Fig. 1. Gene of arylsulfatase A and reported mutations in ASA gene of MLD patients.
Mutations identified in Caucasian patients

436T	Ser96 to Phe, identified in pseudo-deficiency allele	(Gieselmann *et al.* 1991)
609A	exon2 splice donor site mutation	(Polten *et al.* 1991)
2194A	exon 7 splice donor site mutation	(Fluharty *et al.* 1991)
△ 2505-15	11 bp deletion in exon 8	(Bohne *et al.* 1991)
400A	Arg84 to Glu	(Kappler *et al.* 1992)
799G	Ile179 to Ser	(Fluharty *et al.* 1991)
2381T	Pro426 to Leu	(Polten *et al.* 1991)
Mutation identified in a Japanese patient		
445A	Gly99 to Asp	(Kondo *et al.* 1991)

Introduction

Sphingolipidsosis is a group of genetic disorders involving the accumulation of sphingolipids in various tissues, particularly in reticuloendotherial tissues such as liver, spleen and kidney due to a certain lysosomal hydrolytic enzyme. Among these disorders. Tay-Sachs disease, Gaucher disease, MLD, globoid cell leukodystrophy (GLD), Fabry disease, GM1-gangliosidosis, Niemann-Pick disease and Farber disease are included. These disorders are caused by a genetic deficiency of various lysosomal sulfatase or glycosidase. Patients with these disorders shows characteristic cellular changes in organs of storage such as nervous tissue. The signs and symptoms for each syndrome vary with the enzymatic defect and site of lipid accumulation. The amount of true storage depends on the organ involved and on the amount of lysosomal catabolism required for maintenance of normal chemical composition.

Recently, molecular analysis of these sphingolipidoses has rapidly progressed. Many new mutations in each sphingolipidosis has been reported from patients in different races. Furthermore, the careful molecular investigation of these disorders provides an answer for the explanations of clinical heterogeneities. In present communications, we focused the molecular basis of Japanese patients with MLD and Gaucher disease and characterized the mutations in these disorders among Japanese patients.

Metachromatic leukodystrophy (MLD) is characterized by an accumulation of sulfatide in nervous tissues and also kidney due to a deficiency of lysosomal arylsulfatase A or cerebroside sulfatase [1]. Clinically, MLD can be classified into three phenotypes according to the age of onset which are late infantile, juvenile, and adult form. Each clinical form also contains various clinical varieties. All these forms exhibit severe neurological manifestations such as psychomotor retardation, pyramidal tract signs, seizures and dementia. Recently, the nucleotide sequences of a cDNA and the genomic structure of human ASA gene have been identified [2, 3] (Fig. 1). Eight mutations in MLD patients have been reported [4–8]. Most of these mutations were identified in Caucasian people; these are allele 609A (exon 2 splice donar site mutation), allele 2381T (Pro to Leu substitution) and allele 445A (Gly to Asp). The former two mutations show simple genotype-phenotype relationship in Caucasian patients, since the homozygosity of the 609A produced the late infantileform of MLD and also the 2381T mutation occurs in the adult form. Compound heterozygote of these mutations exhibits juvenile form of MLD [4]. Present report concerns the presence of these mutations and also found a novel mutation in Japanese patients with MLD.

On the other hand, Gaucher disease is caused by an accumulation of glucocerebroside in macrophage derived cells such as liver, spleen and others due to a deficiency of lysosomal glucocerebrosidase (GC) [9].

Clinically, Gaucher disease has been classified into three major forms according to clinical symptoms; type I (adult, non-neuronopathic form), type II (acute, neuronopathic form), type III (subacute juvenile form, neuronopathic form). Recently, the isolation of the GC gene has been reported [10–11]. Many mutations including point mutations, deletion and complex alleles have been identified in Gaucher disease patients. Among these mutations, 6433C mutation (C or T substitution, 444 Leu to Pro mutation) originally identified in type 2 patients is a most common among all ethnic group [12]. The 5841 G (A to G, 370 Asn to Ser) and 1035 GG mutations have been found in Ashkenazi Jewish patients [13, 14]. We found a novel mutation, 6433 C and also 3548 A mutations in Japanese Gaucher patients. In this paper, we described molecular characteristics of Japanese patients with Gaucher disease.

Materials and methods

Patients

10 MLD patients were examined in this study; Patient 6 and 9 are juvenile form, while patients 2, 3, 4, 5, 8 and 10 are from late infantile form of MLD. On the other hand, we analyzed 15 Japanese patients with Gaucher disease. Nine patients were classified as type 1, four patients as type II and two patients as type III. Mutations in this paper were designated according to the notation system described by Latham *et al.* [15].

DNA extraction and Oligonucleotides

High molecular genomic DNA was isolated from leukocytes, autopsied tissues and cultured skin fibroblasts according to standard procedure [16]. Oligonucleotide used for PCR amplifications were prepared as described previously [17, 18].

PCR condition and ASO hybridization

PCR conditions and ASO hybridization were carried out described elsewhere [17, 18].

Nucleotide sequence analysis and COS-1 cell expression

Nucleotide sequence by the dideoxy chain termination method and transient expression of mutant by alleles in COS-1 cells were carried out as described previously [17–19].

Table 1. Mutations in Japanese patients with MLD

Patient	Clinical phenotype	Genotype
Patient 1	Late infantile	?
Patient 2	Late infantile	445A/445A
Patient 3	Late infantile	445A/1070A
Patient 4	Late infantile	445A/445A
Patient 5	Juvenile	?/?
Patient 6	Late infantile	?/?
Patient 7	Late infantile	609A/609A
Patient 8	Late infantile	445A/?
Patient 9	Juvenile	1070A/?
Patient 10	Late infantile	445A/445A

Results

(1) Molecular basis of Japanese patients with MLD

The presence of allele 445A and 609A was studied by combining Eco 471 RFLP and ASO hybridization. Allele 445A was found homozygously in three patients with the late infantile type of MLD and heterozygously in two patients with late infantile type of MLD. Allele 609A was found in one patient with the late infantile MLD in a homoallelic state. Allele 2381T was not found in our study. A novel mutation, 1070A was found in one Japanese patient with juvenile MLD in a heterozygote form. To confirm the 1070A mutation is responsible for the deficiency of ASA activity, mutant cDNA (pcD-ASA 1070A) was transiently expressed in COS cells. The activity of ASA in this mutant were decreased whereas the ASA mRNA was increased. Table 1 summerized the mutations of Japanese patients with MLD.

(2) Molecular basis of Japanese patients with Gaucher disease

The presence of 6433C and 3548A mutations were analyzed in 15 Japanese patients with Gaucher disease. We have found five different genotypes in Japanese Gaucher patients by analysis of two known mutations (Table 2). 12 patients out of 15 patients had at least one of these two mutations, comprising 80%. The 6433C mutation was found in 10 patients out of 15. Two type I patients were homozygous for this mutation. No patients with type II or III was homozygous for this mutation. We detected six patients which were heterozygous for the 3548A mutation. No patients were homozygous for this mutation.

Table 2. Genotypes of Japanese patients with Gaucher disease

	6433C/6433C	6433C/?	6433C/3548A	3548A/?	3548A/3548A	?/?
Type 1	2	1	3	0	0	3
Type 2	0	2	0	2	0	0
Type 3	0	1	1	0	0	0
Total (15)	2	4	4	2	0	3

Discussion

We present the molecular basis of Japanese patients with MLD and Gaucher disease. Two disorders are rather common lipid storage disease. The frequency of these disorders can be estimated as one to 70,000 to 100,000 in Japan. Mutation analysis in Japanese patients in these disorders has not been described previously except for one adult case with Gaucher disease [20]. In present study, we identified the mutations in more than 30 to 50% of Japanese patients with Gaucher or MLD.

In MLD gene analysis, we found that allele 445A mutation has also moderately frequent incidence among Japanese patients with MLD (8/16 allele examined) and also homozygosity of this mutation exhibits late infantile MLD in Japanese patient. Allele 609A and 2381T which is relatively common in Caucasian patients were not found in Japanese patients. The novel mutation which we identified is G to A transition at the 1070th nucleotide of the ASA gene (Allele 1070A) was found in one Japanese patient with late infantile form of MLD [17]. This mutation results in a single amino acid substitution of Gly245 to Arg. The substitution from neutral amino acid to basic amino acid in ASA protein molecule is predicted to be significantly changed in the structure or stability of the ASA protein.

The simple genotype-phenotype relationship reported by Polten *et al.* [4] appears to fit in Caucasian MLD patients. However, allele 609A has been found in a patient with adult form combined with other allele. Allele 1070A was found heterozygously in patients with juvenile and late infantile form. The heterozygote combination of allele 1070A and 445A exhibits in the late infantile form of MLD, although the 445A mutation in heterozygote state was originally described in adult form of Japanese MLD patients. Our data strongly indicate that allele 445A and allele 1070A lead to missence mutations which produce no active enzyme and results in the late infantile form of MLD homozygously. Table 3 summarized various ASA gene mutations of MLD patients including our data reported previously.

On the other hand, we analyze the presence of two known mutations in the GC gene in 15 Japanese patients with Gaucher disease by selective amplification

Table 3. Summary of various mutations in arylsulfatase A gene in MLD patients

Genotype	Clinical phenotype	Race
Allele 609A/609A exon 2 splicing donar site	Late infantile	J, C
Allele 2194/609A (2194A: exon 7 splicing site)	Late infantile	C
Allele 2194A/2506 (2506: 11 bp deletion in exon 8)	Late infantile	C
Allele 1070A/445A	Late infantile	J
Allele 445A/445A	Late infantile	J
Allele 799C/2194A	Juvenile	C
Allele 609A/2381T	Juvenile or adult	C
Allele 609A/2381T	Juvenile	C
Allele 400A/2381T	Late juvenile	C
Allele 2381T/2381T	Adult	C
Allele 436T + 1788G + 2723G/436T + 1788G + 2723 G	Late infantile	C
Allele 1788G + 2723G	Pseudodeficiency	C
33 nucleotide insertion in mRNA of SAP-1	SAP-1-deficient MLD	C
A-> T substitution in SAP precursor mRNA	SAP-1, 2 deficiency C	

C: Caucasian, J: Japanese.

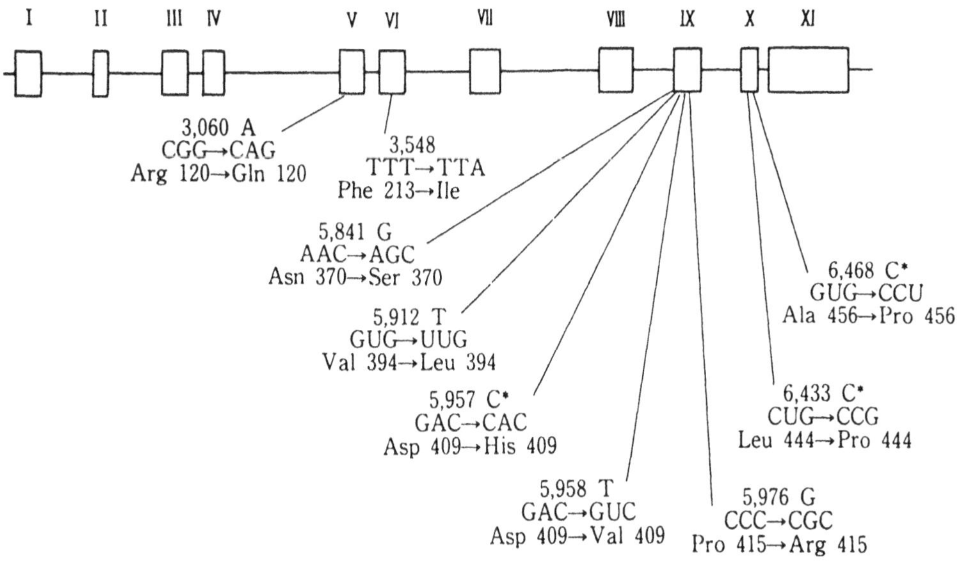

Fig. 2. Summary of mutations in Gaucher disease.

and restriction digestion analysis. The mutation 6433C has been identified in various ethnic origins [10, 11]. This mutation was also high frequency in Japanese Gaucher patients, 10 patients out of 15. The phenotype-genotype relationship for this mutation was found in Caucasian patients, since the true homozygousity of this mutation was identified in both neuronopathic type II and III Gaucher disease [4, 15, 21]. Furthermore, Latham *et al.* [15] reported that the patients with this mutation in homoallelic form showed severe neuronopathic state. However, the homozygousity of this mutation in Japanese Gaucher patients was only found in non-neuronopathic patients. A similar data from Japanese non-neuronopathic Gaucher disease was reported by Masuno *et al.* [22]. Therefore, these findings suggest that ethnic factors including genetic environmental influences may be accounted for phenotypic development of Japanese Gaucher disease. The mutation of the 3548A in Japanese Gaucher patients was found in 20% allele frequency. As shown in Table 2, this mutation was detected in all subtypes of Japanese Gaucher disease. There are still 40% of unknown mutant allele among Japanese Gaucher disease and 10 out of 15 patients were genetically compound heterozygotes. We also found other several mutations in the GC gene by PCR-SSCP method [23]. Fig. 2 illustrated the reported mutations of Gaucher disease. The phenotype-genotype relationship in Japanese patients with Gaucher disease could be included ethnic genetic background or environmental factors. More complete genotypic studies in Japanese Gaucher are currently under investigation.

Acknowledgement

This work is supported by grants of the Japanese Ministry of Education (1991, 1992) and also of the Japanese Ministry of Welfare (1991, 1992). We would like to express our sincere thanks for his encouragement in our research and also to say 'congratulations' on his great success in his cardiology research to Prof. M. Nagano.

References

1. Mehl E, Jatzkewitz H: Evidence for the genetic block in metachromatic leukodystrophy. Biochem Biophys Res Commun 19: 407–411, 1965
2. Stein C, Gieselman V, Kreysing J, Schmidt B, Pohlmann R, Waheed A, Meyer HE, O'Brien JS, Von Figura K: Cloning and expression of human arylsulfatase A. J Biol Chem 264: 1252–1259, 1989
3. Kreysing J, Von Figura K, Gieselmann V: Structure of the arylsulfatase A gene. Europ J Biochem 191: 627–631, 1990
4. Polten A, Fluharty AL, Fluharty CB, Kappler J, Von Figura K, Gieselmann V: Molecular basis of different forms of metachromatic leukodystrophy. N Engl J Med 324: 18–22, 1991
5. Bohne W, Von Figura K, Gieselamm V: An 11 bp deletion in the arylsulfatase A gene of a patient with late infantile metachromatic leukodystrophy. Hum Genet 87: 155–158, 1991
6. Gieselmann V, Fluharty AL, Tonnesen T, Von Figura K: Mutations in the arylsulfatase A pseudodeficiency allele causing metachromatic leukodystrophy. Am J Hum Genet 49: 407–413, 1991
7. Fluharty AL, Fluharty CB, Bohne W, Von Figura K, Gieselmann V: Two new arylsulfatase A (ARSA) mutations in a juvenile metachromatic leukodystrophy (MLD) patient. Am J Hum Genet 49: 1340–1350, 1991

8. Kappler J, Von Figura K, Gieselmann V: Late onset metachromatic leukodystrophy: Molecular pathology in two siblings. Ann Neurol 31: 256–261, 1992
9. Brady RO, Kanfer JN, Shapiro D: Metabolism of glucocerebrosidase. II. Evidence of an enzymatically deficiency in Gaucher's disease. Biochem Biophys Res Commun 18: 221–225, 1965
10. Sorge J, West C, Westwood B, Beutler E: cDNA cloning of human glucocerebrosidase gene. Proc Natl Acad Sci USA 82: 7289–7293, 1985
11. Tsuji S, Choudary PV, Martin BM, Winfield S, Barranger JA, Ginns EI: Nucleotide sequence of cDNA containing the complete coding sequence for human lysosomal glucocerebrosidase. J Biol Chem 261: 50–53, 1985
12. Tsuji S, Choudary PV, Martin BM, Major JA, Barranger JA, Ginns EI: A mutation in the human glucocerebrosidase gene in neuronopathic Gaucher disease. N Engl J Med 316: 570–575, 1987
13. Theophilus BDM, Latham T, Grabowski GA, Smith FI: Gaucher disease: Molecular heterogeneity and phenotype-genotype correlation. Am J Hum Genet 45: 212–225, 1989
14. Beutler E, Gilbart T, Kuhl W, Sorge J, West C: Identification of the second common Jewish Gaucher disease mutation makes possible population-based screening for the heterozygous state. Proc Natl Acad Sci, USA 88: 10544–10547, 1991
15. Latham T, Grabowski GA, Theophilus BDM, Smith FI: Complex allele of the acid beta-glucosidase gene in Gaucher disease. Am J Hum Genet 47: 79–86, 1990
16. Kawame H, Eto Y: A new mutation glucocerebrosidase-gene missense mutation responsible for neuronopathic Gaucher disease in Japanese patients. Am J Hum Genet 49: 1378–1380, 1991
17. Maniatis T, Fritsch EF, Sambrook J: Molecular cloning: A laboratory manual. Cold Spring Harbor Laboratory, Cold Spring Harbor, NY, 1982
18. Hasegawa Y, Kawame H, Eto Y: Mutations in arylsulfatase A (ASA) gene of Japanese patients with metachromatic leukodystrophy (MLD). DNA and Biology, in press, 1992
19. Kawame H, Maekawa K, Eto Y: Molecular screening of Japanese patients with Gaucher disease: Phenotypic variabilities in the same genotypes. Human Mutation, in press, 1992
20. Kondo R, Wakamatsu N, Yoshino H, Fukuhara N, Miyatake T, Tsuji S: Identification of a mutation in the arylsulfatase A gene of a patient with adult-type metachromatic leukodystrophy. Am J Hum Genet 48: 971–978, 1991
21. Sidransky E, Tsuji S, Martin BM, Stubblefield B, Ginns EI: DNA mutation analysis of Gaucher disease. Am J Med Genet 42: 331–336, 1992
22. Masuno M, Tomatsu S, Sukegawa K, Orii T: Non-existence of a tight association between a 444 Leucine to proline mutation and phenotypes of Gaucher disease: High frequency of a Nci I polymorphism in the non-neuronopathic form. Hum Genet 84: 203–206, 1990
23. Kawame H, Hasegawa Y, Maekawa K, Eto Y: Rapid identification of mutations in the glucocerebrosidase gene of Gaucher disease patients by analysis of single strand conformation polymorphism (SSCP) Hum Genet, in press, 1992

Molecular and Cellular Biochemistry **119**: 185–193, 1993.

Characterization of adenylyl cyclase in heart sarcolemma in the absence or presence of alamethicin

Rajat Sethi, Ken S. Dhalla, Kanu R. Shah and Naranjan S. Dhalla
Division of Cardiovascular Sciences, St. Boniface General Hospital Research Centre, and Department of Physiology, Faculty of Medicine, University of Manitoba, Winnipeg, Manitoba, R2H 2A6 Canada

Abstract

Alamethicin is commonly used as an agent for unmasking the latent enzyme activities in vesicular membrane preparations; however, relatively little is known about the effect of this agent on the characteristics of adenylyl cyclase in heart sarcolemma. By employing rat heart sarcolemmal preparation, we observed 5 to 6 fold increase in adenylyl cyclase activity upon treatment with alamethicin. Kinetic experiments using various concentrations of MgATP revealed that the increase in adenylyl cyclase activity in alamethicin treated membranes was associated with an increase in V_{max} as well as affinity of the substrate for the enzyme. Dose-responses of the control and alamethicin-treated preparations to various activators of adenylyl cyclase revealed that the sensitivity of the enzyme to forskolin, NaF and GppNHp, was markedly increased upon treating sarcolemma with alamethicin. The activation of adenylyl cyclase by forskolin was also enhanced by increasing the concentration of alamethicin in the incubation medium. Furthermore, there was a greater increase in adenylyl cyclase activity with different concentrations of Mn^{2+} in the presence of alamethicin. These results suggest that alamethicin treatment alters the characteristics of adenylyl cyclase in addition to unmasking the enzyme activity in the purified sarcolemmal vesicular preparation. (Mol Cell Biochem **119:** 185–193, 1993)

Key words: heart sarcolemma, adenylyl cyclase, signal transduction, alamethicin

Introduction

Sarcolemmal vesicular preparations from the myocardium are widely used to study the biochemical and functional properties of cardiac membranes. The putative enzyme marker activity, ouabain sensitive Na^+-K^+ ATPase activity, has been reported to be predominantly in a latent form in the sarcolemmal vesicles [1], and evidence has been presented to suggest that this latency is probably due to the permeability barrier to ions and substrates or both due to the membranes of the vesicles. It is generally accepted that adenylyl cyclase is also a vectorial enzyme system present in sarcolemma, the catalytic subunit of which is exposed to the cytoplasmic compartment whereas the hormone receptors are exposed to the extracellular environment [2]. In sealed membrane vesicles which retain normal, low passive permeability, the adenylyl cyclase activity would be

Address for offprints: N.S. Dhalla, Division of Cardiovascular Sciences, St. Boniface General Hospital Research Centre, 351 Tache Avenue, Winnipeg, Manitoba, R2H 2A6 Canada

manifested only by the inside-out vesicles. It has been shown in previous studies that it is possible to unmask the latent activity of adenylyl cyclase in parallel to the activity of ouabain sensitive Na^+-K^+ ATPase by eliminating the effect of membrane sideness upon treatment with an ionophore alamethicin [3–5]. Although these effects of alamethicin on membrane enzymes, adenylyl cyclase and Na^+-K^+ ATPase, have been attributed to the formation of pores in the vesicular membranes by alamethicin, the possibility of changes in the characteristics of these enzymes upon treatment with alamethicin has not been explored previously. In the present study we have measured the patent and the latent activities of adenylyl cyclase upon treatment of a purified sarcolemmal preparation from the heart with alamethicin. We also found an increased sensitivity of adenylyl cyclase for various stimulants in alamethicin-treated preparations in addition to unmasking the enzyme activity and changes in its characteristics.

Materials and methods

Isolation of heart sarcolemma

Rats were killed by decapitation and their hearts removed. The atria, connective tissue and right ventricle were trimmed, and the left ventricle was processed for the preparation of the sarcolemmal membranes according to the method described by Pitts [6] with a slight modification [7]. All isolation steps were carried out at 0° to 4° C, and membrane fractions were frozen immediately in liquid N_2 after suspending them in 0.25 M sucrose and 10 mM histidine. The membranes were stored up to 4–6 weeks at − 80° C before carrying out the studies. In some experiments, crude membrane preparation obtained according to the method described by Wagner *et al.* [8] were employed. Protein concentration of all membranes was determined by the method of Lowry *et al.* [9].

Measurement of Na-K ATPase activities

Estimation of Na^+-K^+ ATPase activity was carried out by a previously described method [10] with some modifications. Briefly, sarcolemmal vesicles (10 μg) were incubated at 37° C with 1.0 mM ethylene glycol-bis(B-aminoethyl ether)-N,N,N'N'-tetraacetic acid (EGTA) (Tris), pH 7.4 at 37° C, 5 mM NaN_3, 6 mM $MgCl_2$, 100 mM NaCl and 10 mM KCl, 2.5 mM phosphoenolpyruvate (PEP) and 10 IU/ml pyruvate kinase. The reaction was started by the addition of 0.025 ml of 80 mM Na_2ATP, pH 7.4 and terminated after 10 min with 0.5 ml ice cold trichloroacetic acid. The liberated phosphate was measured by the method of Taussky and Shorr [11]. In some experiments 1 mM ouabain was added to the reaction medium before the addition of the sarcolemmal preparation. Na^+-K^+ ATPase activity was calculated as the difference between activities with and without Na^+ plus K^+. Mg^{2+}-ATPase activity was estimated as the difference between the activities registered with and without Mg^{2+} in the absence of Na^+ and K^+ in the incubation medium.

Determination of adenylyl cyclase activity

Adenylyl cyclase activity was determined by measuring [^{32}P]-cAMP formation from [^{32}P]ATP as described previously [12]. Unless otherwise indicated the incubation assay medium contained 50 mM glycylglycine (pH 7.5), 0.5 mM MgATP, [^{32}P] (1–1.5 × 10^6 cpm), 5 mM $MgCl_2$ (in excess of the ATP concentration), 100 mM NaCl, 0.5 mM cAMP, 0.1 mM EGTA and an ATP regenerating system consisting of 2 mM creatine phosphate, 0.1 mg creatine kinase/ml in a final volume of 200 μl. Incubations were initiated by the addition of membrane (30–70 μg) to the reaction mixture which had been thermally equilibrated for 3 min at 37° C. The reactions (in triplicate) for 10 min at 37° C were terminated by the addition of 0.6 ml of 120 mM zinc acetate. cAMP was determined by co-precipitation of other nucleotides with $ZnCO_3$ and sequent chromatography by the double column system, as described by Salmon *et al.* [13]. Under the assay conditions used, adenylyl cyclase activity was linear with respect to protein concentration and time of incubation used.

Western immunoblotting

Immunoblotting was performed as described by Mymby *et al.* [14]. Briefly membrane proteins were solubilized in Laemmli buffer, resolved on sodium dodecyl sulfate (SDS)-11% polyacrylamide gel [15] and then electrophoretically transfered to PVDF protein sequencing membrane. Blotts were treated with Tris-buffered saline (TBS) containing 5% (wt/vol) non fat dry milk (TBS-bloto) followed by a 14 hr incubation with specific antisera [AS/7 specific for Gia and RM/1 specific for Gsa] (1 : 3000) in TBS bloto and then two 10 min washes with 0.05% Tween 20 and TBS alternately. The antigen antibody complexes were detected by incubating the blots with Protein A for 2 hr at room temperature. The blots were washed with 0.05% Tween-20 and TBS alone. Blots were dried and autoradiographed with Kodak X-)mat AR film and an intensifying screen for overnight at − 70° C. Molecular masses were determined by running a parellel lane of protein standard (14–96 kDa).

Statistical analysis

Paired and unpaired Student's t tests were used to reveal a statistically significant difference between the treated and untreated membrane preparations.

Materials

ATP, cAMP, isoproterenol, forskolin, alamethicin were purchased from Sigma. Creatin kinase, myokinase and GTP were purchased from Boehringer Manheim, Canada. Alpha ^{32}P ATP and anti G proteins (AS/7 and RM/1) were purchased from Du Pont, Wilmington. DE.

Results

The data in Table 1 show that alamethicin at a ratio of 1 : 1 increased the ouabain sensitive Na^+-K^+ ATPase activity by 3 folds and the adenylyl cyclase activity by 4–5 folds in the purified heart sarcolemma. However, the values for Mg^{2+} ATPase, Na^+-K^+ ATPase and digitoxin sensitive Na^+-K^+ ATPase were depressed by about 25% to 50% in the presence of alamethicin compared to their respective control values. These results with alamethicin are consistent with studies from our laboratory [10] and others [16]. In order to gain further insight into the effects of alamethicin on the characteristics of adenylyl cyclase, the enzyme activity was measured using different activators such as NaF, Gpp(NH)p, and forskolin in the presence and absence of alamethicin. Results in Figs 1A, 2A and 3A show a dose dependent increase in adenylyl cyclase activity both in the presence or absence of alamethicin. However the activity of the enzyme was greater in alamethicin treated membranes. Moreover, data in Figs 1B, 2B and 3B showing the adenylyl cyclase activity as % maximal, reveal an increase in sensitivity of the enzyme to these activators in the presence of alamethicin. It should be pointed out that the forskolin response was modified to a greater extent by alamethicin in comparison to those for Gpp (NH)p and NaF. The results in Fig. 4 show that this marked increase in the activation of adenylyl cyclase by forskolin was also evident when different concentrations of alamethicin were used in the incubation medium; the activity of adenylyl cyclase reached maximum at the alamethicin to sarcolemmal protein ratio of 1.0.

To further investigate the effect of alamethicin on the catalytic site of the enzyme, we studied the adenylyl cyclase activity in the presence of different concentrations of Mn^{2+}. It should be noted here that a role for free Mn^{2+} ion as an activator of adenylyl cyclase at the catalytic unit separate from its function as part of the substrate complex has been established [17]. Figure 5 shows that the adenylyl cyclase activity increased in a dose dependent manner with increasing concentrations of Mn^{2+} in the presence or absence of alamethicin. This indicates alterations by alamethicin at the level of catalytic sites of the adenylyl cyclase. In order to gain further insight into the mechanism of actions of alamethicin, the adenylyl cyclase activity was measured in the

Table 1. Enzyme activities in alamethicin (1 : 1) treated and non-treated purified sarcolemmal preparations

	− alamethicin	+ alamethicin
Mg^{2+} ATPase	122 ± 4	83 ± 7*
Na^+-K^+ ATPase	24 ± 3	13 ± 2*
Digitoxin sensitive Na^+-K^+ ATPase	22 ± 2	15 ± 1*
Ouabain sensitive Na^+-K^+ ATPase	5.6 ± 2	15.3 ± 3*
Adenylyl cyclase	65 ± 10	325 ± 27*

Each value is a mean ± SE of 4–6 experiments. * $P < 0.05$ – Significantly different from values in the absence of alamethicin. Na^+-K^+ ATPase and adenylyl cyclase activities are expressed as μmol Pi/mg/hr and Pmol cAMP/mg/min, respectively. The concentrations of digitoxin and ouabain were 10 μM and 1 mM, respectively.

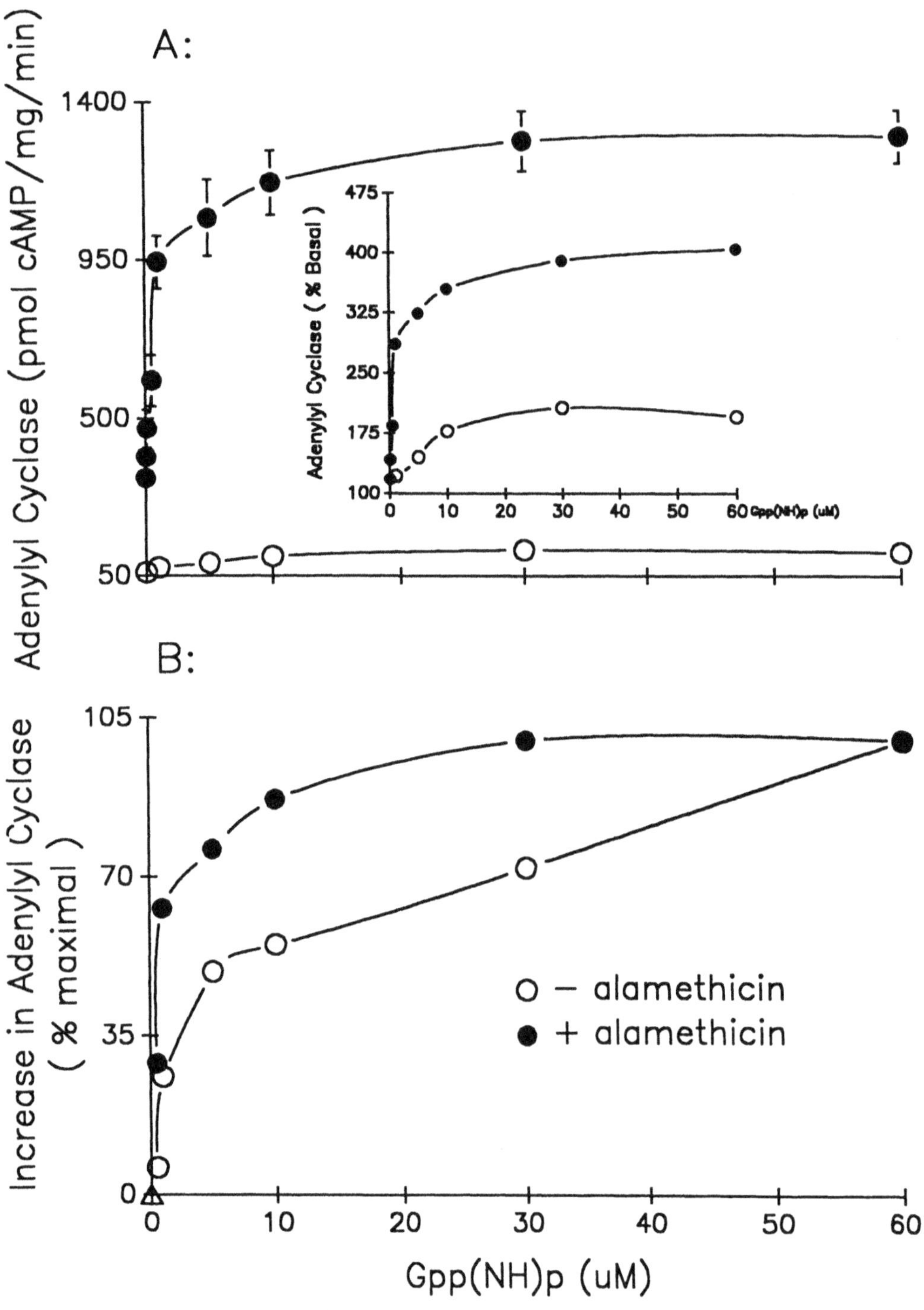

Fig. 1. Effect of different concentrations of Gpp(NH)p on adenylyl cyclase activity in purified sarcolemma in the presence and absence of alamethicin (1 : 1). Each value is a mean ± SE of 6 experiments. Inset shows values expressed as % basal activity whereas B shows values expressed as % maximal activity.

presence of different concentrations of Mg ATP. The results from Fig. 6 show that the enzyme activation exhibited a hyperbolic character. The inset data from Fig. 6 reveal that the kinetic parameters were different for both groups. The V_{max} values for adenylyl cyclase (pmol cAMP/mg/min) in the presence and absence of alamethicin were 270.6 ± 18 and 62.5 ± 2, respectively. Also, alamethicin increased the sensitivity of the

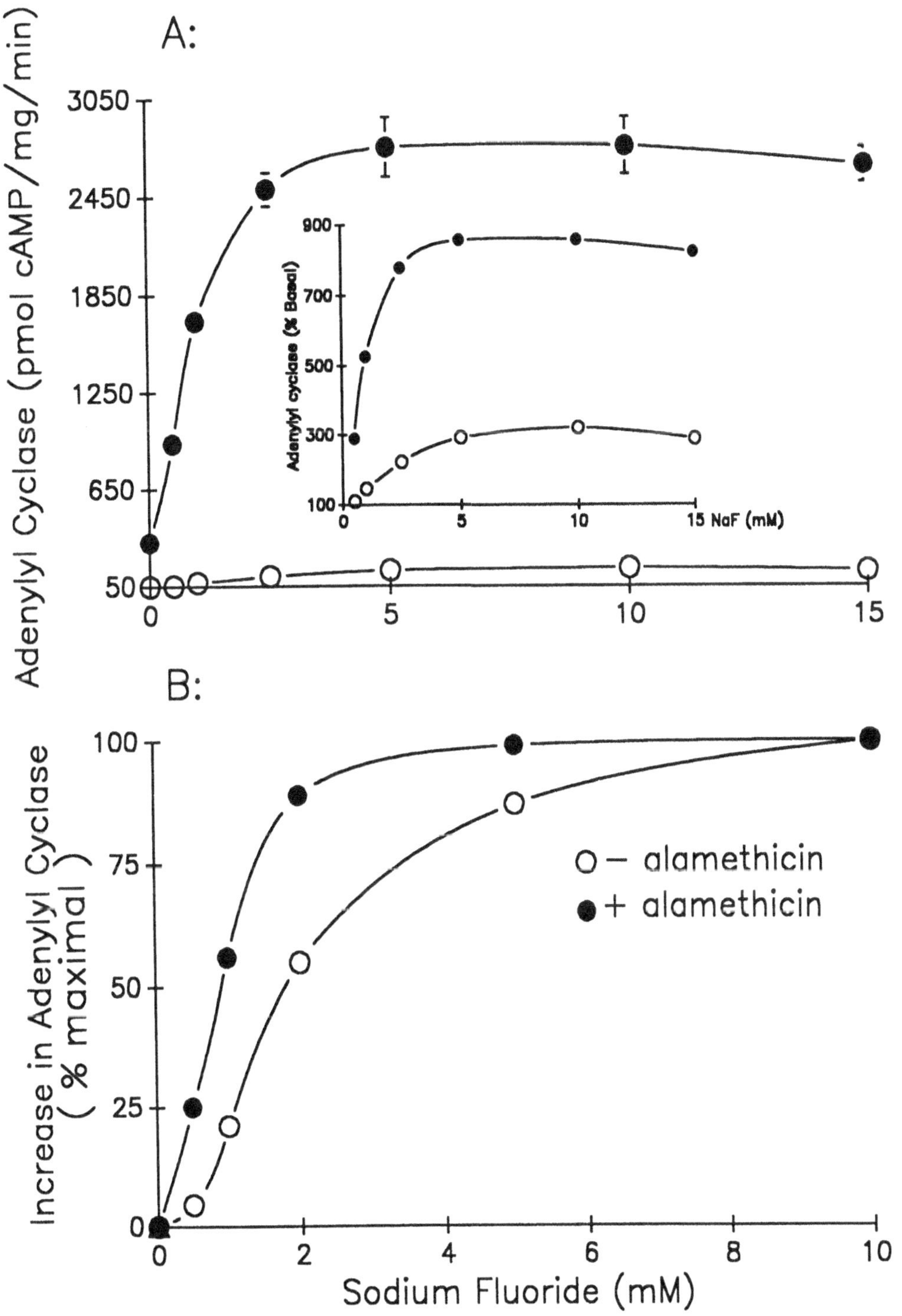

Fig. 2. Effect of different concentrations of NaF on adenylyl cyclase activity in purified sarcolemma in the presence and absence of alamethicin (1 : 1). Each value is a mean ± SE of 4 experiments. Values are expressed as indicated in legend to Fig. 1.

enzyme to MgATP; the K_m values for Mg ATP (mM) in the presence and absence of alamethicin were 0.009 ± 0.003 and 0.019 ± 0.002, respectively.

As adrenergic receptors in heart membrane are shown coupled with adenylyl cyclase through G proteins, we studied the effect of isoproterenol, a beta adrenergic receptor agonist, on the adenylyl cyclase activity in our sarcolemmal preparation. Results in Fig. 7 demonstrate no increase in adenylyl cyclase activity in the purified sarcolemmal preparation in the absence of

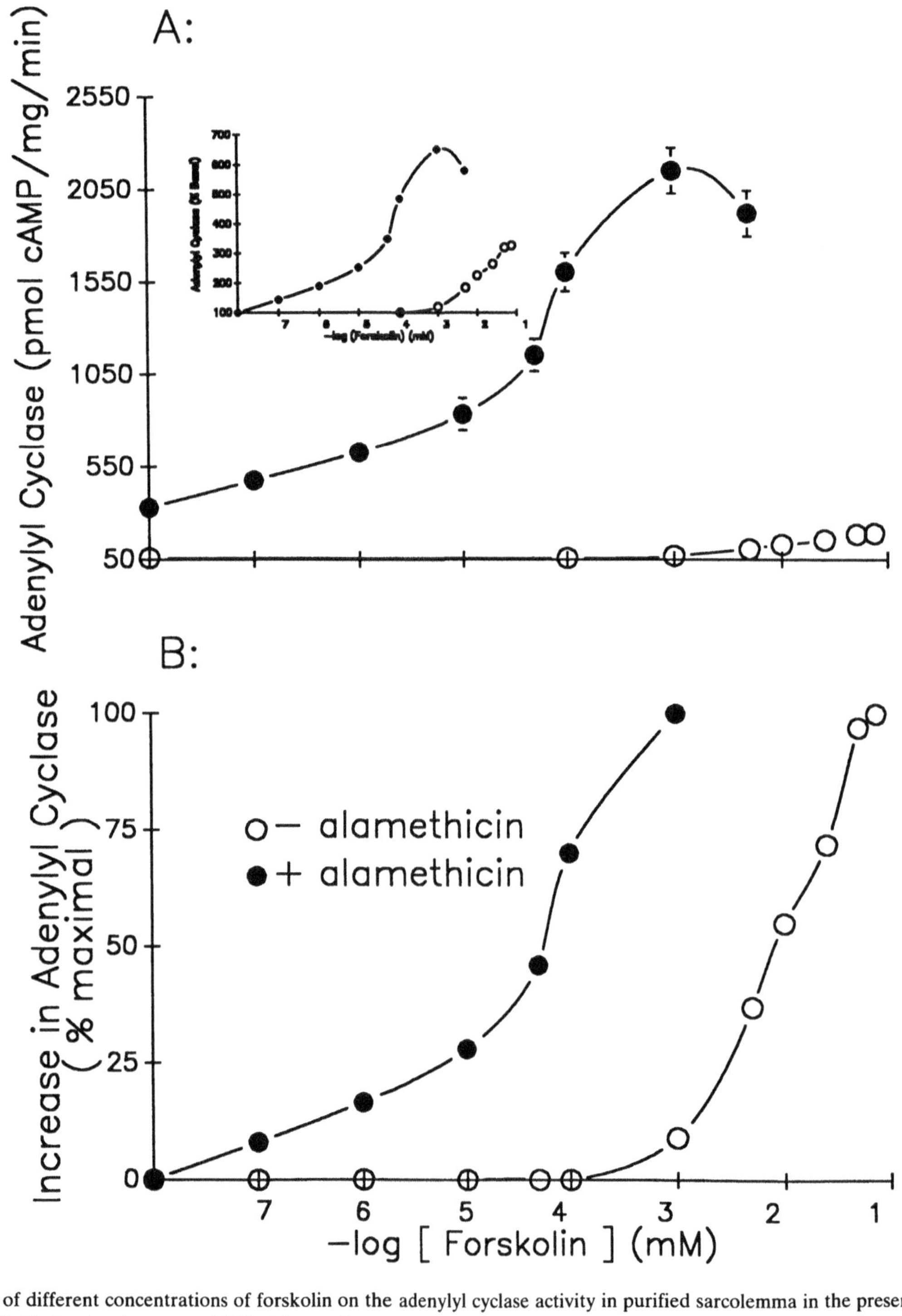

Fig. 3. Influence of different concentrations of forskolin on the adenylyl cyclase activity in purified sarcolemma in the presence and absence of alamethicin (1 : 1). Each value is a mean ± SE of 4 experiments.

alamethicin. On the other hand, there was a 200% increase in adenylyl cyclase activity with a maximal dose of 1 μM isoproterenol in the crude membrane. Results in Table 2 show that the inability of isoproterenol to activate adenylyl cyclase in the purified sarcolemmal preparation was also apparent when alamethicin was present in the incubation medium. In order to rule out the possibility of the absence of G proteins in the purified sarcolemmal preparation, we carried out immunobloting experiments using Giα (AS/7) and Gsα (RM/1) specific antibodies. AS/7 and RM/1 antibody as shown in Fig. 8A and 8B recognized the presence of

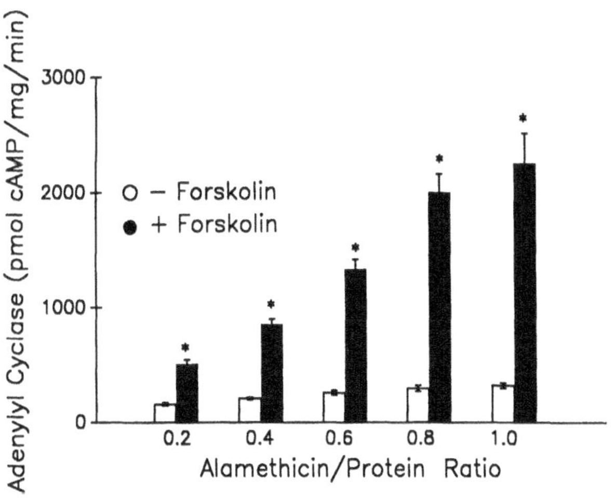

Fig. 4. Effect of different concentrations of alamethicin on adenylyl cyclase activity in purified sarcolemma in the presence and absence of 1 μM forskolin. Each value is a mean ± SE of 4 experiments. * – $P < 0.05$.

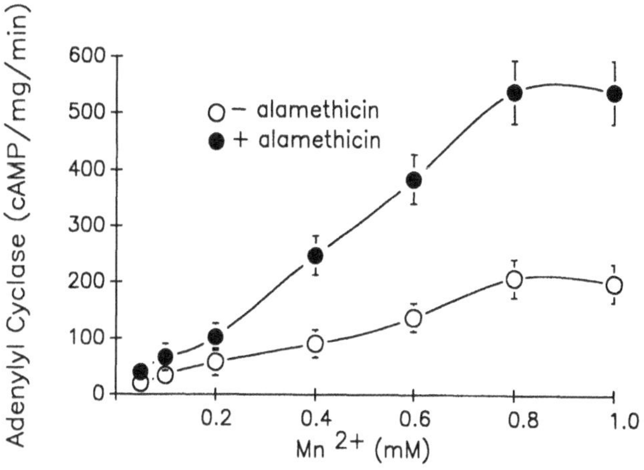

Fig. 5. Influence of different concentrations of Mn^{2+} on the adenylyl cyclase activity in purified sarcolemma in the presence and absence of alamethicin (1 : 1). Each value is a mean ± SE of 5 experiments.

G-proteins of Mr 40000 and 45000 in both purified sarcolemmal preparation and crude membranes obtained from the rat ventricle.

Discussion

The results in this study suggest that alamethicin treatment alters the characteristics of adenylyl cyclase in addition to unmasking the enzyme activity in the rat heart sarcolemma. As treatment of membranes with alamethicin or with detergents is known to solubilize the membrane proteins including enzymes [3], it may be that alamethicin as an amphipathic molecule exerts a detergent-like action on the cell membrane and thus removes phospholipids or proteins and makes the cell membrane permeable [18]. Studies have shown that alamethicin disrupts membrane bariers of sarcolemmal vesicles, which are mostly right side out, giving substrates and activators access to enzymic sites in the interior of the vesicles while preserving functional coupling of enzymes to their effectors [16]. Apparently the same mechanism may underlie the activation of ouabain sensitive Na^+-K^+ APTase by alamethicin in our experiments. The data characterizing Na^+-K^+ ATPase activities provide a basis for an approximate estimation of the orientation of the purified sarcolemmal preparation. Because only 20% of the total Na^+-K^+ ATPase activity in the absence of alamethicin was ouabain sensitive and because ouabain binds on the outer side where as ATP hydrolysis occurs at the inner side of the sarcolemma membranes [4], about 20% of the enzyme activity can be seen to be manifested by the presence of leaky membrane vesicles. It can also be suggested that the rest of the Na^+-K^+ ATPase activity (80%) most likely belonged to the intact vesicles with inside-out orientation. It should also be pointed out that the basal adenylyl

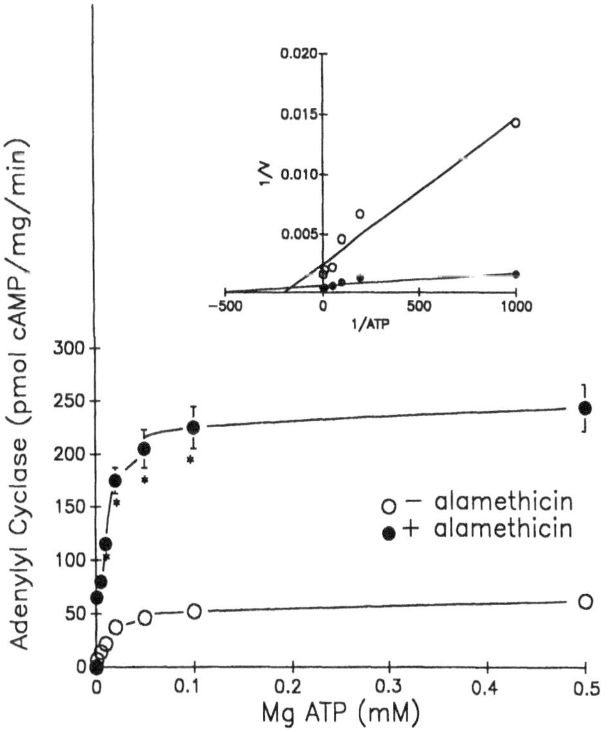

Fig. 6. Modulation of the adenylyl cyclase activity in the purified sarcolemma in the presence of different concentrations of MgATP in the presence and absence of alamethicin (1 : 1). Lineweaver-Burk plot of the data are shown in the upper part of the figure. Each value is a mean ± SE of 5 experiments.

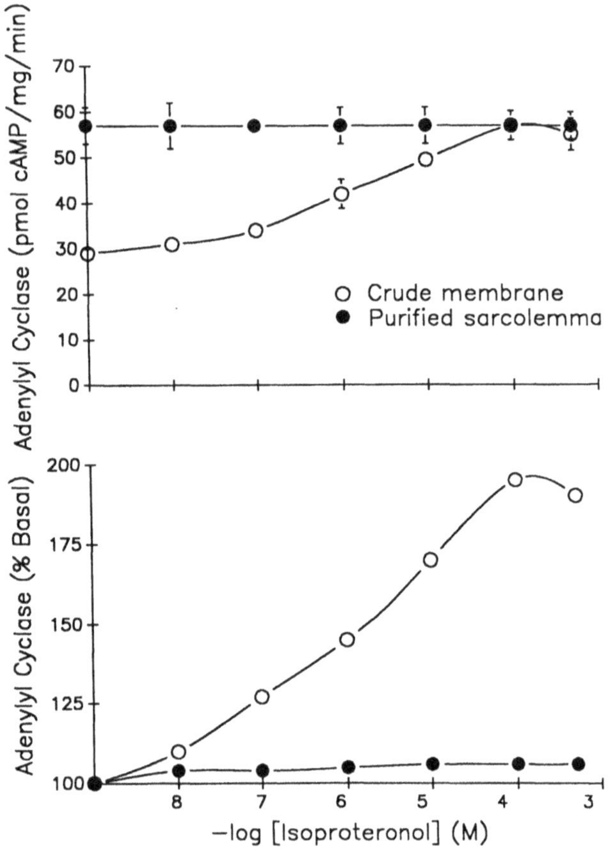

Fig. 7. Effect of different concentrations of isoproterenol on adenylyl cyclase activity in purified sarcolemmal and crude membrane preparations. Each value is a mean ± SE of 4 experiments. Results are given as absolute and % basal values.

cyclase activity as well as activities in the presence of sodium fluoride, forskolin and Gpp(NH)p were significantly increased in the presence of alamethicin. Since active sites of this enzyme are presumably on the inner surface of the sarcolemmal membrane in intact cells, some other mechanism may underlie the 4-5 fold increase in adenylyl cyclase activity by alamethicin as compared to untreated membranes. Moreover, the sensitivity of the enzyme for these activators was remarkably altered in the presence of alamethicin. Since fluoride is an activator of Gs [19] and forskolin potentiates the activation of the catalytic moiety of adenylyl cyclase by Gs [20], this would imply that there is an alteration in the catalytic moiety of adenylyl cyclase by alamethicin. The increase in stimulation by Mn^{2+} is consistent with a change in catalytic unit by alamethicin in the purified sarcolemma. This view is confirmed by our findings that in addition to increase in V_{max} by alamethicin there was an increase in affinity of the substrate MgATP for the adenylyl cyclase, confirming that alamethicin may be

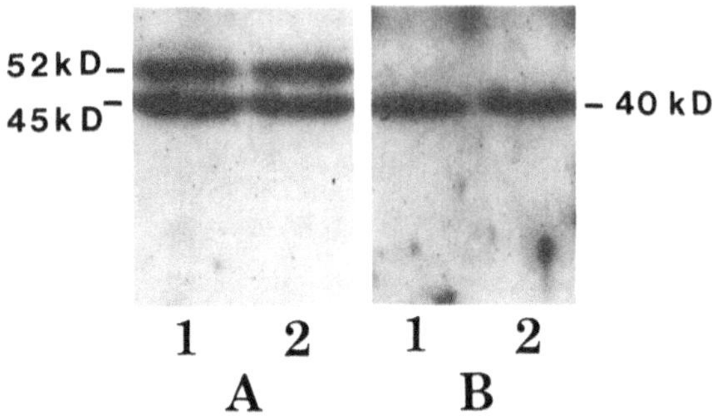

Fig. 8. Autoradiograms of Western immunoblot for G-proteins in purified sarcolemma [1] and crude membrane [2] preparations. A – shows bands for Gsα whereas B – shows band for Giα.

exerting its effect by altering the catalytic moiety of the enzyme.

The present results also provide information on the coupling of the β-adrenergic receptors to adenylyl cyclase via G proteins. The data presented above suggest that β-adrenergic receptors remained functionally uncoupled to the adenylyl cyclase in the purified sarcolemmal preparation in the absence or presence of alamethicin. Such results are not due to methodological problems because an increase in adenylyl cyclase activity with isoproteronol was observed in crude membranes. This absence of isoproterenol response in stimulating adenylyl cyclase can not be attributed to absence of G proteins in the purified sarcolemmal preparation for two reasons: a) an increase in adenylyl cyclase activity was observed on treating the membranes with various activators which are known to act via G proteins and b) respective immunoblots for Gs and Gi showed the presence of G proteins in both crude membranes and purified sarcolemmal preparations. Furthermore, β-adre-

Table 2. Effect of isoproterenol on the adenylyl cyclase activity in the presence of 10 μM Gpp(NH)p in the purified sarcolemmal preparation in the absence or presence of alamethicin

Isoproterenol (μM)	Adenylyl cyclase (pmol cAMP/mg/min)	
	− alamethicin	+ alamethicin
None	105 ± 12	1148 ± 89
1	104 ± 14	1156 ± 98
10	109 ± 18	1143 ± 83
100	105 ± 13	1140 ± 99

Each value is a mean ± SE of 4–6 experiments. Basal values without Gpp(NH)p for adenylyl cyclase in the absence and presence of alamethicin (1 : 1) were 60 ± 8 and 330 ± 27 pmol cAMP/mg/min, respectively.

nergic receptors have been shown to be present in the purified sarcolemmal preparations used in this study [21]. Thus it is concluded that the sarcolemmal preparation employed in this study is uncoupled with respect to β-adrenergic receptors and adenylyl cyclase. Nonetheless, the present study demonstrates that alamethicin is capable of affecting the characteristics of adenylyl cyclase and thus caution should be exercised while interpretting the results with respect to its unmasking effect in vesicular preparations.

Acknowledgements

This work was supported by a grant from the Medical Research Council of Canada. Mr. Rajat Sethi is a predoctoral fellow with the Heart and Stroke Foundation of Canada.

References

1. Besch Jr HR, Jones LR, Fleming JW, Watanabe AM: Parallel unmasking of latent adenylate cyclase and Na^+-K^+ ATPase activities in cardiac sarcolemmal vesicles. A new use of the channel forming ionophore alamethicin. J Biol Chem 252: 7905–7908, 1977
2. Rothman JE, Lenard J: Membrane assymetry. Science 195: 743–753, 1977
3. Besch Jr HS, Jones LR, Watanabe AM: Intact vesicles of canine cardiac sarcolemma: Evidence from vectorial properties of Na^+-K^+ ATPase. Circ Res 39: 589, 1976
4. Seppet EK, Dhalla NS: Characteristics of Ca^{2+}-stimulated ATPase in rat heart sarcolemma in the presence of dithiothreitol and alamethicin. Mol Cell Biochem 91: 137–147, 1989
5. Kato M, Kako KJ: Orientation of vesicles isolated from basolateral membranes of renal cortex. Mol Cell Biochem 78: 9–16, 1987
6. Pitts BJR: Stiochiometry of sodium-calcium exchange in cardiac sarcolemmal vesicles. J Biol Chem 254: 6232–6235, 1979
7. Hata T, Kaneko M, Beamish RE, Dhalla NS: Influence of oxygen free radicals on heart sarcolemmal Na^+-Ca^{2+} exchange. Coronary Artery Dis 2: 397–407, 1991
8. Wagner JA, Reynolds IJ, Weisman HF, Dudeck P, Weisfeldt ML, Snyder SH: Ca^{2+} antagonists receptors in cardiomyopathic hamster: Selective increases in heart, muscle, brain. Science 232: 515–518, 1987
9. Lowry OH, Rosebrough NJ, Farr AL, Randall RJ: Protein measurement with Folin phenol reagent. J Biol Chem 193: 265–275, 1951
10. Dixon IMC, Hata T, Dhalla NS: Sarcolemmal Na^+-K^+ ATPase activity in congestive heart failure due to myocardial infarction. Am J Physiol 262: C664–C671, 1992
11. Taussky H, Shorr E: A microcaloric method for the estimation of inorganic phosphorus. J Biol Chem 202: 678–685, 1953
12. Anand-Srivastava MB: Altered responsiveness of adenylate cyclase to adenosine and other agents in the myocardial sarcolemma and aorta of spontaneously-hypertensive rats. J Biochem Pharmacol 37: 3017–3022, 1988
13. Salmon Y, Londos C, Rodbell M: A highly sensitive adenylate cyclase assay. Anal Biochem 58: 541–548, 1974
14. Mumby SM, Kahn RA, Manning DR, Gilman AG: Antisera of designed specifity for subunits of guanine nucleotide binding regulatory protein. Proc Natl Acad Sci USA 83: 265–269, 1986
15. Laemmli UK: Cleavage of structural proteins during the assembly of the head of bacteriophage T4. Nature London 227: 680–685, 1970
16. Jones L, Maddock SW, Besch Jr HR: Unmasking effect of alamethicin on the Na^+-K^+ ATPase, beta adrenergic receptor coupled adenylate cyclase, and cAMP-dependent protein kinase activities of cardiac sarcolemmal vesicles. J Biol Chem 255: 9971–9980, 1980
17. Steinberg SF, Chow YK, Bilezikian JP: Regulation of rat heart membrane adenylate cyclase by magnesium and manganese. J Pharm Exp Ther 237: 764–772, 1986
18. Bonnafous JC, Dornand J, Mani JC: Detergent like effects of alamethicin on lymphocyte plasma membranes. Biochem Biophys Res Commun 86: 536–544, 1979
19. Downs RW, Speigel AM, Singer M, Reen S, Aurbach GD: Fluoride stimulation of adenylate cyclase is dependent on the guanine nucleotide regulatory protein. J Biol Chem 255: 949–954, 1980
20. Bristow MR, Ginsburg R, Strosberg A, Montgomery W, Minobe W: Pharmacology and inotropic potential of forskolin in the human heart. J Clin Invest 74: 212–223, 1984
21. Dixon IMC, Dhalla NS: Alterations in cardiac adrenoceptors in congestive heart failure secondary to myocardial infarction. Coronary Artery Dis 2: 805–814, 1991

Molecular and Cellular Biochemistry **119**: 195–201, 1993.

Cyclic AMP metabolism in intact rat ventricular cardiac myocytes: Interaction of carbachol with isoproterenol and 3-isobutyl-1-methylxanthine

Yumi Katano and Masao Endoh
Department of Pharmacology, Yamagata University School of Medicine, 990–23 Yamagata, Japan

Abstract

Experiments were carried out to elucidate the characteristics of regulation of cyclic AMP levels in intact myocardial cells. For this purpose, the influence of isoproterenol, a nonselective cyclic nucleotide phosphodiesterase (PDE) inhibitor 3-isobutyl-1-methylxanthine (IBMX) and carbachol on cyclic AMP levels was investigated in isolated rat cardiac myocytes. The extent of cyclic AMP accumulation induced by isoproterenol was much less than that produced by IBMX: submaximal concentrations of isoproterenol and IBMX elevated the cyclic AMP level 2.4- and 4.8-fold of the control level, respectively. Both agents in combination increased the cyclic AMP level markedly 48-fold. Carbachol inhibited the cyclic AMP accumulation induced by isoproterenol, IBMX and their combination by 30%, 60% and 80% of the respective response. The extent of inhibition produced by carbachol of the cyclic AMP accumulation induced by IBMX + isoproterenol was smaller than that caused by propranolol, and carbachol produced only a marginal additional inhibitory action to that of propranolol, implying that carbachol does not affect the process of cyclic AMP degradation. The present findings indicate that in intact cardiac myocytes the rate of cyclic AMP degradation catalyzed by PDE may be a crucial process of cyclic AMP turnover. This view is supported by the observations that the inhibitory action of carbachol on the effect of isoproterenol was less than that on the effect of IBMX, and that the inhibitory action of carbachol was markedly enhanced by the simultaneous presence of IBMX. (Mol Cell Biochem **119:** 195–201, 1993)

Key words: 3-isobutyl-1-methylxanthine; carbachol, propranolol, cyclic AMP accumulation; rat ventricular myocytes

Introduction

In mammalian ventricular muscle the muscarinic receptor stimulation scarcely affects the basal force of contraction, but induces a pronounced negative inotropic effect when the contractile force had been elevated by β-adrenoceptor stimulation [1–3]. This characteristic inhibitory action of muscarinic receptor stimulation in ventricular myocardium has been termed accentuated antagonism [4], and the subcellular mechanism of the inhibitory action has been shown to involve the signal transduction mediated by cyclic adenosine 3′, 5′-monophosphate (cyclic AMP) [5–7].

Biochemical processes involved in the receptor-me-

Address for offprints: M. Endoh, Department of Pharmacology, Yamagata University School of Medicine, 2-2-2 Iida-nishi, 990-23 Yamagata, Japan

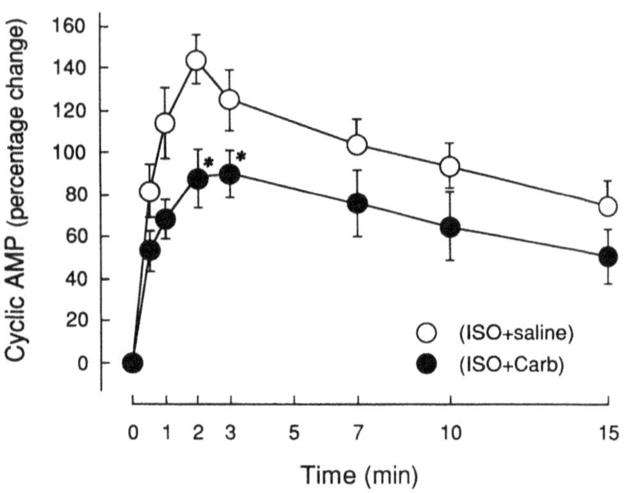

Fig. 1. The inhibitory action of carbachol (Carb; 10^{-4} M) on the cyclic AMP accumulation caused by 10^{-6} M isoproterenol (ISO) in isolated rat cardiac myocytes. Ordinate: percentage changes in cyclic AMP levels. Agents or/and saline in parentheses were added simultaneously into the incubation tubes throughout this and subsequent figures. Control cyclic AMP levels prior to the drug administration: 2.61 ± 0.51 (ISO + saline) and 2.39 ± 0.46 (ISO + Carb) pmol/10^5 cells ($n = 6$, each). Abscissa: time after the administration of ISO + saline or ISO + Carb. * $P < 0.05$; ** $P < 0.01$ vs the corresponding values with ISO + saline.

diated stimulation or inhibition of adenylate cyclase, the enzyme responsible for generation of cyclic AMP, through stimulatory (Gs) or inhibitory (Gi) GTP binding protein [5–6], and degradation of cyclic AMP by cyclic nucleotide phosphodiesterase (PDE) [8, 9] in cell-free system have been well documented. On the other hand, regulation of cyclic AMP turnover in intact myocardial cells has been relatively less understood. While it has long been known that β-adrenoceptor stimulation and/or PDE inhibition increase cyclic AMP content, and muscarinic receptor stimulation antagonizes the cyclic AMP accumulation induced by these interventions in intact myocardium, relative significance of the regulation and characteristics of interaction induced by these interventions has not been studied in detail.

Isolated cardiac myocytes provide an advantage to study the cyclic AMP turnover in intact myocardium in more detail because the contribution of other types of cells such as vascular smooth muscle, endocardial and vascular endothelial cells to the observed changes in cyclic AMP levels could be obviated in isolated single cardiac myocytes.

In order to address the question described above, the effect of isoproterenol, 3-isobutyl-1-methylxanthine (IBMX) and carbachol, and their interaction in regulation of cyclic AMP accumulation were investigated in isolated rat cardiac myocytes.

Materials and methods

The isolated rat cardiac myocytes were used to investigate the interaction of the drugs on cyclic AMP levels. Male Wistar rats (190–230 g) were lightly anesthetized with ether and given heparin (500 U i.p.). Hearts were quickly removed and retrograde perfusion was performed through aorta with Krebs-Henseleit solution containing collagenase (70 U/ml) and 0.1% bovine serum albumin (BSA) in the presence of 50 μM Ca^{2+} according to the method of Powell and Twist [10] with a slight modification. The constituent of the solution (in mM) was: Na^+ 142.9, K^+ 5.9, Mg^{2+} 1.2, Ca^{2+} 1.25, H_2PO^{4-} 1.2, HCO_3^- 24.9, SO_4^{2-} 1.2, Cl^- 127.8, glucose 5.0, and pyruvic acid 2.0; pH 7.4 when bubbled with 95% O_2 and 5% CO_2. The viability of the freshly prepared cells was about 90%. Incubation of cardiac myocytes and extraction of cyclic AMP from the cells were performed according to the method described by Katada and Ui [11] with a slight modification. Briefly, each incubation tube contained 100 μl of cells (5 × 10^5 cells/ml) suspended with Krebs-Henseleit solution containing 2% BSA and 100 μl of physiological saline with drugs. Incubation was carried out at 37° C. At the end of incubation, 200 μl of 0.2 N HCl was quickly added and the tube immersed in boiling water for 3 min to extract cyclic AMP. After centrifugation at 600 × g for 10 min, cyclic AMP was assayed in duplicate by the sensitive radioimmunoassay method (Yamasa Shoyu Co., Choshi, Japan). Submaximal concentration of isoproterenol (10^{-6} M) or IBMX (5 × 10^{-4} M), and the maximal concentration of carbachol (10^{-4} M), which were determined in the preliminary experiment, were used in these experiments. The time course of changes in cyclic AMP levels after the administration of isoproterenol, IBMX or their combination in the absence or presence of carbachol administered simultaneously with these compounds was determined. In other series of experiments carbachol was added when a stable cyclic AMP level in response to cyclic AMP accumulating agents had been reached. In these experiments, IBMX or isoproterenol (in the presence of IBMX administered 5 min before isoproterenol) was allowed to act for 5 min before carbachol was administered. The time of carbachol addition was set at time 0. In another series the

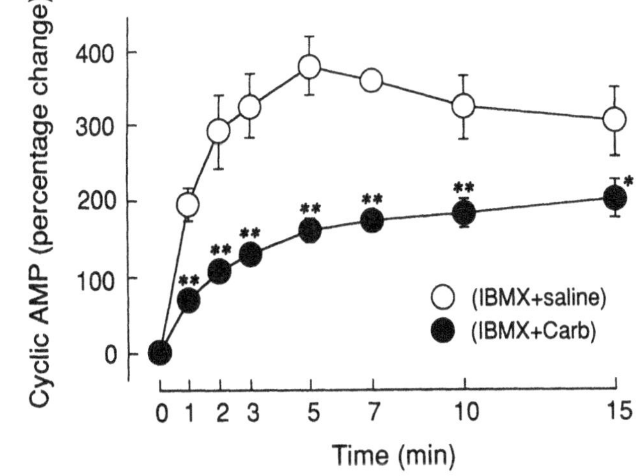

Fig. 2. The inhibitory action of carbachol (Carb; 10^{-4} M) on the cyclic AMP accumulation caused by 5×10^{-4} M IBMX in isolated rat cardiac myocytes. Ordinate: percentage changes in cyclic AMP levels. Control cyclic AMP levels prior to the drug administration: 2.04 ± 0.25 (IBMX + saline) and 2.03 ± 0.34 (IBMX + Carb) pmol/10^5 cells (n = 6, each). Abscissa: time after the administration of IBMX + saline or IBMX + Carb. * P < 0.05; ** P < 0.01; *** P < 0.001 vs the corresponding values with IBMX + saline.

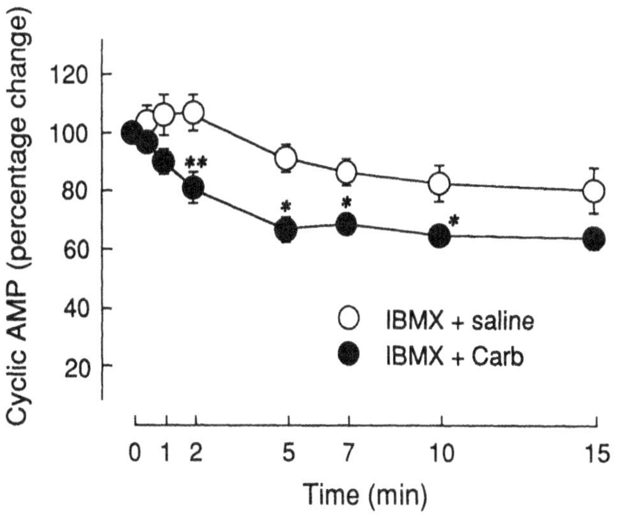

Fig. 3. The inhibitory action of carbachol (Carb; 10^{-4} M) on the cyclic AMP accumulation caused by 5×10^{-4} M IBMX in isolated rat cardiac myocytes. Carbachol was administered 5 min after IBMX. Ordinate: percentage changes in cyclic AMP levels. The cyclic AMP levels at 5 min after the administration of IBMX (immediately before the addition of carbachol or saline: time 0) [7.18 ± 2.28 (IBMX + saline) and 5.89 ± 1.29 (IBMX + Carb) pmol/10^5 cells; n = 4, each] were taken as 100%, respectively. Abscissa: time after the administration of carbachol or saline. * P < 0.05; ** P < 0.01 vs the corresponding values with IBMX + saline.

influence of propranolol (10^{-6} M) was compared with that of carbachol. In these experiments, propranolol was administered alone or with carbachol when the cyclic AMP level reached a stable level in response to IBMX or IBMX + isoproterenol.

Experimental values were presented as means ± S.E.M. Statistical significance between mean values was estimated by Student's t-test. A t-test for the paired comparison was used when it was applicable. A P value smaller than 0.05 was considered to be significant.

The drugs used were obtained from the following sources: 3-isobutyl-1-methylxanthine (IBMX, Aldrich Chemical Co., Milwalkee, WI, USA); (-)-isoproterenol hydrochloride, carbachol (carbamylcholine chloride; Sigma Chemical Co., St. Louis, MO, USA); (±)-propranolol hydrochloride (Sumitomo Seiyaku Co., Osaka, Japan); cyclic AMP assay kit (Yamasa Shoyu Co., Choshi, Japan). IBMX was dissolved in 50% N,N-dimethylformamide in a concentration of 4×10^{-3} M, and diluted to the desired concentration with physiological saline solution. The stock solution of isoproterenol was freshly prepared in 1% ascorbic acid on each experimental day, diluted with distilled water and kept on ice.

Results

Inhibitory action of carbachol on the isoproterenol-induced response

The time course of changes in cyclic AMP levels in cardiac myocytes after administration of isoproterenol alone (ISO + saline), and isoproterenol administered simultaneously with carbachol (ISO + Carb) is shown in Fig. 1. Isoproterenol elevated rapidly the cyclic AMP level within 1 min with the time to a half maximum increase at 30 sec. The cyclic AMP accumulation reached a peak level of 140% higher than the control at 2 min, and declined to a level of 80% up to 15 min. The cyclic AMP accumulation induced by isoproterenol + carbachol was significantly less than that induced by isoproterenol alone. Statistically significant attenuation was achieved at 2 and 3 min. However, the extent of attenuation by carbachol of the isoproterenol-induced cyclic AMP accumulation was relatively small: it amounted to only 25–30% at the maximum, which was comparable to that induced by choline esters in the cell-free system in the early study [12].

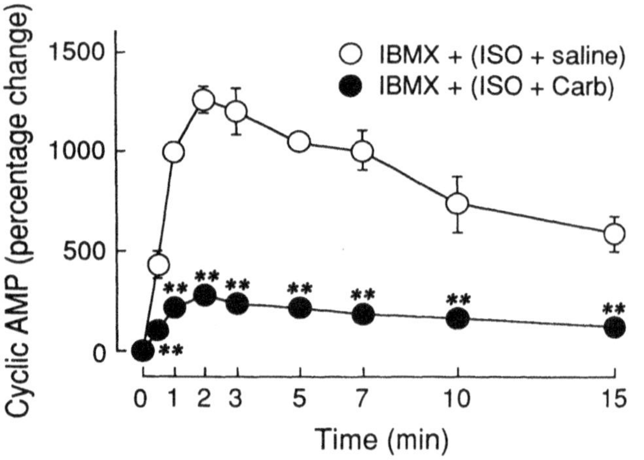

Fig. 4. The inhibitory action of carbachol (Carb: 10^{-4} M) on the cyclic AMP accumulation caused by 10^{-6} M ISO in the presence of 5×10^{-4} M IBMX in isolated rat cardiac myocytes. Isoproterenol (ISO) was administered 5 min after the administration of IBMX. Ordinate: percentage changes in cyclic AMP levels. The cyclic AMP levels 5 min after IBMX [6.60 ± 0.54 (IBMX + ISO + saline) and 6.52 ± 0.78 (IBMX + ISO + Carb) pmol/10^5 cells; $n = 5$, each] were taken as 100%. Abscissa: time after the administration of ISO + saline or ISO + Carb. ** $P < 0.01$ vs the corresponding values with ISO + saline.

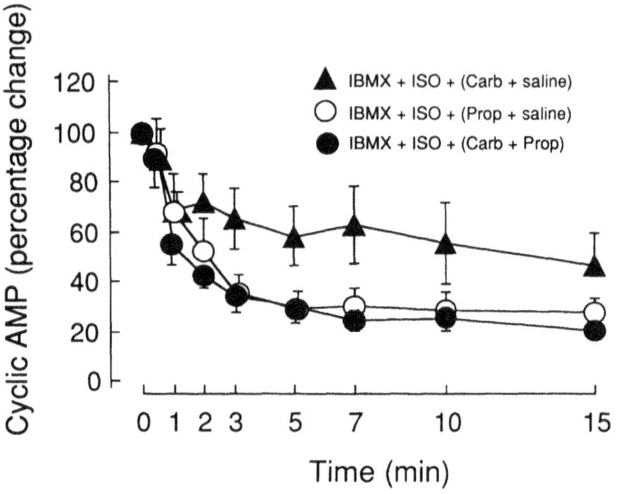

Fig. 5. The inhibitory action of propranolol (Prop; 10^{-6} M), carbachol (Carb; 10^{-4} M) or both on the cyclic AMP accumulation caused by 10^{-6} M ISO + 5×10^{-4} M IBMX in isolated rat cardiac myocytes. Prop, Carb or both were administered 3 min after isoproterenol (ISO; time 0) in the presence of IBMX (administered 5 min prior to ISO). Ordinate: percentage changes in cyclic AMP levels. The cyclic AMP levels prior to the drug administration [62.4 ± 12.48 (Prop + saline), 86.3 ± 15.24 (Carb + saline) and 62.8 ± 9.00 (Prop + Carb) pmol/10^5 cells; $n = 6$, each] were taken as 100%. Abscissa: time after the administration of Prop, Carb or both.

It is noteworthy that the time course of cyclic AMP accumulation induced by isoproterenol was scarcely affected by the simultaneous administration of carbachol (Fig. 1).

Inhibitory action of carbachol on the IBMX-induced response

The IBMX-induced accumulation of cyclic AMP developed more slowly than that of isoproterenol with the time to a half maximum increase at 1 min (Fig. 2). The peak of cyclic AMP accumulation in response to IBMX reached a level 380% higher than the control level at 5 min, and the cyclic AMP level declined slightly at 15 min. By simultaneous administration of carbachol, the IBMX-induced cyclic AMP accumulation was markedly attenuated, and the maximum accumulation amounted to 200% (the maximum accumulation at 5 min was attenuated by 60%). A marked retardation of the time course of IBMX-induced cyclic AMP accumulation was induced by carbachol: the time to a half maximum increase was prolonged twofold to 2 min. The peak of cyclic AMP accumulation was not achieved within 15 min (Fig. 2). These results indicate that the rate of cyclic AMP generation may play a crucial role for the cyclic AMP accumulation induced by IBMX in the presence of carbachol.

In order to examine further the contribution of cyclic AMP generation during the interaction of carbachol with IBMX, the effect of carbachol was examined when

Table 1. The inhibitory action of propranolol (10^{-6} M), carbachol (10^{-4} M), and propranolol and carbachol in addition on the cyclic AMP level elevated by 5×10^{-4} M IBMX and 10^{-6} M isoproterenol in the isolated rat cardiac myocytes

Time after Prop or Carb (sec)	Cyclic AMP level (%)		
	Prop (10^{-6} M)	Carb (10^{-4} M)	Prop + Carb
0	100	100	100
20	95.1 ± 0.7	78.6 ± 4.7	81.8 ± 9.8
40	81.4 ± 3.0	63.0 ± 2.4	51.4 ± 7.1**
60	56.4 ± 6.4	54.8 ± 4.6	36.5 ± 3.7*
90	39.7 ± 4.8	44.3 ± 3.4	26.6 ± 5.2
120	32.8 ± 5.0	40.5 ± 5.9	18.4 ± 1.2
150	26.8 ± 3.3	35.8 ± 4.3	14.3 ± 0.3

Prop or/and Carb were administered 3 min after ISO (time 0) in the presence of IBMX (administered 5 min prior to ISO). The cyclic AMP levels prior to addition of Prop or/and Carb were taken as 100%: 68.9 ± 4.76 pmol/10^5 cells (IBMX 5×10^{-4} M + ISO 10^{-6} M; $n = 12$). Numbers of experiments (n) = 4, each. * $P < 0.05$; ** $P < 0.01$ vs. the respective values with Prop alone.

a stable cyclic AMP level had been achieved in response to IBMX. Whereas carbachol reduced significantly the IBMX-induced accumulation of cyclic AMP (Fig. 3), the inhibitory action of carbachol on the IBMX-induced response was less than that when carbachol was administered simultaneously with IBMX (Fig. 2).

Inhibitory action of carbachol on the response induced by IBMX and isoproterenol in combination

Figure 4 shows the time course of changes in cyclic AMP levels of cardiac myocytes after the administration of isoproterenol and carbachol in the presence of IBMX. IBMX was allowed to act for 5 min, and then isoproterenol and saline, or isoproterenol and carbachol was administered in the presence of IBMX. Isoproterenol produced an enormous cyclic AMP accumulation in the presence of IBMX (Fig. 4). The cyclic AMP accumulation reached a peak level of 13-fold the level with IBMX alone at 2 min of isoproterenol administration, and declined slowly to a level 6-fold higher than that with IBMX alone at 15 min. Since IBMX elevated the cyclic AMP level by itself, the accumulation induced by IBMX and isoproterenol amounted to 48-fold the baseline level prior to the administration of IBMX.

Carbachol administered simultaneously with isoproterenol attenuated markedly the cyclic AMP accumulation: isoproterenol administered with carbachol elevated the cyclic AMP level only 2.8-fold (Fig. 4), indicating that the inhibitory action of carbachol on the isoproterenol-induced cyclic AMP was more pronounced in the presence of IBMX.

Comparison of carbachol- and propranolol-induced inhibitory action

The inhibitory action of carbachol on cyclic AMP accumulation was compared with the action of propranolol that has been used to produce a selective cessation of cyclic AMP generation by β-adrenoceptor blockade [13].

The influence of carbachol, propranolol and their combination on the cyclic AMP accumulation induced by IBMX + isoproterenol is shown in Fig. 5. Carbachol decreased the cyclic AMP level rapidly by approximately 30% within 1 min, but the extent of the carbachol-induced inhibition was smaller than that caused by propranolol. Propranolol reduced the cyclic AMP level 70% at 5 min (Fig. 5). Carbachol administered with propranolol dit not produce an obvious additional inhibitory action on the time course and extent of the propranolol-induced inhibition. However, since a close inspection of the changes revealed that in the early period (up to 2 min) carbachol appear to produce a small additional action, we examined the changes in cyclic AMP levels within 2 min after the administration of propranolol and carbachol in more detail.

In the presence of 5×10^{-4} M IBMX and 10^{-6} M isoproterenol, carbachol elicited a small but significant additional inhibitory action compared with propranolol alone up to 2 min (Table 1).

Discussion

An important finding in the present study in the isolated intact rat ventricular cardiac myocytes is that carbachol attenuated the IBMX-induced cyclic AMP accumulation more effectively than the isoproterenol-induced accumulation: isoproterenol at 10^{-6} M elevated the cyclic AMP level 2.4-fold, and carbachol at 10^{-4} M attenuated the accumulation less than by 30%, while IBMX (5×10^{-4} M) increased the level 4.8-fold, and carbachol inhibited the accumulation maximally by 60%.

Although the inhibitory action of muscarinic receptor agonist on the IBMX-induced cyclic AMP accumulation has been demonstrated in chick embryonic heart [1], rat atria [14] and dog ventricular trabeculae [15], the effectiveness of the inhibitory action of muscarinic stimulation on β-adrenoceptor- and IBMX-induced cyclic AMP accumulation has not been compared in detail. IBMX is generally accepted to be a pure and most effective nonselective PDE inhibitor among methylxanthines and employed as a prototype of PDE inhibitors [16–18].

While our expectation before the experiments was that the cyclic AMP accumulation induced by isoproterenol through activation of adenylate cyclase may be more effectively antagonized by carbachol through muscarinic receptors coupled to Gi than the IBMX-induced cyclic AMP accumulation, the results obtained in intact rat cardiac myocytes were opposite. We consider that these results can be explained reasonably well by the characteristics of cyclic AMP turnover in intact cardiac myocytes that the rate of degradation of cyclic AMP catalyzed by PDE may play a crucial role in regulation cyclic AMP levels. The following observations support the view described above: (1) inhibition of

PDE by IBMX (5×10^{-4} M) elevated the cyclic AMP content to a level about three times as high as that reached by isoproterenol (10^{-6} M); (2) isoproterenol elevated the cyclic AMP level only 2.4-fold, while in the presence of IBMX it increased the cyclic AMP level by 48-fold of the baseline value supports this view. In intact myocardial cells, the PDE may provide an ample reserve capacity for cyclic AMP hydrolysis, and, therefore, the time course and extent of cyclic AMP accumulation induced by isoproterenol in the absence of IBMX may be essentially determined by the rate of degradation, though cyclic AMP generation is facilitated by β-stimulation. The observations that carbachol decreased markedly the isoproterenol-induced cyclic AMP accumulation in the presence of IBMX imply the actual inhibitory action of carbachol may have been masked by the effective degradation of cyclic AMP produced by PDE in the absence of IBMX.

We suspected whether other mechanism such as stimulation of PDE in addition to inhibition of adenylate cyclase through Gi may be also involved in the carbachol-induced lowering of cyclic AMP accumulation. Activation of phosphodiesterase mediated by muscarinic receptor stimulation has been reported in noncardiac cells [19]. However, since carbachol produced only a marginal additional action to the propranolol-induced decrease in cyclic AMP accumulation, such mechanism may not contribute to the carbachol-induced inhibitory action. The findings that carbachol was more effective when administered simultaneously with IBMX than when IBMX-induced response reached a steady level are also consistent with the view that carbachol may act on cyclic AMP generation.

IBMX accumulated cyclic AMP more slowly than isoproterenol with the time to a half maximum increase at 1 min, and the peak of cyclic AMP accumulation by IBMX reached a level 380% higher than the control at 5 min; carbachol markedly decreased the rate of cyclic AMP accumulation induced by IBMX, indicating that the rate of cyclic AMP generation becomes critical in determining the time course in the presence of IBMX and carbachol. This indicates that cyclic AMP is actively metabolized in the baseline state, and the basal rate of cyclic AMP generation may be high in intact myocardial cells.

In addition the possibility has also to be taken into account that IBMX elicits an effect more than a pure PDE inhibition. It has recently been shown that IBMX acts on G_i to lead to inhibition of its function, and thereby facilitates cyclic AMP generation in rat adipocytes [20]. Although it is unknown whether G_i plays a role in the regulation of basal cyclic AMP turnover in myocardial cells, such mechanism may be likely in rat cardiac myocytes. The observation that the cyclic AMP level was rapidly lowered by carbachol even in the presence of IBMX suggests that the nucleotide is hydrolyzed actively in the presence of IBMX. The PDE isoenzyme that is resistant to IBMX has been reported in hepatocytes [21].

In summary the present results indicate that the cyclic AMP turnover rate in intact cardiac myocytes may be high involving a complex interaction of generating and hydrolyzing systems. Drugs that interact with the system can modulate markedly the cyclic AMP turnover rate and thereby the cell function in intact myocardial cells.

Acknowledgements

This work was partly supported by Grant-in-Aid for Scientific Research on Priority Areas (No. 62624004, 63641002 and 01641003) from the Ministry of Education, Science, and Culture, Japan.

References

1. Biegon RL, Epstein PM, Pappano AJ: Muscarinic antagonism of the effects of a phosphodiesterase inhibitor (methylisobutylxanthine) in embryonic chick ventricle. J Pharmacol Exp Ther 215: 348–356, 1980
2. Hollenberg M, Carriere S, Barger AC: Biphasic action of acetylcholine on ventricular myocardium. Circ Res 16: 527–536, 1965
3. Löffelholz K, Pappano AJ: The parasympathetic neuroeffector junction of the heart. Pharmacol Rev 37: 1–24, 1985
4. Levy MN: Sympathetic-parasympathetic interactions in the heart. Circ Res 29: 437–445, 1971
5. Birnbaumer L, Codina J, Sunyer T, Hildebrandt JD, Sekura RD, Cerione RA, Lefkowitz RJ, Caron MG: Dual coupling mechanisms in the regulation of cAMP formation. In: AM Gotto Jr and BW O'Malley (eds). The role of receptors in biology and medicine. Raven Press, New York, 1986, p 193–211
6. Gilman AG: G proteins: transducers of receptor-regulated signals. Ann Rev Biochem 56: 615–649, 1987
7. Ui M, Katada T, Maruyama T, Kurose H, Yajima M, Tamura M, Nakamura T, Nogimori K: Islet-activating protein, pertussis toxin: a specific uncoupler of receptor-mediated inhibition of adenylate cyclase. Adv Cyclic Nucleotide Protein Phosphorylation Res 17: 145–151, 1984
8. Thompson WJ, Pratt ML, Strada SJ: Biochemical properties of

high-affinity cyclic AMP phosphodiesterases. Adv Cyclic Protein Phosphorylation Res 16: 137–148, 1984

9. Weishaar RE, Kobylarz-Singer D, Keiser JA, Wright CD, Cornicelli J, Panek R: Cyclic nucleotide phosphodiesterases in the circulatory system. Biochemical, pharmacological, and functional characteristics. Adv Cyclic Protein Phosphorylation Res 25: 249–269, 1992
10. Powell T, Twist VW: A rapid technique for the isolation and purification of adult cardiac muscle cells having respiratory control and a tolerance to calcium. Biochem Biophys Res Commun 72: 327–333, 1976
11. Katada T, Ui M: Islet activating protein: enhanced insulin secretion and cyclic AMP accumulation in pancreatic islets due to activation of native calcium ionophores. J Biol Chem 254: 469–479, 1979
12. Murad F, Chi YM, Rall TW, Sutherland EW: Adenyl cyclase: II. The effect of catecholamines and choline esters on the formation of adenosine-3′-5′-monophosphate by preparations from cardiac muscle and liver. J Biol Chem 237: 1233–1238, 1962
13. Lindemann JP, Watanabe AM: Muscarinic cholinergic inhibition of β-adrenergic stimulation of phospholamban phosphorylation and Ca^{2+} transport in guinea pig ventricles. J Biol Chem 260: 13122–13129, 1985
14. Brown BS, Polson JB, Krzanowski JJ, Wiggins JA: Influence of isoproterenol and methylisobutylxanthine on the contractile and cyclic nucleotide effects of methacholine in isolated rat atria. J Pharmacol Exp Ther 212: 325–332, 1980
15. Endoh M: Dual inhibition of myocardial function through muscarinic and adenosine receptors in the mammalian heart. J Appl Cardiol 2: 213–230, 1987
16. Chasin M, Harris DN: Inhibitors and activators of cyclic nucleotide phosphodiesterase. Adv Cyclic Nucleotide Res 7: 225–264, 1976
17. Korth M: Effects of several phosphodiesterase-inhibitors on guinea-pig myocardium. Naunyn-Schmiedeberg's Arch Pharmacol 302: 77–86, 1978
18. Weishaar RE, Cain MH, Bristol JA: A new generation of phosphodiesterase inhibitors: Multiple molecular forms of phosphodiesterase and the potential for drug selectivity. J Med Chem 28: 537–545, 1985
19. Meeker RB, Harden TK: Muscarinic cholinergic receptor-mediated activation of phosphodiesterase. Mol Pharmacol 22: 310–319, 1982
20. Parsons WJ, Ramkumar V, Stiles GL: Isobutylmethylxanthine stimulates adenylate cyclase by blocking the inhibitory regulatory protein, G_i. Mol Pharmacol 34: 37–41, 1988
21. Lavan BE, Lakey T, Houselay MD: Resolution of soluble cyclic nucleotide phosphodiesterase isoenzymes, from liver and hepatocytes, identifies a novel IBMX-insensitive form. Biochem Pharmacol 38: 4123–4136, 1989

Molecular and Cellular Biochemistry **119**: 203–207, 1993.

Purification and properties of pyridoxal kinase from bovine brain

Tae Hirakawa-Sakurai, Kiyoshi Ohkawa and Makoto Matsuda
Department of Biochemistry, Jikei University School of Medicine, 3-25-8, Nishi-Shinbashi, Minato-Ku, Tokyo 105, Japan

Abstract

A 27,000-fold purification of pyridoxal kinase from bovine brain tissue has been achieved by a combination of ammonium sulfate fractionation, DEAE-cellulose chromatography, hydroxyapatite chromatography, Sephadex G-150 gel filtration, Blue Sepharose CL-6B chromatography, and Phenyl-Superose chromatography. The final chromatography step yields a homogeneous preparation of high specific activity (2105 nmol/min/mg protein). The molecular mass of the native enzyme was estimated to be approximately 80,000 on gel filtration. The subunit molecular mass was determined by sodium dodecyl sulfate polyacrylamide gel electrophoresis to be approximately 39,500. This indicates that pyridoxal kinase is a dimeric enzyme. (Mol Cell Biochem **119:** 203–207, 1993)

Key words: pyridoxal kinase, bovine brain, vitamin B_6

Introduction

Pyridoxal phosphate dependent enzymes synthesize some neurotransmitters and are, therefore, important in the maintenance of the integrity of the physiological homeostasis in the central nervous system. As a coenzyme of L-amino acid decarboxylase, pyridoxal phosphate is essential for the synthesis of dopamine, norepinephrine, serotonine, and 4-aminobutyric acid, respectively.

The formation of pyridoxal phosphate from ATP, pyridoxal, and divalent metal ion, such as Zn^{2+}, Co^{2+} or Mn^{2+}, is catalyzed by pyridoxal kinase (EC 2.7.1.35.), an enzyme which has been detected in virtually all mammalian species [1, 2], and in many micro-organisms [3]. The pyridoxal kinases have been purified from brain [4–7] and liver [8] of several mammalian species. However, there has been no report about the purification of the enzyme to homogeneity from bovine brain [4].

Recently our studies [9–12] have shown that when pyridoxine or pyridoxamine was given orally to mice, every one was entirely transformed to pyridoxal in the intestine, which can serve as the only source of pyridoxal phosphate for most tissues including neural tissue. Thus, pyridoxal kinase appears to play an important role in pyridoxal phosphate synthesis.

In this paper, we describe the purification of pyridoxal kinase from bovine brain and report the kinetic data for several substrates.

Materials and methods

Materials

DEAE-cellulose was purchased from Whatman (Clifton, NJ, U.S.A.). Hydroxyapatite was from Bio-Rad (Richmond, CA, U.S.A.). Sephadex G-150, Blue Se-

Address for offprints: K. Ohkawa, Dept. Biochem. Jikei Univ. Sch. Med. 3-25-8, Nishi-Shinbashi, Minato-Ku, Tokyo 105, Japan

pharose CL-6B, EAH-Sepharose 4B, and Phenyl-Superose (HR 5/5) were from Pharmacia (Uppsala, Sweden). Vitamin B_6 and all other reagents were of analytical grade from Sigma (St. Louis, MO, U.S.A.), except where specifically noted.

Pyridoxyl Sepharose was prepared by a coupling EAH-Sepharose 4B gel with pyridoxal according to the procedure of Kerry *et al.* [7].

Enzyme assays and determination of kinetic constants

Enzyme samples were incubated at 37° C for 30 min with 1.5 mM ATP, 0.15 mM $ZnCl_2$, and 0.3 mM pyridoxal in 70 mM potassium phosphate (pH 6.5) in a total volume of 1 ml. The reaction was terminated by the addition of 0.1 ml $HClO_4$. The amount of pyridoxal phosphate produced was determined by reaction with phenylhydrazine according to the method of Wada and Snell [13], except that the color development reaction was allowed to proceed for only 15 min. Enzyme activity was determined spectrophotometrically at 410 nm on ice.

Pyridoxal kinase activities, using either pyridoxine or pyridoxamine as substrates, were measured by HPLC according to the method of Kurioka *et al.* [14].

Michaelis constant (K_m) values and maximum velocities (V_{max}) for pyridoxal, pyridoxine and pyridoxamine were estimated from double reciprocal plots, respectively, and inhibition constant (K_i) value from Dixon plots.

Protein determination

Protein concentration was determined by Bio-Rad protein assay kit (Bio-Rad) using bovine serum albumin as standard.

Purification procedure

All procedures were carried out at 4° C unless otherwise stated. In each step, pyridoxal kinase activity in the fractions was assayed by the method described above.

Extraction
To prepare a 50% (wt./vol.) homogenate, 5 Kg fresh bovine brains, obtained from the slaughter house, were mixed with ice-cold 5 l of 10 mM potassium phosphate, pH 7.0, containing 0.05 mM pyridoxal, 0.02 mM $ZnCl_2$, 0.1 mM glutathione, 0.1 mM phenylmethylsulphonilfluoride (PMSF) (Buffer A) and were homogenized using Porytron homogenizer at maximum speed for 120 sec. The homogenate was then centrifuged at 10,000 × g for 30 min and a supernatant was obtained.

Ammonium sulfate fractionation
The supernatant was treated with ammonium sulfate and centrifuged at 10,000 × g for 30 min. The precipitate obtained at 40 to 60% saturation was dissolved in 500 ml of Buffer A. The solution was dialyzed against Buffer A extensively.

DEAE-cellulose chromatography
The dialyzed solution was applied to a DEAE-cellulose column (6.2 × 35 cm) equilibrated with Buffer A. After washing the column with Buffer A, it was eluted with linear gradient of 500 ml of 0–250 mM KCl in buffer A, followed by an additional 500 ml of 250 mM KCl in buffer A. The enzymatically active fractions were collected.

Hydroxyapatite chromatography
The active fraction from DEAE-cellulose column was applied directly to a hydroxyapatite column (4 × 25 cm), which was previously equilibrated with Buffer A containing 250 mM KCl. After washing the column, pyridoxal kinase was eluted by a linear gradient of 10–300 mM potassium phosphate, pH 7.0, and the active fractions were combined.

Sephadex G-150 filtration
The hydroxyapatite fractions containing enzyme activities were concentrated and dialyzed against 150 mM potassium phosphate, pH 7.0, containing 0.05 mM pyridoxal, 0.02 mM $ZnCl_2$, 0.1 mM glutathione, 0.1 mM PMSF, using Centriprep concentrator (Amicon, Beverly, MA, U.S.A.). Approximately 10 ml sample was applied on Sephadex G-150 (5 × 100 cm) preequilibrated and eluted with the same buffer.

Blue Sepharose CL-6B chromatography
The eluate containing enzyme activity was dialyzed against 20 mM Tris-HCl, pH 7.0, containing 0.05 mM pyridoxal, 0.1 mM dithiothreitol (DTT), 0.02 mM $ZnCl_2$, and 0.1 mM PMSF (Buffer B). The dialyzed solution was applied to a Blue Sepharose CL-6B column (2.5 × 7 cm) equilibrated with Buffer B. The column was eluted with the same buffer. The flow-

through fractions with enzyme activity were collected, since the enzyme was not retained by Blue Sepharose CL-6B.

Phenyl-Superose chromatography

The active fractions from Blue Sepharose CL-6B were dialyzed against Buffer B containing 3 M KCl and applied on a column of Phenyl-Superose (HR 5/5) using a FPLC system (Pharmacia) preequilibrated with the same buffer. Followed by washing with Buffer B, the enzyme was eluted from the column by decreasing linearly the concentration of KCl from 3 to 0 M in the buffer B. Fractions containing enzyme activity were pooled and used for kinetic studies. Purified enzyme was routinely stored at − 80° C with addition of 0.05 mM pyridoxal, 0.02 mM $ZnCl_2$, 0.1 mM DTT, 0.1 mM PMSF and 10% glycerol.

Determination of molecular mass and isoelectric point (pI) of enzyme

The molecular mass of native enzyme was determined by comparing its mobility on a Sephadex G-150 column equilibrated with 150 mM potassium phosphate, pH 7.0 with that of known standards (Sigma). The subunit molecular mass under reducing condition was estimated by sodium dodecyl sulfate polyacrylamide gel electrophoresis (SDS-PAGE) [15] using standard proteins (Bio-Rad).

To determine pI of the enzyme, isoelectric focusing gel electrophresis (IEF) was performed using commercially available ampholine PAGE plate and pH calibration kit (Pharmacia) under recommended condition. The gels were stained by Silver Staining Kit (Kanto Kagaku, Tokyo, Japan).

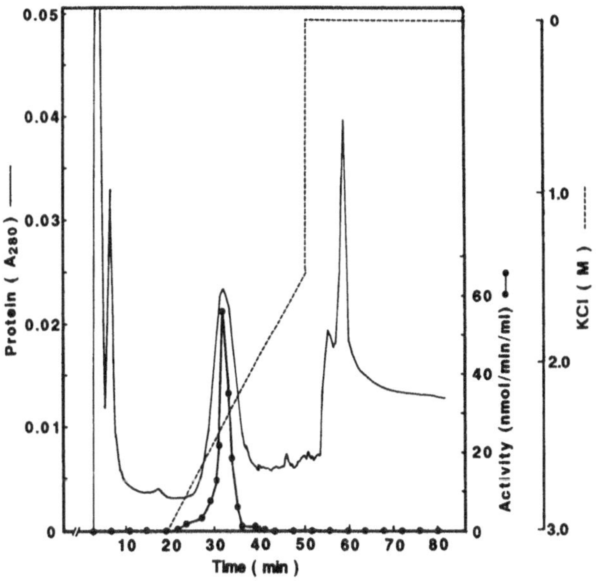

Fig. 1. Elution profile of pyridoxal kinase from Phenyl-Superose column using FPLC system. Enzymatically active fractions were collected. See details in Materials and methods.

Results

Purification of enzyme

Pyridoxal kinase from bovine brain was purified by ammonium sulfate fractionation and subsequent chromatography on DEAE-cellulose, hydroxyapatite, Sephadex G-150, Blue Sepharose CL-6B and Phenyl-Superose (Fig. 1). A summary of the purification is shown in Table 1. At the last step in the purification procedure the homogeneity of the enzyme was assessed by SDS-PAGE, on which the enzyme migrated as a single protein band. The purification fold was approximately

Table 1. Purification of pyridoxal kinase from bovine brain

Treatment	Volume	Total protein	Protein	Total activity	Activity	Specific activity
	ml	mg	mg/ml	nmol/min	nmol/min/ml	nmol/min/mg
Homogenate[1]	10000	369800	36.98	28700	2.87	0.078
40 to 60% $(NH_4)_2SO_4$ fraction	930	15186	16.3	23668	25.5	1.56
DEAE fraction	1037	1787	1.7	17889	17.25	10.0
Hydroxyapatite fraction	100	327	3.3	13000	130.4	39.9
Sephadex G-150 fraction	71	82	1.2	9102	128.2	111.0
Blue Sepharose CL-6B fraction	75	36	0.48	9054	120.7	253.6
Phenyl-Superose fraction	30	1.2	0.039	2457	81.9	2105

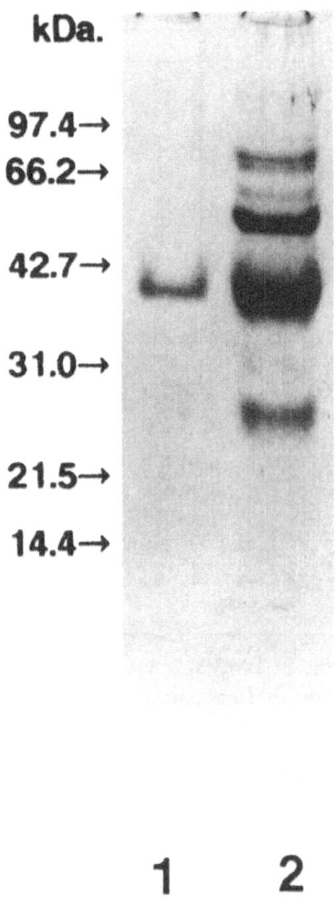

Fig. 2. SDS-PAGE of purified pyridoxal kinase from bovine brain. Lane 1, purified pyridoxal kinase; Lane 2, Blue Sepharose flowthrough fraction; Molecular size markers, 97.4 kDa., phosphorylase b; 66.2 kDa., bovine serum albumin; 42.7 kDa., egg albumin; 31.0 kDa., carbonic anhydrase; 21.5 kDa., soybean trypsin inhibitor; 14.4 kDa., lysozyme.

27,000-fold, and the final yield of the enzyme was 8.5% of the homogenate activity.

The active fractions from Blue Sepharose CL-6B column were tried to be further purified by an affinity chromatography on pyridoxyl Sepharose according to the procedures of Kwok *et al.* [5, 6]. However, the enzyme could hardly be further purified by the procedures probably because of low specific affinity of the enzyme for pyridoxyl Sepharose.

Molecular mass and pI determinations

The subunit molecular mass of the purified pyridoxal kinase, determined by SDS-PAGE was estimated to be 39,500 (Fig. 2). Gel filtration on Sephadex G-150 gave an estimated molecular mass for the native enzyme of 80,000. IEF revealed a pI of 5.0 for the purified enzyme.

Michaelis constants of the enzyme

Table 2 shows the kinetic properties of the purified enzyme for several substrates. The K_m value for pyridoxamine was higher than those for pyridoxine and pyridoxal. The V_{max} obtained for pyridoxal and pyridoxine were only one-fifth of that obtained for pyridoxamine. Thus, the rate of phosphorylation catalyzed by the kinase may be dependent upon the chemical structure of the substrates.

4-Deoxypyridoxine was known to be an effective competitive inhibitor of pyridoxal kinase. As shown in Table 2, the K_i value obtained for 4-deoxypyridoxine was 39 μM with respect to pyridoxal as substrate.

Discussion

Pyridoxal kinase from bovine brain was first purified 27,000-fold to apparent homogeneity with high activity (2105 nmol/min/mg protein). The effectiveness of the hydrophobic column chromatography for purification of pyridoxal kinase was also demonstrated in the present study. We employed the hydrophobic (Phenyl) gel column in the final procedure of purification because the pyridoxyl Sepharose chromatography, usually used in the isolation of the enzyme from pig brain [5], sheep brain [7], rat brain [6] and pig liver [13], was not applicable to the case of pyridoxal kinase from bovine brain.

Purified pyridoxal kinase from bovine brain is a protein of 80,000 molecular mass, which dissociates into subunits of 39,500 molecular mass under reducing condition, suggesting that the enzyme may be a dimer. The pyridoxal kinase from bovine brain is almost similar in molecular mass to those from other mammalian species [7, 8].

Table 2. Kinetic data[1] of pyridoxal kinase from bovine brain

Compounds	K_m (μM)	V_{max} (μ moles/ min·mg)	K_i (μM)
Pyridoxal	91	4.2	
Pyridoxine	67	3.7	
Pyridoxamine	2100	21.4	
ATP	210	4.5	
4-Deoxypyridoxine			39

[1] Samples (360 ng) of pyridoxal kinase (1.78 μmol/min·mg protein) in 70 mM potassium phosphate (pH 6.0) containing 1 mM ATP and 0.05 mM $ZnCl_2$ were incubated with various concentrations of substrates or inhibitor for 20 min at 37° C.

Table 3. Kinetic values of various pyridoxal kinases

Compounds	Bovine[b] brain	Bovine[c] brain	Pig[d] brain	Sheep[e] brain	Rat[c] liver	Pig[f] liver
Pyridoxal	91[a]	50	25	40	15	125
Pyridoxine	67	200	17	26	25	
Pyridoxamine	2100	500		6	150	
4-Deoxypyridoxine	39[a]	90	10		15	
ATP	210		12	20		59

[a] K_m and K_i values were expressed as μM.
[b] present study.
[c] reference No. 2.
[d] reference No. 5.
[e] reference No. 7.
[f] reference No. 8.

The kinetic properties of the enzyme indicated that the affinity for pyridoxal was lower than that for pyridoxine. This tendency was also reported in the pyridoxal kinases purified from pig [5] and sheep brain [7] (Table 3). K_m values of the enzyme from bovine brain, in the present study, for pyridoxal and pyridoxine were 3.6-fold and 3.9-fold larger than those from pig brain, respectively, and 2.3-fold and 2.6-fold larger than those from sheep brain, respectively. The pyridoxal kinase from bovine brain showed somewhat lower affinity to vitamin B_6 as compared with those from other mammalian species. This fact is one of the reasons to be insufficient activity of the pyridoxyl Sepharose column to isolate the pyridoxal kinase from bovine brain.

This purified preparation is useful to confirm the amino acid sequences in order to deduce the nucleotide sequences of the pyridoxal kinase and to produce the specific antibody against pyridoxal kinase. These materials will act as powerful probes to determine the possible regulation of pyridoxal phosphate biosynthesis in brain.

References

1. McCormick DB, Gregory ME, Snell EE: Pyridoxal phosphokinases. I. Assay, distribution, purification, and properties. J Biol Chem 236: 2076–2084, 1961
2. McCormick DB, Snell EE: Pyridoxal phosphokinases II. Effects of inhibitors. J Biol Chem 236: 2085–2088, 1961
3. White RS, Dempsey WB: Purification and properties of vitamin B_6 kinase from Escherichia coli B. Biochemistry 9: 4057–4064, 1970
4. Neary JT, Diven WD: Purification, properties, and a possible mechanism for pyridoxal kinase from bovine brain. J Biol Chem 245: 5585–5593, 1970
5. Kwok F, Churchich JE: Brain pyridoxal kinase: Purification, substrate specificities, and sensitized photodestruction of an essential histidine. J Biol Chem 254: 6489–6495, 1979
6. Cash CD, Maitre M, Rumigny JF, Mandel P: Rapid purification by affinity chromatography of rat brain pyridoxal kinase and pyridoxamine-5-phosphate oxidase. Biochem Biophys Res Commun 96: 1755–1760, 1980
7. Kerry JA, Rohde M, Kwok F: Brain pyridoxal kinase: Purification and characterization. Eur J Biochem 158: 581–585, 1986
8. Tagaya M, Yamano K, Fukui T: Kinetic studies of the pyridoxal kinase from pig liver: Slow-binding inhibition by adenosine tetraphosphopyridoxal. Biochemistry 28: 4670–4675, 1989
9. Sakurai T, Asakura T, Matsuda M: Transport and metabolism of pyridoxine and pyridoxal in mice. J Nutr Sci Vitamimol 33: 11–19, 1987
10. Sakurai T, Asakura T, Matsuda M: Transport and metabolism of pyridoxine in the intestine of the mouse. J Nutr Sci Vitaminol 34: 179–187, 1988
11. Sakurai T, Asakura T, Mizuno A, Matsuda M: Absorption and metabolism of pyridoxamine in mice I. Pyridoxal as the only form of transport in blood. J Nutr Sci Vitaminol 37: 341–348, 1991
12. Sakurai T, Asakura T, Mizuno A, Matsuda M: Absorption and metabolism of pyridoxamine in mice II. Transformation of pyridoxamine to pyridoxal in intestinal tissues. J Nutr Sci Vitaminol 38: 227–233, 1992
13. Wada H, Snell EE: The enzymatic oxidation of pyridoxine and pyridoxamine phosphates. J Biol Chem 236: 2089–2095, 1961
14. Kurioka S, Ishioka N, Sato J, Nakamura J, Ohkubo T, Matsuda M: Assay of vitamin B_6 in human plasma with graphitic carbon column. Anal Biochem, submitted
15. Weber K, Osborn M: The reliability of molecular weight determinations by dodecyl sulfate-polyacrylamide gel electrophoresis. J Biol Chem 244: 4406–4412, 1969.

Molecular and Cellular Biochemistry **119**: 209–215, 1993.

Modulation of the effect of glibenclamide on K_{ATP} channels by ATP and ADP

László Virág,[1] Tetsushi Furukawa and Masayasu Hiraoka
Department of Cardiovascular Diseases, Medical Research Institute, Tokyo Medical and Dental University, 1-5-45 Yushima, Bunkyo-ku, Tokyo-113, Japan
([1] Present address: Department of Pharmacology, Albert Szent-Györgyi Medical University, Dóm tér 12., P.O. Box: 115, H-6701 Szeged, Hungary)

Abstract

The inside-out configuration of the patch-clamp technique was used to study the effect of glibenclamide on the ATP-sensitive K^+ channel current in isolated guinea-pig ventricular myocytes. The inhibitory effect of glibenclamide was tested in the bath solution containing two different concentrations of ATP (100 μM and 200 μM). It was found that the effect of the drug on the K_{ATP} current was stronger in the presence of the higher concentration of ATP. The blocking effect of glibenclamide on the channels was weaker if, in addition to ATP, ADP was applied in the intracellular solution. Similarly, the inhibitory effect of the drug was not pronounced for the channels reactivated by ADP after run-down. As application of the drug in the presence and absence of Mg^{2+} did not show different effects on the channel inhibition, we concluded that the effect of glibenclamide may not depend on the phosphorylation of the channel protein. These results suggest that in addition of the previously described effect of ADP, ATP also has some modulatory effect on inhibition of the K_{ATP} channel by glibenclamide. (Mol Cell Biochem **119:** 209–215, 1993)

Key words: ATP-sensitive K^+ channel, glibenclamide, ventricular myocytes

Introduction

Potassium channels which are inhibited by intracellular ATP were first described by Noma [1] in cardiac cells. Later, these channels were detected in pancreatic β-cells [2], in skeletal muscle [3] and in vascular smooth muscle [4]. The K_{ATP} channels are mainly responsible for the shortening of the action potential and the decreased myocardial contractility in ischemic cardiac cell [5–7]. However in intact hearts the normal cytosolic ATP decreases only by 25–50% in the first 10 min. of ischemia which looks insufficient to open the channels [8]. As one of a possible explanation for this discrepancy, it was suggested that ADP can reduce the ATP-sensitivity of K_{ATP} channels, so the $[ADP]_i/[ATP]_i$ ratio determines the open probability of the channels better than the $[ATP]_i$ itself [9]. The most potent inhibitor of this channel is the sulfonylurea glibenclamide. The limited effectiveness of this drug at reducing cellular K^+ loss in ischemic conditions [10–12] turned the attention to the glibenclamide – nucleotides interactions. The aim of this study was to gain more insight into the processes which control the glibenclamide-sensitivity of K_{ATP} channels, in particular view of ADP and ATP.

Address for offprints: Masayasu Hiraoka, MD, Department of Cardiovascular Diseases, Medical Research Institute, Tokyo Medical and Dental University, 1-5-45 Yushima, Bunkyo-ku, Tokyo-113, Japan

Methods

Single ventricular myocytes were obtained from guinea-pigs hearts using enzymatic dissociation technique, as described previously [13].

The inside-out configuration of the patch-clamp technique was used to record single channel current through K_{ATP} channels using a patch-clamp amplifier (model 8900, Dagan Corp., USA). Current signals were recorded at a fixed membrane potential of -60 mV with videocassette recorder (HR-S5800, Victor Corp., Japan) through a pulse code modulation converter (RP-880, NF Instruments, Japan). The recorded signals were filtered off line by and eight-pole Bessel low pass filter (FV-665, NF Instruments) at 500 Hz and digitized at 2.5 kHz using an analog to digital converter (TL-1, Axon Instruments, USA) to store on disc of a computer (IBM PC/AT) for analysis.

The intracellular (bath) solution contained (mM): KCl 140, HEPES 5, EGTA 2, Glucose 5.5. The composition of the pipette (extracellular) solution in mM: KCl 140, $CaCl_2$ 1.8, $MgCl_2$ 0.53, HEPES 5, Glucose 5.5. The pH of both solutions was adjusted to 7.3 by addition of KOH. The concentration of free Ca^{2+} and Mg^{2+} in the bath solution was estimated as 2×10^{-10} M and 5×10^{-6} M, respectively, using a computer program for calculating metal ion/ligand binding. In the solution containing Mg^{2+}, Mg^{2+} was added to the bath solution to obtain the desired concentration of free Mg^{2+}. The experiments were done at room temperature (23–25° C).

The open probability (P_o) was calculated as

$$P_o = \frac{\sum n_i t_i}{NT}$$

where t_i is the duration of an n_ith number of open channels, N is the number of available channels in the patch, the total recording time (T) was usually 30 second. The number of available channels was estimated as the maximum open channels in the bath solution containing no ATP. The run-down of the channels highly influenced our experimental results. In order to minimize the influence of run-down, we monitored the activity of K_{ATP} channels in the ATP-free solution for 2 min after establishment of inside-out patches. If the P_o did not decrease more than 10% of the control value, we then used these patches for further experiments. Furthermore, we measured the P_o of the channels at the end of each experiment after washing out of the ATP, ADP, and/or glibenclamide. If the fraction of the P_o at the end of experiment was less than 0.8 of the initial value, the experiment was then discarded from the analysis.

All results are presented as means ± S.E. Statistical significance was evaluated using Student's t-test. A value of $p < 0.05$ was considered significant.

Results

Effect of ATP on the glibenclamide-inhibition of K_{ATP} channels

After making inside-out patches in ATP-free internal solution, a population of K^+ channels with unitary conductance of 75 pS appeared. The activity of these channels were inhibited by the presence of ATP in a dose-dependent manner, indicating these channels as the K_{ATP} channels. First, in order to test if the presence of ATP in the intracellular side of the membrane affected glibenclamide induced inhibition of K_{ATP} channel activity, we studied the inhibitory effect of glibenclamide in the bath solution containing two different concentrations of ATP (100 μM and 200 μM). Figure 1A shows representative tracings and Fig. 1B displays amplitude histograms. Panels *a* and *b* are for experiments in 100 μM ATP solution, and panels *e* and *f* are for experiments in 200 μM ATP solution; panels *a* and *e* for control tracings and panels *b* and *f* for tracings in the solution containing 10 nM glibenclamide. It appeared that the inhibition by 10 nM glibenclamide was greater at a 200 μM ATP than at a 100 μM ATP.

In Fig. 2, in order to make a more quantitative analysis, we tested the effects of glibenclamide at three different concentrations (1 nM, 10 nM, and 100 nM) in the presence of 100 μM ATP (filled reverse triangles) and 200 μM ATP (filled squares). Ten nM glibenclamide inhibited the K_{ATP} channels to a greater extent when the internal solution contained 200 μM ATP than 100 μM ATP. The P_o in the presence of 100 nM glibenclamide relative to the control values was 0.674 ± 0.037 at a 100 μM ATP and 0.335 ± 0.058 at a 200 μM ATP ($p < 0.05$). Similarly, one nM glibenclamide tended to inhibit the channel activity with a greater magnitude at a 200 μM ATP than a 100 μM ATP; the difference, however, did not reach a statistically significant level (0.944 ± 0.028 for 100 μM ATP versus and 0.734 ± 0.115 for 200 μM ATP [NS]). The inhibition of K_{ATP} channel activity by 100 nM glibenclamide was very

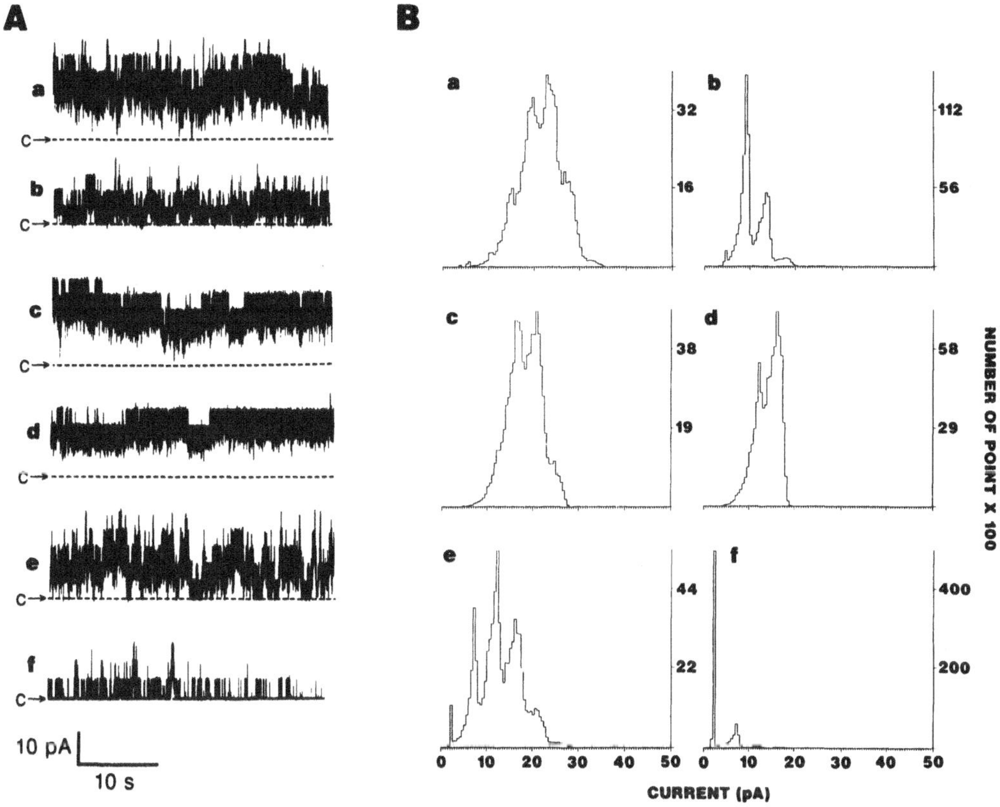

Fig. 1. Effect of 10 nM glibenclamide on the K_{ATP} current in the presence of 100 μM ATP (*a* for control, *b* for test), 100 μM ATP and 100 μM ADP (*c* for control, *d* for test), and 200 μM ATP (*e* for control, *f* for test). A: representative tracings, B: amplitude hitograms from the same patches as A. In each panel, *c* and an arrow represent the closed level. The holding potential of the patches was − 60 mV.

strong, so that we did not find any significant different effect of the drug between at 100 and 200 μM ATP (0.105 ± 0.033 versus 0.090 ± 0.039, [NS]).

Effect of ADP on the glibenclamide-inhibition of K_{ATP} channels

We next studied if the presence of ADP at the intracellular aspect of membrane affected the effect of glibenclamide. In order to test this hypothesis, we compared the effect of glibenclamide between in the presence and absence of 100 μM ADP. In Fig. 1, panels *c* and *d* show the representative data displaying the effect of 10 nM glibenclamide in the presence of 100 μM ATP and 100 μM ADP; panel A for representative tracings and panel B for amplitude histograms. It appeared that, in the solution containing both 100 μM ATP and 100 μM ADP, the inhibition of K_{ATP} channel activity was not strong.

We studied the inhibitory effect of glibenclamide at three different concentrations (1 nM, 10 nM, and 100 nM) in the presence of 100 μM ATP and 100 μM ADP, and the results are shown in Fig. 2. Comparing the inhibitory effects of glibenclamide on the K_{ATP} channel activity between in the presence of 100 μM ATP alone and in the presence of both 100 μM ATP and 100 μM ADP, 100 nM glibenclamide blocked the channels significantly stronger if the intracellular solution contained only ATP, compared to that containing both ATP and ADP (0.105 ± 0.033 versus 0.527 ± 0.081, $p < 0.05$). Although, at concentrations of 1 nM and 10 nM, glibenclamide tended to inhibit K_{ATP} channels to a lesser degree in the presence of ADP than in its absence, these differences were not statistically significant.

Effect of glibenclamide on run down K_{ATP} channels reactivated by ADP

It is known that various nucleotide diphosphates, such as GDP, UDP or ADP, can induce openings of the K_{ATP} channels after channel run-down [14]. Figure 3 shows the effect of glibenclamide on K_{ATP} channels after run-down channels were reactivated by 150 μM ADP. After

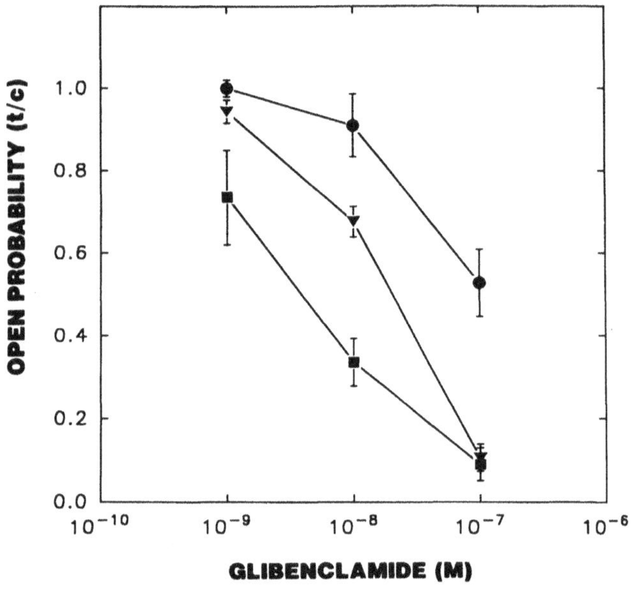

Fig. 2. Effect of glibenclamide on the relative open probability of the K_{ATP} channels in the presence of 100 μM ATP (reverse triangles), 100 μM ATP and 100 μM ADP (circles), and 200 μM ATP (squares). The relative open probability was defined as the ratio of the channel activity with glibenclamide to that without, n = 2–5 for each point.

formation of inside-out patches the openings of the K_{ATP} channels decayed to a minimal level within 2–3 min. under continuous perfusion of the ATP-free internal solution containing 1 mM free Mg^{2+}, mostly due to dephosphorylation of the channels. After run-down of the K_{ATP} channels occurred, in the assumed dephosphorylated state, 150 μM ADP was applied in the presence of 1 mM free Mg^{2+} to the internal side of the patch membranes. Reactivation effect of ADP varied considerably among patches, sometimes it was very low. The only experiments were chosen for the analysis, in which the reactivation of the channels was well pronounced and stable. It is interesting to note that in the experiments when the rate of run-down was slow, i.e. it took 10 or more minutes until the complete run-down, we could never reactivate the K_{ATP} channels by ADP.

Application of 100 nM glibenclamide in the intracellular solution decreased the activity of the reopened K_{ATP} channels to 0.445 ± 0.081 (n = 3) of the value before addition of glibenclamide. This magnitude of inhibition by glibenclamide was significantly weaker than in the presence of ATP alone, and was similar to the value in the presence of both ATP and ADP using phosphorylated K_{ATP} channels before run-down.

Effects of glibenclamide in the presence and absence of Mg^{2+}

Figure 4A shows an experiment, in which we examined whether the phosphorylated state of the K_{ATP} channels was equivalent to the dephosphorylated state from the point of view of glibenclamide-sensitivity. After forming inside-out patches in Mg^{2+} free internal solution, the effect of glibenclamide on the activity of K_{ATP} channels was tested in the presence of MgATP, i.e. in phosphorylated state, then the drug and MgATP were washed out using Mg^{2+} free bath solution and the effect of the drug was tested again in the presence of K_2ATP. We assumed that in the absence of MgATP the channels were partly dephosphorylated. Ten nM glibenclamide decreased the activity of the channels to 0.603 ± 0.152 in the presence of 100 μM MgATP and to 0.598 ± 0.103 in the presence of 100 μM K_2ATP (3 patches).

In other 3 experiments the concentration of 200 μM was used for Mg- and K_2ATP. In these cases the corresponding values of P_o by glibenclamide were 0.266 ± 0.016 and 0.219 ± 0.077 respectively. We concluded, therefore, that the glibenclamide sensitivity of the K_{ATP} channels was not changed between in dephosphorylated and phosphorylated state.

It has been reported in biochemical studies that ATP, added as the Mg^{2+} salt, can inhibit the glibenclamide binding, but only in the presence of dithiothreitol (DTT) [15]. In order to examine this possibility, experiments were done using similar protocol as above, in which effects of glibenclamide were tested in the presence of 0.1 mM DTT (Fig. 4B). Results obtained were essentially the same as those in the absence of DTT. Ten nM glibenclamide blocked the activity of the channels to 0.587 ± 0.118 in the presence of 100 μM MgATP and to 0.614 ± 0.084 in the presence of 100 μM K_2ATP. In the case of 100 nM glibenclamide, the corresponding values were 0.306 ± 0.050 (n = 4) and 0.325 ± 0.094 (n = 4) respectively.

Discussion

Our results suggest that there is some positive co-operativity between the binding of ATP to its binding site and the inhibitory effect of glibenclamide, i.e. applying higher concentration of ATP in the intracellular solution can make the block of the drug on the K_{ATP} channels stronger. Some evidences seem to support this finding. It was reported that using inside-out membrane

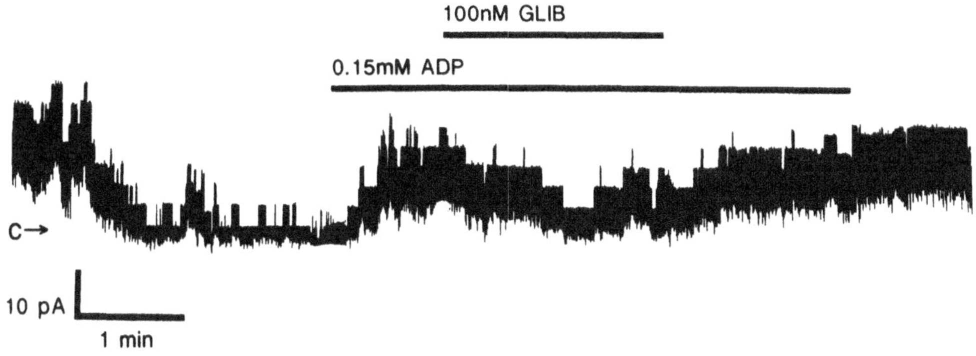

Fig. 3. Effect of 100 nM glibenclamide on K_{ATP} channels after run-down reactivated by 150 μM ADP. *c* and an arrow represent the closed level. The holding potential of the patch was − 60 mV.

patches in the absence of nucleotides, 50% channel inhibition requires very high, more than 100 μM concentration of glibenclamide, though the same report could not find any ATP effect on the dose-response curve to glibenclamide [16]. Under similar conditions another research group reported a K_d value of 0.5 μM for glibenclamide inhibition, the Hill coefficient was 0.5 [17]. The estimated value of Hill coefficient is about 1.2 according to our experiments in the presence of 100 μM ATP which is similar to that measured by Findaly [18], though the experimental conditions were different between his study and ours. The difference in the Hill coefficient in the presence and absence of nucleotides and the high K_d value for nucleotide-free conditions may suggest that the ATP has some role in the regulation of glibenclamide sensitivity of K_{ATP} channels.

It is known that the blocking effect of ATP on K_{ATP} channels is stronger in the presence of Mg^{2+} [19]. Therefore it may be interesting to examine the possibility that the inhibitory effect of glibenclamide is also depend on the presence of Mg^{2+}. In our experiments glibenclamide blocked the K_{ATP} channels to the same extent, regardless of the presence or absence of Mg^{2+} in the internal solution. Thus, the effect of glibenclamide may not depend on the phosphorylation of the channel protein. It was reported using radiolabelled glibenclamide binding assay that MgATP inhibited the glibenclamide binding but only in the presence of dithiothreitol (DTT) suggesting that an important sulfhydryl group was present at or near the nucleotide binding site [15]. In the presence of DTT, however, we failed to show any difference in the effect of glibenclamide with or without

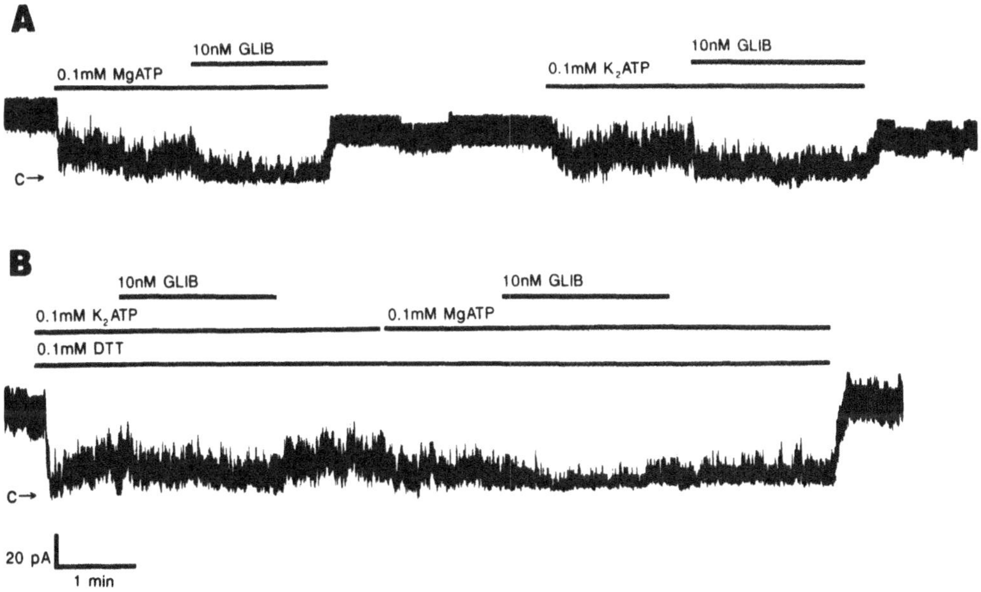

Fig. 4. Lack of effect of Mg^{2+} on the inhibitory effect of glibenclamide on K_{ATP} channels in the absence (A) and presence of DTT (B). *c* and an arrow represent the closed level. The holding potential of the patches was − 60 mV.

Mg^{2+} in the bath solution. It could be possible that, in the glibenclamide binding assay, the inhibition of glibenclamide binding in the presence of MgATP came from the hydrolysis of ATP to ADP.

It has been shown previously that the sensitivity of K_{ATP} channels to ATP for inhibitory action is weaker in the presence of ADP and other nucleotide diphosphates [9, 20–22]. According to these results Lederer and Nichols hypothesized [22] that the K_{ATP} channel has two ATP binding sites which interact in a positive cooperative way, and proposed that in the presence of Mg^{2+}, ADP can bind to one of the ATP binding inactivation sites to prevent the ATP binding to the other site, hereby restoring the channel activity. Tung and Kurachi suggested another model to explain their result that nucleotide diphosphates could reactivate the K_{ATP} channels after run-down in the presence of Mg^{2+} [14]. Consequently, the K_{ATP} channel would have a high affinity and Mg^{2+}-dependent binding site for ADP and a high affinity but Mg^{2+}-independent binding site for ATP. Ligand binding to the ATP site causes channel closure, while NDP binding to the ADP binding site can open the channels in spite of the possible dephosphorylated state.

In our experiments the blocking effect of glibenclamide on the K_{ATP} channels was significantly weaker if, in addition to ATP, ADP was applied to the intracellular side of the patch membrane. Similarly, the inhibitory effect of glibenclamide was not pronounced for the channels reactivated by ADP after run-down, assumed in dephosphorylated state. These findings appear to be in consistent with other reports using radiolabelled glibenclamide binding and electrophysiological techniques [9, 15, 17, 20–24]. The mechanism by which the ADP antagonize the inhibitory effect of glibenclamide on K_{ATP} channels is still unclear. Venkatesh *et al.* [17] have proposed a two-binding site model which is similar to the explanation of Tung and Kurachi. Notably the first binding site, which may be the same as the ATP binding site, promotes channel closure when occupied by a ligand and has high affinity for glibenclamide and low affinity for ADP. The second binding site, which is analogous to the nucleotide diphosphate binding site in the Tung and Kurachi model [14], promotes channel openings when occupied by a ligand, has high affinity for ADP and low affinity for glibenclamide.

Our experimental results may be explained in the line of this two-binding site model with minor modifications. Binding of a ligand to the high affinity ATP binding site, which has low affinity for ADP, promotes channel closure and makes the glibenclamide effect stronger by elevating the affinity of sulfonylurea binding site. Binding of a ligand to the high affinity ADP binding site promotes opening even if the ATP and glibenclamide binding sites are reserved by the ligand or in dephosphorylated channel state. Nevertheless, our results did not exclude the possibility of the regulatory model proposed by Lederer and Nickols, and further studies are needed to clarify the validity of this model and to describe the sulfonylurea – nucleotide interactions in detail.

Acknowledgements

V.L. is supported by the Foreign Special Scholarship from the Japanese Government (Mombusho). He thanks Professor J. Gy. Papp and Professor L. Szekeres for giving chance to perform this study. This work was supported by grants from the Japanese Government (Mombusho) to M.H. and T.F.

References

1. Noma A: ATP-regulated K^+ channels in cardiac muscle. Nature 305: 147–148, 1983
2. Cook DL, Hales CN: Intracellular ATP directly blocks K^+ channels in pancreatic B-cells. Nature 311: 271–273, 1984
3. Spruce AE, Standen NB, Stanfield PR: Voltage-dependent, ATP-sensitive potassium channels of skeletal muscle membrane. Nature 316: 736–738, 1985
4. Standen NB, Quayle JM, Davies NW, Brayden JE, Huang Y, Nelson MT: Hyperpolarizing vasodilatators activate ATP-sensitive K^+ channels in arterial smooth muscle. Science 245: 177–180, 1989
5. Noma A, Shibasaki T: Membrane current through adenosine-triphosphate-regulated potassium channels in guinea-pig ventricular cells. J Physiol (Lond) 363: 463–480, 1985
6. Weiss JN, Lamp ST: Glycolysis preferentially inhibits ATP-sensitive K^+ current in isolated gluinea-pig cardiac myocytes. Science 238: 67–69, 1987
7. Cole WC, McPherson CD, Sontag D: ATP-regulated K^+ channels protect the myocardium against ischemia/reperfusion demage. Circ Res 69: 571–581, 1991
8. Rovetto MJ, Whitmer JT, Neely JR: Comparison of the effects of anoxia and whole heart ischaemia on carbohydrate utilization in isolated working rat hearts. Circ Res 32: 699–711, 1973
9. Weiss JN, Venkatesh N, Lamp ST: ATP-sensitive K^+ channels and cellular K^+ loss in hypoxic and ischaemic mammalian ventricle. J Physiol 447: 649–673, 1992
10. Kantor PR, Coetzee WA, Carmeliet EE, Dennis SC, Opie LH: Reduction of ischemic K^+ loss and arrhythmias in rat hearts: Effect of glibenclamide, a sulfonylurea. Circ Res 66: 478–485, 1990.

11. Venkatesh N, Lamp ST, Weiss JN: Effects of sulfonylureas on K^+ loss during myocardial ischemia and metabolic inhibition (abstract). Circulation 80 (suppl II): II-607, 1989
12. Wilde AAM, Escande D, Schumacher CA, Thuringer D, Mestre M, Fiolet JWT, Janse MJ: Potassium accumulation in the globally ischemic mammalian heart: A role for the ATP-sensitive K^+ channel. Circ Res 67: 835–843, 1990
13. Hirano Y, Hiraoka M: Barium-induced automatic activity in isolated ventricular myocytes from guinea-pig hearts. J Physiol 395: 455–472, 1988
14. Tung RT, Kurachi Y: On the mechanism of nucleotide diphosphate activation of the ATP-sensitive K^+ channel in ventricular cell of guinea-pig. J Physiol 437: 239–256, 1991
15. Gopalakrishnan M, Johnson DE, Janis RA, Triggle DJ: Characterization of binding of the ATP-sensitive potassium channel ligand, [^{3}H]glyburide, to neuronal and muscle preparations. J Pharmacol Exp Ther 257: 1162–1171, 1991
16. Ripoll C, Lederer WJ, Nichols CG: On the mechanism of inhibition of K_{ATP} channels by glibenclamide in rat ventricular cell membranes (abstract). Circulation 82 (suppl IV): III–12, 1990
17. Venkatesh N, Lamp ST, Weiss JN: Sulfonylureas, ATP-sensitive K^+ channels, and cellular K^+ loss during hypoxia, ischemia, and metabolic inhibition in mammalian ventricle. Circ Res 69: 623–637, 1991
18. Findlay I: Inhibition of ATP-sensitive K^+ channels in cardiac muscle by the sulfonylurea drug glibenclamide. J Pharmacol Exp Ther 261: 540–545, 1992
19. Findlay I: ATP^{4-} and ATP · Mg inhibit the ATP-sensitive K^+ channel of rat ventricular myocytes. Pflugers Arch 412: 37–41, 1988
20. Nichols CG, Lederer WJ: The regulation of ATP-sensitive K^+ channel activity in intact and permeabilized rat ventricular myocytes. J Physiol (London) 423: 91–110, 1990
21. Nichols CG, Ripoll C, Lederer WJ: K_{ATP} channel modulation of the guinea-pig ventricular action potential and contraction. Circ Res 68: 280–287, 1991
22. Lederer WJ, Nichols CG: Nucleotide modulation of the activity of rat heart ATP-sensitive K^+ channels in isolated membrane patches. J Physiol 419: 193–211, 1989
23. Niki I, Nicks JL, Ashcroft SJH: The beta cell glibenclamide receptor is an ADP-binding protein. Biochem J 268: 713–718, 1990
24. Bernardi H, Fosset M, Lazdunski M: ATP/ADP binding sites are present in the sulfonylurea binding protein associated with brain ATP-sensitive K^+ channels. Biochemistry 31: 6328–6332, 1992

Molecular and Cellular Biochemistry **119**: 217–220, 1993.

Index

GPSR Compliance

The European Union's (EU) General Product Safety Regulation (GPSR) is a set of rules that requires consumer products to be safe and our obligations to ensure this.

If you have any concerns about our products, you can contact us on ProductSafety@springernature.com

In case Publisher is established outside the EU, the EU authorized representative is:

Springer Nature Customer Service Center GmbH
Europaplatz 3
69115 Heidelberg, Germany

Batch number: 10755910

Printed by Printforce, the Netherlands